32.95

INTRODUCTION TO FINITE GEOMETRIES

NORTH-HOLLAND TEXTS
IN ADVANCED MATHEMATICS

INTRODUCTION TO FINITE GEOMETRIES

by

F. KÁRTESZI

PROFESSOR AT THE EÖTVÖS LORÁND UNIVERSITY
BUDAPEST

1976
NORTH-HOLLAND PUBLISHING COMPANY
AMSTERDAM · OXFORD
AMERICAN ELSEVIER PUBLISHING COMPANY, INC.
NEW YORK

This monograph was published in Hungary
as Vol. 7 in the series of
DISQUISITIONES MATHEMATICAE HUNGARICAE

Translated by
L. Vekerdi

English translation supervised by
Minerva Translations Ltd.,
5 Burntwood Grange Road, London S.W.18 3JY

Library of Congress Catalog Card Number: 74—83729
North-Holland ISBN: 0 7204 2832 7
American Elsevier ISBN: 0 444 10855 6

© Akadémiai Kiadó · Budapest 1976

All Rights Reserved. No part of this publication may be reproduced, stored in a retrieval system, or transmitted,
in any form or by any means, electronic, mechanical, photocopying, recording or otherwise, without the prior
permission of the Copyright owner.
Joint edition published by North-Holland Publishing Company,
Amsterdam—Oxford
and
Akadémiai Kiadó,
Publishing House of the Hungarian Academy of Sciences, Budapest
Sole distributors for the U.S.A. and Canada:
American Elsevier Publishing Company, Inc., 52, Vanderbilt Avenue,
New York, N.Y. 10017

Printed in Hungary

*To my dear Master,
Professor Beniamino Segre*

PREFACE

This book contains the material elaborated during my courses held at the Eötvös Loránd University of Budapest under the title "projective geometry" since 1948. In the beginning I briefly mentioned the notation of finite projective planes in connection with the classical projective geometry. Over the years I continuously enlarged the share of finite geometries at the expense of the classical themes. As the preliminary training of my students varied widely, I had to begin at a quite elementary level.

More than two decades experience in teaching as well as the rapid development of the subject over the last three decades gave me a didactical point of view which was mirrored in my lectures. Encouraged by some former students of mine I have presented these lectures in the form of an introductory textbook of a didactical character.

As to the treatment, my book is presented in a rather unusual form in comparison with traditional and modern textbooks alike. First of all, I do not strive to give a complete account of the subject up to the present day but rather the ways and means which led to its development. I consider this book to be somewhat experimental — as indeed my lectures were — and I await the opinion of the readers who begin to learn about the theory of finite geometries through the pages of this book.

As to the terminology — mainly because of the shortage of relevant publications in Hungarian — I was often obliged to introduce new notations or to deviate from the usual ones; of course in such cases I have always explained the exact meaning of the terms. Similarly, certain illustrations, such as tables, deformed figures, etc., are presented in an unusual form.

I should like to express my indebtedness to Professors Gy. Strommer and G. Szász for their useful advice and careful work in supervising the manuscript of this book, to M. Frigyesi and G. Vidéki for the clarity and neatness of the figures.

Moreover, I owe thanks to the Akadémiai Kiadó (Publishing House of the Hungarian Academy of Sciences) for the publication of the book.

F. Kárteszi

CONTENTS

Preface vii
Common notation xii

Chapter 1

Basic concepts concerning finite geometries 1
   1.1 The finite plane 1
   1.2 Isomorphic planes, incidence tables 4
   1.3 Construction of finite planes, cyclic planes 8
   1.4 The Γ-table of a finite projective plane 12
   1.5 Coordinate systems on the finite plane 19
   1.6 The concepts of Galois planes and Galois fields 22
   1.7 Closed subplane of a finite projective plane 27
   1.8 The notion of the finite affine plane 30
   1.9 Different kinds of finite hyperbolic planes 32
   1.10 Galois planes and the theorem of Desargues 36
   1.11 A non-Desarguesian plane 43
   1.12 Collineations and groups of collineations of finite planes 47
   1.13 Line preserving mappings of a finite affine plane and of a finite regular hyperbolic plane 54
   1.14 Finite projective planes and complete orthogonal systems of Latin squares 58
   1.15 The composition of the linear functions and the $D(X, Y)$ plane 68
   1.16 Problems and exercises to Chapter 1 78

Chapter 2

Galois geometries 84
   2.1 The notion of Galois spaces 84
   2.2 The Galois space as a configuration of its subspaces 87
   2.3 The generalization of Pappus' theorem on the Galois plane 94
   2.4 Coordinates on a Galois plane 98
   2.5 Mappings determined by linear transformations 106
   2.6 Linear mapping of a given quadrangle onto another given quadrangle 107
   2.7 The concept of an oval on a finite plane 110
   2.8 Conics on a Galois plane 113
   2.9 Point configurations of order 2 on a Galois plane of even order 119
   2.10 The canonical equation of curves of the second order on the Galois planes of even order 122

2.11 Point configurations of order 2 on a Galois plane of odd order 125
2.12 Correspondences between two pencils of lines 128
2.13 A theorem of Segre 133
2.14 Supplementary notes concerning the construction of Galois planes 137
2.15 Collineations and homographies on Galois planes........................... 142
2.16 The characteristic of a finite projective plane 143
2.17 The set of collineations mapping a Galois plane onto itself 146
2.18 Desarguesian finite planes 150
2.19 Problems and exercises to Chapter 2 162

Chapter 3

Geometrical configurations and nets 164
   3.1 The concept of geometrical configurations 164
   3.2 Two pentagons inscribed into each other 166
   3.3 The pentagon theorem and the Desarguesian configuration 170
   3.4 The concept of geometrical nets 177
   3.5 Groups and R nets 181
   3.6 Problems and exercises to Chapter 3 183

Chapter 4

Some combinatorial applications of finite geometries 185
   4.1 A theorem of closure of the hyperbolic space 185
   4.2 Some fundamental facts concerning graphs 187
   4.3 Generalizations of the Petersen graph 191
   4.4 A combinatorial extremal problem 195
   4.5 The graph of the Desargues configuration 198
   4.6 Problems and exercises to Chapter 4 199

Chapter 5

Combinatorics and finite geometries 201
   5.1 Basic notions of combinatorics 201
   5.2 Two fundamental theorems of inversive geometry 205
   5.3 Finite inversive geometry and the t-(v, k, λ) block design 209
   5.4 General theorems concerning the Möbius plane 213
   5.5 Incidence structure and the t-block design 216
   5.6 Problems and exercises to Chapter 5 218

Chapter 6

Some additional themes in the theory of finite geometries 219
   6.1 The Fano plane and the theorem of Gleason 219
   6.2 The derivation of new planes from the Galois plane 232
   6.3 A generalization of the concept of the affine plane 241
   6.4 Problems and exercises to Chapter 6 244

7. Appendix

7.1 Notes concerning algebraic structures in general 250
7.2 Notes concerning finite fields and number theory 257
7.3 Notes concerning planar ternary structures 261

Bibliographical notes 263
Index 265

COMMON NOTATION

$\|\mathbf{H}\|$	number of the elements of a set $\mathbf{H}$.
$\{a, b, ..., h\}$	the set consisting of the elements $a, b, ..., h$.
$\{...\|...\}$	the set of the elements to the left of the dividing line which all have the property denoted on the right hand side of the dividing line. e.g. $\{x \in G \| x^2 = 1\}$.
$x \in \mathbf{H}$	the element x belongs to the set $\mathbf{H}$.
$x \notin \mathbf{H}$	x does not belong to $\mathbf{H}$.
$\mathbf{U} \subseteq \mathbf{H}$	every element of set $\mathbf{U}$ belongs to $\mathbf{H}$.
$\mathbf{U} \subset \mathbf{H}$	every element of $\mathbf{U}$ belongs to $\mathbf{H}$ but not every element of $\mathbf{H}$ belongs to $\mathbf{U}$.
$\mathbf{A} \cap \mathbf{B}$	$\{x \| x \in \mathbf{A}$ and $x \in \mathbf{B}\}$, greatest common subset, intersection.
$\mathbf{A} \cup \mathbf{B}$	$\{x \| x \in \mathbf{A}$ or $x \in \mathbf{B}\}$, union of $\mathbf{A}$ and $\mathbf{B}$.
$\mathbf{A} \otimes \mathbf{B}$	the set of all ordered pairs (a, b) where $a \in \mathbf{A}$ and $b \in \mathbf{B}$.
$\mathbf{A} - \mathbf{B}$ or $\mathbf{A} \setminus \mathbf{B}$	If $\mathbf{B} \subseteq \mathbf{A}$, $\{x \| x \in \mathbf{A}$ and $x \notin \mathbf{B}\}$.
$\circ, *, \perp, \curlywedge$	signs of operations (with meaning to be specified from case to case).
$(A, \circ)$	structure with one operation: written $(A, +)$ or $(A, \times)$ when the operation is addition or multiplication, respectively.
$(A, +, \cdot), (A, \perp, \curlywedge)$	structures with two operations.
$F(xmb)$	ternary operation.
$GF(q)$	Galois field of q elements.
(x_1, x_2, x_3)	point-coordinates.
$[u_1, u_2, u_3]$	line-coordinates.
$\bullet$	sign of incidence (on an incidence table).
Σ	(arbitrary) incidence table.
Ω	cyclic incidence table.
$C^{m,x}$	parcel (on an incidence table).
$\Gamma(q), \Gamma$	the parcelled incidence table of order q.
$\vec{AB}; \mathbf{a}, ..., \mathbf{x}$	vectors.
$\lambda \mathbf{x}$	vector multiplied by a scalar.

(x) and $[u]$	point and line determined by their coordinates.
$[u](x)$	$u_1 x_1 + u_2 x_2 + u_3 x_3$.
$S_{n,q}$	the n-dimensional (arithmetic) space over the coordinate field $GF(q)$; Galois space.
t-(v, k, λ)	the t-(v, k, λ) block system.
I	the axiom system consisting of the axioms I_1, I_2, I_3, I_4.
$D-, P-, R-,$	Desarguesian, Pascalian, Reidemeister.
$MD, MP,$	micro-Desarguesian, micro-Pascalian,
MR	micro-Reidemeister.

CHAPTER 1

BASIC CONCEPTS CONCERNING FINITE GEOMETRIES

In this chapter a simple and elementary treatment will be given to some basic concepts, problems and methods of finite geometries. To begin with we shall outline what this new branch of mathematics is about and how it developed out of several classical subjects.

1.1 The finite plane

The projective plane obtained by extending the Euclidean plane by ideal points and by the ideal line is called the *classical projective plane*. On this plane the following basic facts concerning the incidence relation of points and lines are known:

(1) *Given any two distinct points there exists just one line incident with both of them* (called the *connecting line* of the two points).

(2) *Given any two distinct lines there exists just one point incident with both of them* (called the *point of intersection* of the two lines).

(3) *There exist four points such that a line incident with any two of them is not incident with either of the remaining two.* (We shall express this by saying that there exists a real quadrangle.)

Let us now consider the basic facts enumerated above as axioms. Every theorem which is a consequence of these axioms is also a theorem on the classical projective plane. There are, however, other systems consisting of objects called points and objects called lines together with a relation called incidence — which either does or does not hold for any pair consisting of one point and one line — and satisfying the requirements of (1), (2), and (3). The oldest and most simple example of this is the *Fano figure*. (See Fig. 2.)

Let us consider an uncoloured chessboard of 7 squares by 7 squares and upon it 21 chessmen in the arrangement that can be seen in Fig. 1. Let the columns of the chessboard be called "points" $P_1, P_2, P_3, P_4, P_5, P_6$, and P_7 and let the rows be called "*lines*" $l_1, l_2, l_3, l_4, l_5, l_6$ and l_7. Let the "*incidence*" of a point and a line be interpreted as the fact that the square at the intersection of the corresponding row and the column is occupied by a chessman — i.e. on the figure by a sign o — and let an empty square be interpreted as "non-incidence". It is easy to see that this model in fact fulfils the requirements of (1), (2) and (3). — Thus, for instance, the only line connecting points P_4 and P_6 is l_6. The only point incident with both of the lines l_3 and l_5 is P_7. A real quadrangle is given by $P_1 P_2 P_3 P_4$, the six lines connecting pairs of points of this quadrangle are l_1, l_2, l_3, l_5, l_6 and l_7 so the system satisfies (3). In what follows, this system will be called a *Fano plane*.

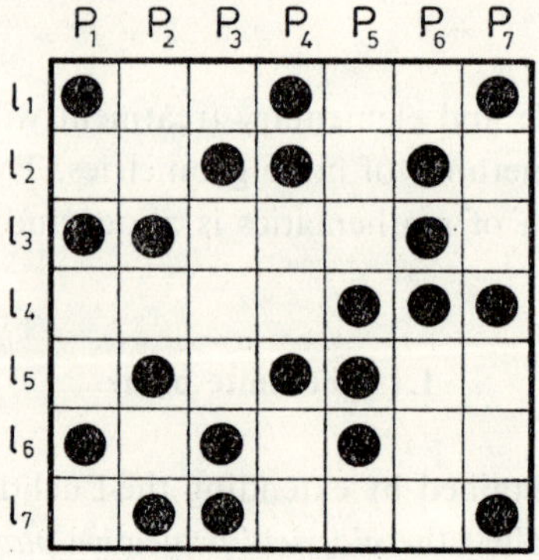

Figure 1

We shall principally be concerned with the investigation of finite planes; the treatment of a plane as a point set has many advantages. Thus the line appears as a subset defined by certain properties. So the plane is a point set, say, $\Sigma = \{P_1, P_2, ...\}$ and the lines of the plane are certain subsets of this set Σ. We shall denote points by upper case Latin letters, lines by lower case Latin letters. The projective plane will be defined by the following axiom system:

I_1 *If $P \in \Sigma$ and $Q \in \Sigma$ and $P \neq Q$, then there exists uniquely a line l for which $P \in l$ and $Q \in l$.*

I_2 *If $g \subset \Sigma$ and $l \subset \Sigma$ and $g \neq l$, then there exists P for which $P \in g$ and $P \in l$.*

I_3 *There exist four points determining two by two six distinct lines according to I_1.*

We know already that the abstract projective plane defined by these three axioms is not an empty concept; we have the examples of the classical projective plane and the Fano plane.

(a) *It follows already from axioms I_1 and I_2 that any two distinct lines have one and only one common point.*

Namely according to I_2 there exists one and according to I_1 there cannot exist more than one common point of two distinct lines.

(b) There exist four lines of which no three have a common point. (That is, there exist real quadrilaterals.)

Namely, if P_1, P_2, P_3 and P_4 are four points satisfying axiom I_3, then the lines P_1P_2, P_2P_3, P_3P_4 and P_4P_1 correspond to the statement (b).

The Principle of Duality, as in classical projective geometry, is valid on the abstract projective plane as well. This follows from the fact that I_1 and I_2 and, furthermore, I_3 and Theorem (b) are the duals of each other.

(c) Any line contains at least three points; each point is contained in at least three lines.

The second part of the statement is the dual of the first, therefore it suffices to prove the first part.

There exist real quadrangles; let one of them be the quadrangle $P_1P_2P_3P_4$. An arbitrary line l of the plane may contain either two, one or none of the vertices of its quadrangle. In the first case, suppose $P_3 \in l$ and $P_4 \in l$. The lines P_1P_2 and P_3P_4 have a unique common point which is distinct from the vertices of the quadrangle. This point, however, is already a third point of line l. In the second case suppose l only passes through the vertex P_4 then the lines P_1P_2, P_2P_3 each meet l in one point. In the third case the lines P_4P_1, P_4P_2 and P_4P_3 each meet l in one point. Thus our statement (c) is true.

These theorems, which reflect the basic properties of the classical projective plane, define, if taken as axioms, a much too general notion of the abstract projective plane. We introduce a fourth axiom in order to restrict this concept and obtain the notion of the finite projective plane of order q.

I_4 *There exists a line which consists of $q+1$ points, where $q(>1)$ is a (suitable) positive integer.*

In fact, this axiom can be considered a reasonable one, since it is satisfied by the Fano plane namely with $q=2$. Furthermore, it follows from the former three axioms that $q>1$. (Theorem (c).)

The axiom system consisting of I_1, I_2, I_3 and I_4 will be denoted by **I**. We shall now discuss some simple but significant consequences of the axiom system **I**.

(d) Every line consists of $q+1$ points.
(e) There are $q+1$ lines through every point.
(f) The plane consists of q^2+q+1 points.
(g) The plane contains q^2+q+1 lines.

In the example given in Fig. 1, where $q=2$, these theorems are exhibited as the following properties:

$2+1=3$ signs occur in each row of the table.
3 signs occur in each column of the table.

The table consists of $2^2+2+1=7$ (non-empty) columns.

The table consists of 7 (non-empty) rows.

And now we shall prove statements *(d)* to *(g)* successively. The perspective correlation of ranges and pencils on the classical projective plane is a notion which can also be extended to the plane defined by the axiom system I and the proof of our statements is based upon this one-to-one perspective correspondence.

(d) Let $l=\{P_1, P_2, \ldots, P_{q+1}\}$ be a line satisfying I_4 and let l' be an arbitrary line distinct from l, i.e. $l' \neq l$. According to I_2, the lines l and l' have a common point; let their common point be P_j. Line l' contains — according to Theorem *(c)* — a point P_r distinct from P_j; furthermore, the line connecting the point P_k of l, distinct from P_j, with P_r also contains a point P_s which is distinct from both P_r and P_k. Now the lines connecting the point P_s with the points of the line l exhaust the totality of the lines passing through the point P_s — as can be seen from axioms I_1 and I_2 — and each of them meets l' in a point; thus the line l' also contains $q+1$ points. Clearly, l' cannot contain more than $q+1$ points because otherwise, by connecting them with the point P_s, we would obtain more than $q+1$ lines in the pencil through P_s.

(e) By dualizing the proof of Theorem *(d)* we obtain the proof of Theorem *(e)*, since we know already that the number of lines passing through point P_s is $q+1$.

(f) Every point of the plane is, according to I_1, a point on a line of the pencil through P_s. On each of the $q+1$ lines of this pencil — according to *(d)* — there are q points, distinct from P_s. Thus the total number of the points is: $(q+1)q+1=$
$=q^2+q+1$.

(g) The proof of Theorem *(g)* is given by dualizing the proof of Theorem *(f)*.

Definition: *The figure corresponding to the axiom system* I *is said to be a finite projective "plane of order* q*".*

Where no misunderstanding can occur we shall abbreviate this to "plane of order q".

The following question remains to be answered: Does a plane of order q exist if q is an integer greater than 2?

1.2 Isomorphic planes, incidence tables

Let us consider two finite projective planes of order q, i.e. two sets having the same number, $q^2+q+1=n$, of elements in which there can be found n subsets each having $q+1$ elements so that the structure of the two sets generated by these subsets — we shall call it the *combinatorial structure* — should satisfy the axiom system I. As both sets have the same number of elements, the same names, i.e. indices, can be used to denote the elements in each set.

For example let us consider Figs 1 and 2. Figure 1 has already been examined; Fig. 2 also satisfies the requirements of **I**, as can readily be checked. We established a one-to-one correspondence between the elements of the two different planes of order 2 by using the same indexing set. This correspondence, however, gives little insight into the structural similarity of the two kinds of sign pattern.

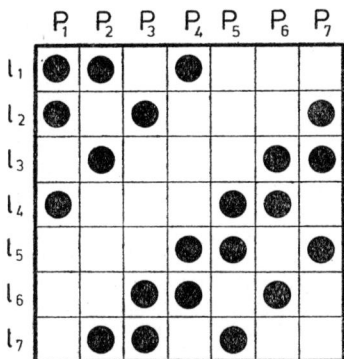

Figure 2

We can obtain more revealing information by the following simple operations. In what follows, we shall call Fig. 1 the "Σ_1 table", and Fig. 2 the "Σ_2 table".

1° We decompose table Σ_1 into columns. Next we perform the interchanges expressed by the scheme

$$\omega = \begin{pmatrix} 1 & 2 & 3 & 4 & 5 & 6 & 7 \\ 1 & 7 & 2 & 4 & 3 & 5 & 6 \end{pmatrix}$$

referring to the indices of the columns. This means that we do not move the first column, we replaced the second column by the seventh, the third by the second and so on; of course together with the columns, the sign patterns contained in them will be interchanged as well. We then unite the rearranged columns. An operation performed according to prescription will be called an *ω-transformation*.

2° We now decompose the table into rows and after performing e.g. the interchanges given by

$$\sigma = \begin{pmatrix} 1 & 2 & 3 & 4 & 5 & 6 & 7 \\ 1 & 3 & 4 & 6 & 2 & 5 & 7 \end{pmatrix},$$

we unite the rearranged rows. An operation like this one will be called a *σ-transformation*.

3° Starting with a given table let us perform first an ω-transformation and then, starting with the transformed table, a σ-transformation. Conversely, let us now first perform the σ-transformation and then the ω-transformation. Any square of the table will be moved to the same place by either operation (together with the sign in it). We have composed two translations that are perpendicular to each other and two such translations can be interchanged. This fact is expressed by writing $\omega\sigma = \sigma\omega = \tau$; the composed operation will be called a τ-*transformation*. Clearly, an ω-transformation is also a τ-transformation being the composite of itself with the σ-transformation that fixes all the rows. Similarly, a σ-transformation is also a τ-transformation.

If we now apply the particular ω- and σ-transformations given above to our example Σ_1 it is readily seen that the sign pattern of Σ_1 is transformed into the sign pattern of Σ_2. This is expressed by writing

$$\omega\sigma(\Sigma_1) = \sigma\omega(\Sigma_1) = \tau(\Sigma_1) = \tau(\Sigma_1) = \Sigma_2.$$

Thus, we arrive at the fact that Figs 1 and 2 are in essence two different forms (realizations) of the same combinatorial structure; i.e. Fig. 2 realizes a Fano plane as well.

Definition: *If an axiom system* **I** *holds for two incidence tables and one table can be derived by a τ-transformation from the other, then we say that the two tables (planes) are tables that are not isomorphic but are said to be heteromorphic.*

The definition refers to incidence tables that satisfy **I**. In general an incidence table is an $m \times n$ table consisting of squares which are either empty or contain a sign •. Later on, we shall see incidence tables (e.g. Figs 72 and 73, pp. 180—181), which do not represent finite planes. In such cases we also say two tables are isomorphic if one can be derived from the other by a τ-*transformation* (e.g. the two incidence tables in Fig. 71 p.179). We also give examples of heteromorphic tables as well, for instance the two tables occurring in Figs 17, 72 and 73, pp. 59, 180—181.

Obviously, two incidence tables representing finite planes can only be isomorphic if they are of the same order q. Also, two planes of the same order can be heteromorphic to each other, this will be dealt with more fully later on — but we mention here that the smallest size of table for which this can occur is 91 squares$\times$91 squares where 7371 of the squares are empty and 910 are occupied by an incidence sign. Given two tables like this, it would be far from easy to show that they are heteromorphic.

Do two heteromorphic planes of order 2 exist? Before dealing with this question, we add the following results to the theorems enumerated in **1.1**.

(h) An incidence table representing a finite plane cannot contain an incidence sign in every one of the 4 squares that form the intersection of two rows and two columns. (I.e. it cannot contain a pattern corresponding to Fig. 3.)

For if the incidence table contained such a pattern, it could not satisfy the requirements either of axiom I_1 or of I_2 because there would exist two points connected by two distinct lines and two lines intersecting in two distinct points.

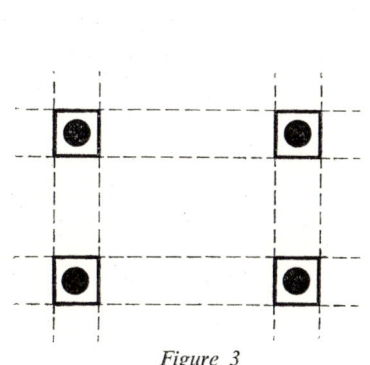

Figure 3

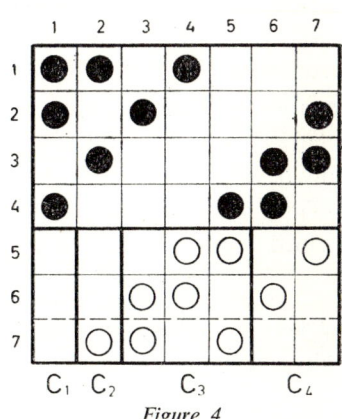

Figure 4

Let us now consider a Σ table representing a plane of order 2. The arrangement shown in Fig. 4 either holds for the signs occurring in the first row of Σ or it can be obtained by an appropriate ω-transformation. Let us take from the transformed table $\omega(\Sigma) = \Sigma_1$ a row containing in its first square the sign •. This is either the second row or it can be made the second row by a suitable σ-transformation leaving the first row invariant. By applying an appropriate ω-transformation to the table $\sigma(\Sigma_1) = \Sigma_2$ we can, in view of Theorem *(h)*, obtain a table Σ_3 which has the same sign pattern in the first two rows as that in Fig. 4. Thus it is easy to see that we can obtain by ω- and σ-transformations from our original table a sign pattern agreeing in its first four rows with Fig. 4. But the sign pattern of the last three rows will also be transformed in some way. Let us separate the upper part formed by the dark signs • from the lower part consisting of 7×3 squares and let us divide the latter, as can be seen in the figure by vertical lines into the parcels C_1, C_2, C_3 and C_4. Let us use the light sign ○ to denote incidence in the lower part of the table.

Now, the upper part of the first column of the table contains three signs • so C_1, being the lower part of this column, can contain no sign ○. The parcel C_2 contains only one sign ○ and this can be moved by a suitable σ-transformation which leaves the first four rows invariant, into the last row. In this row two further signs o occur and these can be situated, according to Theorem *(h)*, only in the way shown in Fig. 4. Let us now darken the sign of row 7, indicating the fixation of the sign pattern in this row.

Now we have only to examine the arrangement of the $7 \cdot 3 - 5 \cdot 3 = 6$ signs distributed over the $2 \cdot 5 = 10$ squares of the 5th and 6th rows of C_3 and C_4. According to Theorem *(h)* C_4 cannot contain two adjacent signs, but it must still contain one sign o in each column — because of the two fixed signs ● in each. These two signs o are either situated like those in Fig. 4, or they can be brought in such a position by interchanging rows 5 and 6. Let us now darken these signs as well.

We have yet to determine the arrangement of the 4 signs falling into the 6 squares of the 5th and 6th rows of C_3. It is, however, easy to see by virtue of Theorem *(h)* and the arrangement of the signs already fixed that these 4 signs can only be arranged as in Fig. 4.

Thus, any incidence table representing a plane of order 2 can indeed be transformed into the form given in Fig. 2 by means of interchanging rows and columns. In other words, we have the following

Theorem: *Planes of order 2 are indentical up to an isomorphism.* (I.e., essentially, there exists only one kind of second order plane.)

1.3 Construction of finite planes, cyclic planes

For the time being we can realize only the classical projective plane and the plane of order two among the finite projective planes; in fact, we can only prove the existence of these two by the construction of models satisfying the axiom system **I**. Planes of order q are known to exist for certain integers $q(>1)$; however, the questions of existence in general and the representation of all finite planes are unanswered. These are, perhaps, among the most difficult unsolved problems of mathematics today.

In what follows, we shall discuss some simple constructions giving rise to certain known finite planes. The method will be introduced for the case $q=4$, but in such a way that, following its example we shall easily be able to construct finite planes in the cases of $q=2, 3, 4, 5$ as well. (Moreover, by using a certain algebraical procedure, our method can be applied to the construction of finite planes for an infinity of positive integers q.)

Let us take

$$n = q^2 + q + 1 = 2\binom{q+1}{2} + 1$$

points on a circle k, centre O, which divide the circle into equal arcs; these n points are the vertices of a regular n-gon $P_1 P_2 ... P_n = \Sigma(n)$, say. Let ω denote the angle subtended at O by a side of the polygon, thus $n\omega = 2\pi$. Figure 5 shows the 21-gon, $\Sigma(21)$, corresponding to the case $q=4$. The chords obtained by joining distinct vertices of the polygon will be of $\binom{q+1}{2}$ different lengths, this being the number of chords of different lengths meeting at a single vertex. The number of the point-pairs which can be formed from $q+1$ points is also $\binom{q+1}{2}$. There is an obvious question, namely:

Is it possible to choose $q+1$ of the vertices of $\Sigma(n)$ so that all the chords obtained by joining pairs of these points are of different lengths?

Obviously, this requirement cannot be fulfilled if more than $(q+1)$ vertices are chosen, i.e. there will certainly occur, among the chords spanned by them, two of equal length. If $q+1$ vertices can be chosen, fulfilling our requirements, then the convex $(q+1)$-gon determined by them will be denoted by Λ. In view of the definition and the remark, above, we shall call the subpolygon Λ, and from the definition that in the case of when it exists, the *completely irregular subpolygon with the greatest number of vertices of the regular polygon $\Sigma(n)$*.

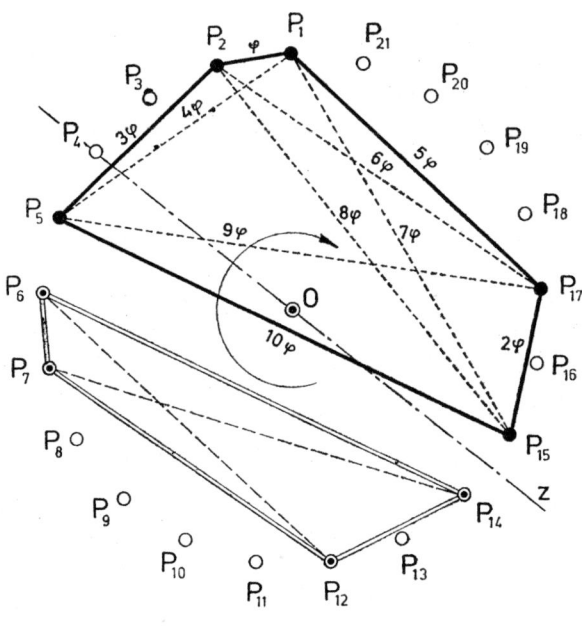

Figure 5

We give an example to show that the concept of the polygon Λ is not an empty one. In the figure we choose the pentagon $\Lambda = P_1 P_2 P_5 P_{15} P_{17}$ from the polygon $\Sigma(21)$; the vertices of this pentagon span 10 chords of different lengths. It is known that the subpolygon Λ, corresponding to our definition, does not exist for every integer q; indeed, the smallest number for which a subpolygon like this does not exist is $q = 6$. For the case of $q = 6$, we can also express this fact as follows: *Among the 21 chords spanned by the vertices of any of the sub-heptagons of the regular 43-gon there exist always two of equal length.* The proof of this statement will not be given here. It is comparatively easy to find a subpolygon Λ in $\Sigma(n)$, if $q = 2, 3, 4, 5$; trial and error, the most elementary method that exists, leads to the choices of the vertices of such a subpolygon. But, as q increases, the number of cases to be tested increases too rapidly for this method to be practical. Certain algebraical procedures make trial and error unnecessary; these, however, cannot be dealt with in this introductory chapter.

Suppose, for some n, the subpolygon Λ of $\Sigma(n)$ exists. Consider the following chain of polygons $\Lambda_1, \Lambda_2, ..., \Lambda_n$; where $\Lambda_1 = \Lambda$ and Λ_k is the polygon derived from Λ_1 by a rotation about the point O through an angle of $(k-1)\omega$. (The sense of the rotation being as on the figure, opposite to the ascending order of the subscripts.) The relation of the point set $\{P_1, P_2, ..., P_n\}$ of n elements to polygon set $\{\Lambda_1, \Lambda_2, ..., \Lambda_n\}$ of the same number of elements is characterized by the following statements:

A$_1$ *If $1 \leq j, k \leq n$ and $j \neq k$, then there exists one and only one polygon Λ_r, which has both the points P_j and P_k as vertices.*

A$_2$ *If $1 \leq j, k \leq n$ and $j \neq k$, then there exists one and only one point P_s which is vertex of both the subpolygons Λ_j and Λ_k.*

A$_3$ *There exists a subquadrangle $\Pi = P_i P_j P_h P_k$ of the polygon $\Sigma(n)$ of which no more than two vertices can also be vertices of any one of the subpolygons $\Lambda_1, ..., \Lambda_n$.*

The basis of the proof of these statements is the fact that each of the chord lengths occurring in the polygon $\Sigma(n)$ occurs also in each the polygons Λ_k as the length of a single chord. Since

$$n = q^2 + q + 1 = q(q+1) + 1$$

is an odd number the rotation, in the sense indicated above, which carries the element of index j into that of index k, $j \neq k$, will be through a different angle to that which carries the element of index k into that of index j. We may assume, without loss of generality, that the angle of the rotation carrying the element of index j into that of index k is the smaller one.

Now in order to prove statement A_1 let us consider the chord of the polygon Λ_1 which has a length equal to $P_j P_k$. It is easily seen that this chord is carried into the chord $P_j P_k$ only by a rotation, in the given sense, through an angle of $l\omega < 2\pi$, where $0 \leq l \leq n-1$. Therefore Λ_{1+l} is the only polygon fulfilling the requirements of A_1.

In order to prove statement A_2 let us consider the angle $l\omega$, the angle of the rotation carrying Λ_j into Λ_k. The polygon Λ_j has only one chord which subtends the angle $l\omega$ at O. The rotation in question carries one endpoint of this chord into the other, but this second endpoint, because of the equality of the angle of rotation and the angle subtended at O is also a vertex of Λ_k, thus it is a common vertex of Λ_j and Λ_k.

For the proof of statement A_3 let us choose any polygon Λ_s and reflect it in a diameter passing through any of the points P_i. Any subtriangle of the image, Λ'_i so obtained is of opposite orientation to the triangle from which it was derived by the reflection, thus there is only a single side of the original triangle which can be carried into its mirror image by a rotation. From this it this follows that every subquadrangle of the polygon Λ'_s satisfies the statement A_3. — Of course, if $q=2$, the polygon Λ_s is only a triangle and our proof no longer holds. But, then take $\Lambda_s = \{P_1 P_2 P_4\}$ and so $\Pi = \{P_1 P_2 P_5 P_7\}$ satisfies the statement A_3.

Now, let us call the polygon $\Sigma(n)$ a "plane". The vertices $P_1, P_2, ..., P_n$ of the polygon are the "points" of the plane; the subpolygons (subsets) $\Lambda_1, \Lambda_2, ..., \Lambda_n$ are the "lines" of the plane. The system of A_1, A_2 and A_3 is a transcription of the axioms l_1, l_2 and l_3 in our model. Thus, it can be seen from this that we have given a geometrical-combinatorical construction for the realization of the finite plane of order q. The value of this method is not lessened by the fact that the systematical trials involved become difficult, even practically impossible, when q is large; however, this urges us to look for a complementary mathematical idea which makes the trials unnecessary. We shall return to this problem later on.

We shall give here a list of the vertices of a polygon Λ_1 for some values of q (for brevity, we have written 1, 2, 4 for $P_1 P_2 P_4$, etc.):

if $q = 2$, then $n = 7$ and $\Lambda_1 = 1, 2, 4$;
if $q = 3$, then $n = 13$ and $\Lambda_1 = 1, 2, 5, 7$;
if $q = 4$, then $n = 21$ and $\Lambda_1 = 1, 2, 5, 15, 17$;
if $q = 5$, then $n = 31$ and $\Lambda_1 = 1, 2, 4, 9, 13, 19$;
if $q = 6$, then $n = 43$ and $\Lambda_1 =$ non-existent;
if $q = 7$, then $n = 57$ and $\Lambda_1 = 1, 2, 4, 14, 33, 37, 44, 53$;
if $q = 8$, then $n = 73$ and $\Lambda_1 = \{1, 2, 4, 8, 16, 32, 37, 55, 64\}$;
if $q = 9$, then $n = 91$ and $\Lambda_1 = \{1, 2, 4, 10, 28, 50, 57, 62, 78, 82\}$.

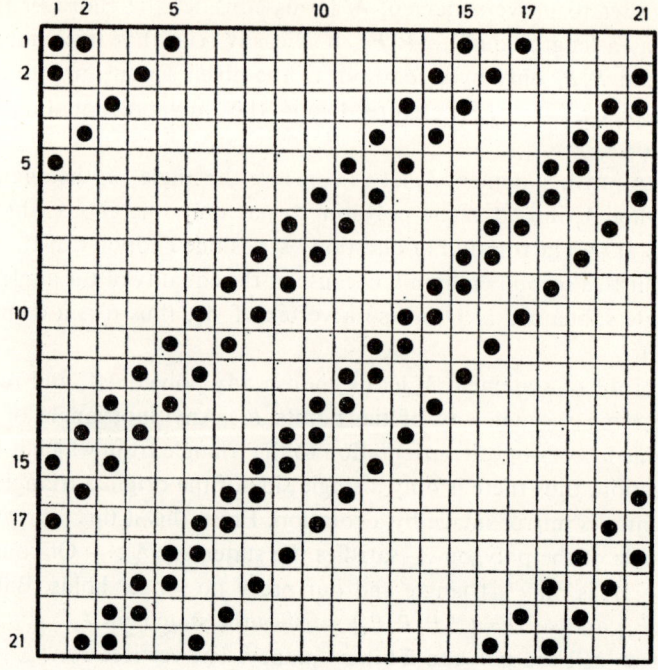

Figure 6

Thus, in possession of this list, we are already able to study seven kinds of finite plane geometries. We can easily compile the incidence table in each case. If we take, for instance the case of $q=4$, i.e. Fig. 5, we obtain the table given in Fig. 6.

This table — to use expressions borrowed from the theory of matrices — is *cyclic* and *symmetrical* (with respect to the principal diagonal). The cyclic incidence table representing a finite plane will, henceforth, be called an Ω-*table*, or briefly $\Omega(q)$. Thus Fig. 2 is just the table Ω (2).

1.4 The Γ-table of a finite projective plane

Among the (isomorphic) incidence tables which represent a finite plane some are more suitable than others for the clear expression of the incidence structure of the plane, as can be seen by comparing Figs 1 and 2, and this is more forcibly the case when q is large. We have shown, when $q=2$, that any incidence table can be transformed into a cyclic table (Fig. 1). However, there are incidence tables

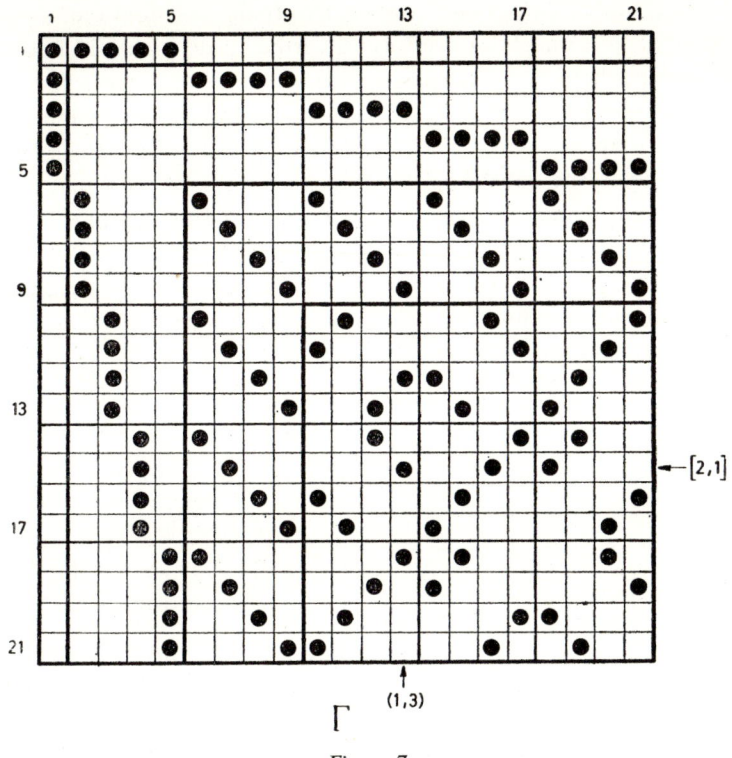

Figure 7

which cannot be transformed into cyclic ones. — We omit here the figure demonstrating an example of this since the smallest table of this kind is a 91×91 table. — Of course, there are other forms of tables, expressing clearly the structural properties of a plane, into which any incidence table of the plane can be rearranged by means of ω- and σ-transformations. We shall now present a scheme of this type.

Consider an empty $n \times n$ table, where $n = 1 + q + q^2$. Let us decompose the table by emphasizing certain dividing lines as is shown in Fig. 7. The first row and the first column are separated from the remainder by a thick line. Thus there emerges a narrow hook or margin or border, having the shape of a Γ. Consider the part of the incidence table filling the "hollow" of this Γ-hook and consisting of $(n-1)^2$ squares; this may be divided into $q+1$ columns each q squares wide, called *columnbands*. Similarly, it may be divided into $q+1$ rows each q squares high, called *rowbands*. By intersecting each columnband with each rowband we obtain $(q+1)^2$ *parcels* of q^2 squares each (emphasized by the thick lines in the figure). Now we form a wide Γ-shaped hook consisting of the first rowband and the first columnband of the subtable filling the hollow of the original Γ-hook and, similarly, in the hollow of this second hook we form a further wide Γ-shaped hook. In our figure the 3 nesting hooks, separated from each other by thick lines, form together a *complete Γ-hook,* having a width of $(2q+1)$ squares. The subtable filling the hollow of this complete Γ-hook and consisting of $(q-1)^2$ parcels will be denoted by Θ. Moving from the outside to the inside, we speak of the *outer, middle* and *inner Γ-hooks* and of the Θ-*hollow*; the part consisting of the inner Γ-hook and of the Θ-hollow is said to be the $\Gamma\Theta$-*hollow*.

An empty table, divided as above, will be called a Γ-*lattice*. This lattice reflects the decomposition of the number n according to the formula $n=1+q+q+(q-1)q$.

Now by rearranging the incidence table $\Omega(4)$ of Fig. 6 by the following ω and σ-transformations:

$$\omega = \begin{pmatrix} 1 & 2 & 3 & 4 & 5 & 6 & 7 & 8 & 9 & 10 & 11 & 12 & 13 & 14 & 15 & 16 & 17 & 18 & 19 & 20 & 21 \\ 1 & 2 & 5 & 15 & 17 & 21 & 16 & 14 & 4 & 7 & 6 & 20 & 10 & 11 & 18 & 19 & 13 & 8 & 3 & 12 & 9 \end{pmatrix}$$

and

$$\sigma = \begin{pmatrix} 1 & 2 & 3 & 4 & 5 & 6 & 7 & 8 & 9 & 10 & 11 & 12 & 13 & 14 & 15 & 16 & 17 & 18 & 19 & 20 & 21 \\ 1 & 2 & 17 & 5 & 15 & 16 & 21 & 4 & 14 & 18 & 11 & 13 & 19 & 3 & 8 & 9 & 12 & 6 & 7 & 10 & 20 \end{pmatrix}$$

it is easy to check that we obtain the table, given on a Γ-lattice, in Fig. 7. Let us observe in Fig. 7 how the incidence signs • are distributed with respect to the Γ-lattice.

In the upper left hand corner of the outermost hook a Γ-shape is formed by $1+2q$ signs. Consequently, the upper left hand parcel of the middle Γ-hook is empty and in the remainder of this hook the signs form a stepped pattern, as shown in the figure. In the parcels of the inner hook the incidence signs fill the squares of the main diagonal. In each parcel belonging to Θ the signs form a diagonal pattern and $q-1$ different patterns occur. By a "diagonal pattern" we mean that there is one and only one sign in every row and in every column of the parcel.

An incidence table so arranged is said to be a Γ-table and is denoted by $\Gamma(q)$.

This description is, of course, not a definition; it is given merely to emphasize a particular structure. The requirements defining a Γ-table are the following:

1° *In the outer and the middle hooks the sign pattern is guided by the Γ-lattice and is of (sectionally)-stepped nature.*

2° *The same sign pattern occurs in each parcel of the inner hook and this pattern is one of the $q!$ different diagonal types.*

Clearly, $\Gamma(q)$ is by no means unique, since, for example, a τ-transformation of the table leaving invariant all the rows and columns with the exception of those passing through an arbitrary parcel preserves the defining properties 1° and 2°. The example above shows that an $\Omega(q)$ table can be transformed into a $\Gamma(q)$ table, but not every $\Gamma(q)$ can be transformed into an $\Omega(q)$ table and we have the following

Theorem: *An incidence table representing a finite plane can be rearranged by appropriate ω- and σ-transformations — i.e. by an appropriate τ-transformation — into a $\Gamma(q)$ table.*

Our proof rests mainly upon the fact that a finite plane cannot contain an incidence sign quadruple of the kind shown in Fig. 3. For the sake of brevity this excluded configuration will be called a *sign-rectangle*.

First we rearrange the signs in the outer and middle Γ-hooks, by appropriate ω and σ-transformations, so that we obtain the (sectionally)-stepped form shown in Fig. 7 and also in picture *i* of Fig. 8. That such a rearrangement is possible follows easily from the properties l_1, l_2 Q, s of the incidence table, and furthermore from the fact that there is no sign rectangle (because of l_1, l_2).

We now introduce once and for all the notations

$$\begin{array}{cccc} C^{00}, & C^{01}, & \ldots, & C^{0,q-1}; \\ C^{10}, & C^{11}, & \ldots, & C^{1,q-1}; \\ \ldots & \ldots & \ldots & \ldots \\ C^{q-1,0} & C^{q-1,1} & \ldots, & C^{q-1,q-1} \end{array}$$

for the q^2 parcels filling the hollow of the middle Γ-hook, i.e. the $\Gamma\Theta$-hollow.

In picture *i* of Fig. 8 we leave empty all the parcels C^{rs} to illustrate the first phase of the rearranging. Of course, certain patterns of incidence signs are also formed in the empty parcels of the figure by the first phase of the reordering in fact, the • signs became already diagonally arranged, in any C^{rs}, by the first reordering i.e. there is a single • sign in any row or column of the parcel.

This follows because two • signs cannot occur in one row (or column respectively of a parcel, since otherwise these together with the two signs in the corresponding row column) of the middle Γ-hook would form a sign rectangle, which is impossible. However, the aforementioned row (column) in the parcel cannot be empty, either; since otherwise it would form together with a row (column) of the Γ-hook a pair of rows (pair of columns), which would not satisfy l_2 (respectively l_1) and this is impossible. Thus, there is indeed one and only one • sign in any of the rows and in any of the columns of the C^{rs}.

Now in order to satisfy property 2 of our definition for the inner hook we continue the reordering of the table, leaving invariant the middle Γ-hooks.

Under the initial transformation the sign patterns (cf. picture *i* of Fig. 8) of the parcels of the inner Γ-hook may not necessarily be alike; but, we can make

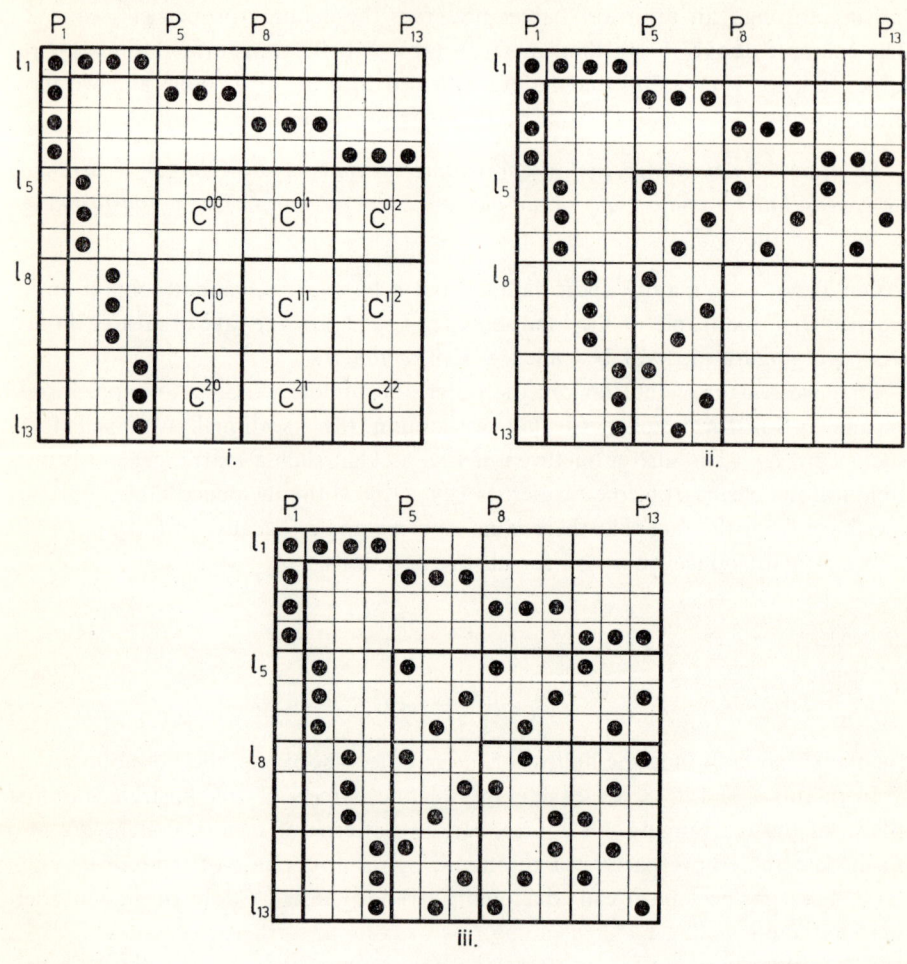

Figure 8

a rearrangement so that the sign pattern in each parcel is the same and, furthermore, this pattern can be arbitrarily chosen from the $q!$ different diagonal forms. Let us, for example, choose the form shown in picture *ii* of Fig. 8.

Consider first any one of the parcels

$$C^{00}, \quad C^{01}, \quad \ldots, \quad C^{0,q-1}$$

say the parcel C^{0s}. The diagonal pattern sign of the parcel is carried into the prescribed form by an appropriate ω-transformation restricted to the columns of the complete table which pass through this parcel. Thus, we arrive at the prescribed sign pattern in each parcel of the upper arm of the inner Γ-hook.

Now, we have still to transform the sign pattern of the parcels

$$C^{10}, \quad C^{20}, \quad \ldots, \quad C^{q-1,0}.$$

into the prescribed form. Consider the parcel C^{r0}; the diagonal sign pattern in this can be rearranged into the required form by a σ-transformation restricted to the rows passing through this parcel, similarly for the other parcels above. With this second phase of the reordering we have finished the proof of our theorem.

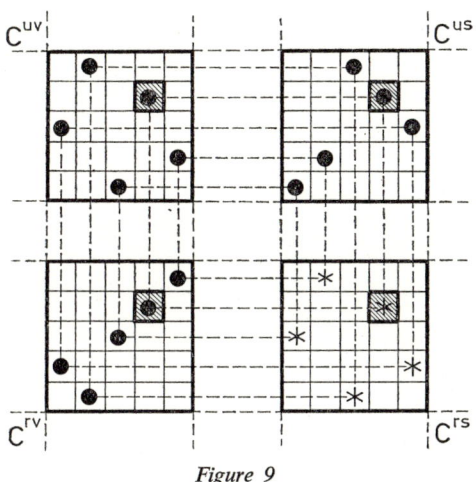

Figure 9

Under the second transformation (picture *ii* of Fig. 8) the sign patterns of the parcels in Θ were also changed into a certain diagonal pattern. We can only make a few general remarks about this. Before discussing these, however, we add a remark related to Fig. 9.

The *homologous elements* (homologous associates) of a square in the $\Gamma\Theta$-part of an incidence table are those squares which can be derived by a translation of the parcel containing the square into another parcel. Thus, for example, the dark squares in Fig. 9 are homologous associates. The figure consisting of four parcels, of the $\Gamma\Theta$-part, which form the intersection of two rowbands with two columnbands, will be called a *parcel rectangle* — by analogy with a sign rectangle. Suppose that the same sign pattern is given in each of the parcels C^{uv}, C^{us} and C^{rv}; then, we certainly know that there are q squares in C^{rs} that contain no sign, since otherwise sign rectangles would occur. In our figure these necessarily empty squares are denoted by $*$.

We make now the following general statements concerning the parcels of the part $\Gamma\Theta$:

D_1 Each of the sign patterns of the parcels C^{rs} is of a diagonal form

D_2 If the first or the second indices of two parcels agree and are non-zero, then their sign patterns differ

D_3 If a rowband or a columnband does not contain the parcel C^{00} and if the sign patterns of the parcels contained in it are superimposed onto one parcel then they will exactly fill this parcel

D_1 was already verified above, D_2 and D_3 are obvious on the basis of our remarks concerning Fig. 9.

We encounter now the following natural question: The number of the parcels C^{rs} is q^2, let the number of the different sign diagonals occurring in the q^2 parcels be denoted by δ, obviously $\delta < q^2 + 1$. What else can we say about the relation of q to δ? We know that $q = \delta$ is a possible case; samples of it are Fig. 7 and picture *iii* of Fig. 8. Clearly, $q > \delta$ since otherwise there would occur two parcels of the same sign pattern in any rowband; which, according to D_2, is impossible. Another example is also known, we do not give it here having, as it does an incidence table of 91×91 squares, for which $q = 9$ and $\delta = 27$. Since the parcels of the inner Γ-hook can be arranged with the same sign pattern in each parcel, our statements can be summarized as follows:

D_4: $q \leq \delta \leq (q-1)^2 + 1$.

We shall illustrate the utility of the $\Gamma(q)$ table by proving the following

Theorem: *There exists only one kind of the plane of order 3.*

The proof of the theorem consists of rearranging the incidence table of any plane of order 3 into a unique form. This can be achieved by first transforming the sign pattern of the outer and the middle Γ-hooks into the form given in picture *ii* of Fig. 8. In each of the parcels C^{11}, C^{12}, C^{21} and C^{22}, according to the properties $D_1 - D_4$, 3 incidence signs must be distributed among the six free squares in a diagonal arrangement. There are only two ways in which we can form a diagonal arrangement in the six free fields of a parcel. Let these be denoted by A and B. These occur in the four parcels as follows:

either $\begin{pmatrix} A & B \\ B & A \end{pmatrix}$ or $\begin{pmatrix} B & A \\ A & B \end{pmatrix}$

and the second of these can be transformed into the first, namely by

ω: 8, 9, 10, 11, 12, 13 → 11, 12, 13, 8, 9, 10

and

$$\sigma: \qquad 3, 4 \to 4, 3$$

referring to the columns and to the rows of the complete table, respectively.* Thus, we see that any incidence table representing a plane of order 3 can be changed by a τ-transformation into the form occurring in picture *iii* of Fig. 8.

1.5 Coordinate systems on the finite plane

We introduce now a new notation for the squares of $\Gamma\Theta$, composed of the Γ-hook and of the Θ-part of the incidence table $\Gamma(q)$; which will distinguish each square and also the parcel to which each belongs.

In the beginning we used the notation $(n=1, 2, \ldots, q^2+q+1)$ for the rows $l_1, l_2, \ldots, l_n$ and the columns $P_1, P_2, \ldots, P_n$ of a Γ-table. What follows applies only to the rows $l_{q+2}, \ldots, l_n$ and the columns $P_{q+2}, \ldots, P_n$. If we denote the rows which pass through a parcel by $0, \ldots, q-1$ (where 0 denotes the top row, 1 the second row, etc.); then the bth row passing through the parcel $C^{m,0}$ will be denoted by $[m, b]$. Similarly, if we denote the columns which pass through a parcel by $0, \ldots, q-1$ (running from the left hand column to the right hand column); then the yth column passing through C^{0x} will be denoted by (x, y). The square at the intersection of the row $[m, b]$ with the column (x, y) will be denoted by

$$\langle m, b; x, y \rangle.$$

The Γ-table of a finite plane establishes a *ternary operation* on the set $0, 1, \ldots, q-1 = \mathbf{Q}$, as follows. The table assigns an index y to every ordered triple (x, m, b) of indices: namely, by specifying x, m and b we restrict our attention to the segment of the row $[m, b]$, of $\Gamma\Theta$, falling into the parcel C^{mx} and this segment can contain one and only one incidence sign $\bullet$ (by D_1,) which occurs in the yth row, say. This operation is defined for every triple of elements (x, m, b) of the set $\mathbf{Q}$. In this sense we may speak of a ternary operation

$$F(x, m, b) = y$$

defined on the set of indices $\mathbf{Q}$. Thus, for instance, in Fig. 7 we can say that

$$F(1, 2, 1) = 3,$$

since there is a sign $\bullet$, in the square at the intersection of the row $[2, 1]$ with the column $(1, 3)$.

---

* The notation $a, b, c, \ldots k, l, m, \ldots$ means the same thing as $\begin{pmatrix} a, b, c, \ldots \\ k, l, m, \ldots \end{pmatrix}$.

As the rows of the table represent lines and the columns points, we may speak of the point (x, y) and of the line $[m, b]$, furthermore of *point coordinates* x, y and of *line coordinates* m, b. Obviously, the points $P_1, \ldots, P_{q+1}$ and the lines $l_1, l_2, \ldots, l_{q+1}$ are outside this coordinatization. However, a coordinatization like this is still useful, as will be seen by the discussion of an analogy.

Let us draw a comparison between the Cartesian point coordinates of the Euclidean plane and the point coordinates of the finite plane introduced above. On the Euclidean plane we can assign an ordered pair of coordinates to every line, except those parallel to the y-axis, in the following way. We assign the pair $[m, b]$ to the line of slope m passing through the point $(0, -b)$. Thus the Euclidean plane establishes a ternary operation on the set of the real numbers by means of the lines not parallel to the y-axis, i.e. by subsets of the point plane. The role of the index set **Q** is fulfilled in this case by the set of real numbers and the role of the ternary operation $F(x, m, b)$ is taken over by the arithmetic operation

$$F(x, m, b) = y: \quad xm - b = y.$$

Geometrically, we assign to the triple of numbers (x, m, b) the ordinate of the point of the line $[m, b]$ having x for its abscissa. — A line parallel to the y-axis can be included in this description by saying that it is the set of the points having the property $x = c$. The Euclidean plane can be extended into the projective plane, if we augment it by the ideal points, i.e. by the ideal line formed by the ideal points. Every line is extended by a single ideal point in such a way that parallel lines are extended by the same point.

The analytic geometry of a finite plane as obtained by means of the index set **Q** is closely analogous to the analytic geometry of the Euclidean plane. Upon this basis we can extend our discussion to the point set $P_1, P_2, \ldots, P_{q+1}$ and to the line set $l_1, l_2, \ldots, l_{q+1}$ omitted thus far.

On the finite plane the role of the ideal points is taken over by the points $P_1, P_2, \ldots, P_{q+1}$ i.e. by the points to which we did not assign a pair of coordinates (x, y). These are the points forming set $l_1 = \{P_1, P_2, \ldots, P_{q+1}\}$ which plays the role of the ideal line. This analogy has the following consequences:

1° The role of the y-axis is taken over by $l_2 = \{P_1, P_{q+2}, P_{q+3}, \ldots, P_{2q+1}\}$
2° The role of parallel lines defining the ideal point P_r is taken over, on the finite plane, by the lines corresponding to the rows which pass through the parcel $C^{r-2, 0}$ $(r = 1, 2, \ldots, q+1)$.*

---

* We extend the notation $C^{r, s}$ to the parcels outside $\Gamma\Theta$ by allowing the indices to take the value -1.

3° The role of the x-axis is taken over by the line $l_{q+2} = \{P_2, P_{q+2}, P_{q+3}, \ldots \ldots, P_{2q+1}\}$

4° The points P_{q+2}, P_{2q+3} correspond to the origin and the unit point $(1, 1)$, of the Euclidean plane, respectively.

On the Euclidean plane we can also determine the set of (ordinary) points incident with the line $[m, b]$ by the linear equation

$$y = mx - b.$$

On the finite plane there corresponds to this the equation

$$y = F(x, m, b)$$

defined by the ternary operation $F(x, m, b)$.

The analogy in question leads also to a type of a graphical model of a finite plane, which often occurs in the literature (mostly without any explanation as to what it represents).

We shall use the analogy discussed above to obtain a model corresponding to Fig. 7. This model is shown in Fig. 10 (not completely drawn). The q^2 (ordinary) points labelled by a pair of indices (x, y) are represented by lattice points on the Euclidean plane. These lattice points are produced as the points of intersection of the equidistant lines $l_2, l_3, \ldots, l_{q+1}$ and the equidistant lines $l_{q+2}, l_{q+3}, \ldots, l_{2q+1}$ perpendicular to them. Thus we represented $2q$ of the lines of the finite plane by

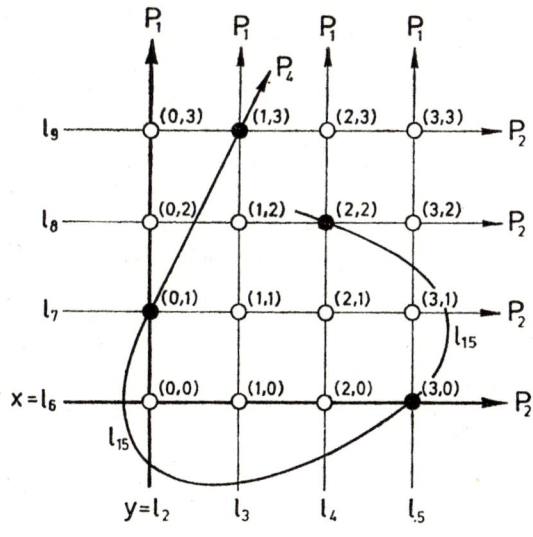

Figure 10

lines on the Euclidean plane. However, the other lines of the finite plane cannot, in general, be visualized by lines of the Euclidean plane. In our figure, for example, the curve l_{15} represents the row l_{15} of Fig. 7 and P_4 written at the point of the arrow indicates that the ideal point P_4 is to be adjoined to this curve.

If we add, to the two families of parameter lines drawn already, as a third family of lines, the curves with the common ideal point P_{k+1} which represent the rows

$$l_{kq+2}, \quad l_{kq+3}, \quad \ldots, \quad l_{(k+1)q+1}$$

we shall obtain a figure resembling a curvilinear net. Moreover, if we consider only those points of the curves that are images of points in the finite plane we obtain a figure of q^2 points; a "curvilinear nomogram" of the *family of functions* $y = F(x, k-1, b)$, one function for each pair of values parameters b and $m = k-1$. If we draw the image of every line of the finite plane (of course, with the exception of l_1), then the figure obtained can be considered as $q-1$ individual curvilinear nets.

The complete figure, even for the case $q = 4$, is highly complicated. Therefore we shall prefer to use the Γ-table, as illustration is easier for this model.

1.6 The concepts of Galois planes and Galois fields

The close connection between the analytic geometry of finite planes and ordinary analytic geometry leads to the notion of a special kind of finite plane. We need here first of all the notions of groups and fields. If the number of elements in a field is finite, then we speak of a *finite field* or *Galois field*.* We shall now discuss the following example of a finite field.

Consider the Galois field of 4 elements; it is denoted by $GF(4)$. We shall specify this field by giving its operation tables. Let its four elements be denoted by a_0, a_1, a_2 and a_3 respectively, furthermore, let a_0 be the additive indentity (zero element) and let a_1 be the multiplicative identity (unity element) of the field. The addition table (A) and the multiplication table (M) are (writing 0 for a_0, etc.):

+	0	1	2	3		×	0	1	2	3
0	0	1	2	3		0	0	0	0	0
1	1	0	3	2		1	0	1	2	3
2	2	3	0	1		2	0	2	3	1
3	3	2	1	0		3	0	3	1	2

$$A \qquad\qquad M$$

---

* Cf. Appendix: 7.1, 4° and the Fundamental Theorem in 7.2.

It is easy to check by means of these tables that the addition and the multiplication so defined satisfy all the requirements of the field axioms. Of course verification is tedious, even in the case of a field of so few elements. For instance in order to check whether the distributive law holds in the single case $(2+3)3 = = 2 \cdot 3 + 3 \cdot 3$ we have to perform the following work. According to table A $2+3=1$, according to table M $1 \cdot 3 = 3$, thus on the left hand side we have the element a_3. According to table M $2 \cdot 3 = 1$ and $3 \cdot 3 = 2$, according to table A $1 + 2 = = 3$, hence we also have 3 on the right hand side. Thus the distributive law holds for the triple of elements a_2, A_3, A_5.

We shall discuss later how tables A and M can be produced for any Galois field. Here we shall only note that the number of elements in a Galois field must be a *power of a prime* and there belongs to any number q of this kind essentially only one field. (In our example we dealt with the case $q=2^2$.) The notation for the Galois field consisting of q elements is $GF(y)$.

Let us return to the example given in Fig. 7. In the parcels

$$C^{10}, \quad C^{11}, \quad C^{12}, \quad C^{13}$$

replace each sign • by the second index of the parcel containing it. Then by superposing the four parcels we obtain a 4×4 index table and this is none other than table A belonging to $GF(4)$.

Consider, in Fig. 7, the 4×4 array of parcels forming $\Gamma\Theta$. In the second row let us denote C^{1i} by i, $i=0, \ldots, 3$. Take any other parcel C^{rs}, of $\Gamma\Theta$; this must agree, in sign pattern, with one of the C^{1i}. Let C^{rs} also be denoted by the index of the parcel with which it agrees. In this way we obtain a 4×4 index table the first row and the first column of which contain only zeros. In fact, this table is none other than the table M belonging to $GF(4)$.

These observations suggest the following questions. *Is it possible to construct from the operation tables of any* GF(q) *the Γ-table of a finite plane, or more precisely the $\Gamma\Theta$-part of it?* By answering this question affirmatively we shall gain access to as many finite planes as there are Galois fields (or prime powers).

Let $q=p^r$ be a given prime power and assume that we have the operation tables A and M of the corresponding field $GF(q)$. Let the elements of the field be denoted by the indices $0, 1, 2, \ldots, q-1$. An essential requirement is that 0 and 1 should denote the zero element and the unity element of the field, respectively; but the choice of the indices of the remaining elements is arbitrary. We enter signs • into the Γ-lattice of $\Gamma(a)$ as follows.

1° In the outer and the middle Γ-hooks form the segmental (sectional) and stepwise sign patterns corresponding to the definition of the Γ-table.

2° We shall establish a correspondence between the parcels of $\Gamma\Theta$ and the squares of the table M in the following way: let the parcel C^{rs} correspond to the square at the intersection of the row of index r and the column of index s in the table M. Let the index occupying this particular square be: t ($rs=t$). Now we locate in table A the squares containing the index t and put a sign • into the corresponding squares of the parcel C^{rs}.

The incidence table constructed in this manner will be denoted by $\Gamma(GF(q))$. The complete sign pattern of this table is symmetrical with respect to the principal diagonal. This follows immediately for the outer and middle Γ-hooks and is true for the sign pattern in $\Gamma\Theta$ because, as this is derived from the tables A and M, it will be symmetrical if and only if both the operation tables A and M are symmetrical. But this is so, as both of the operations are commutative.

We have now to prove that the incidence table $\Gamma(GF(q))$ actually represents a finite projective plane of order q.

Clearly, axiom I_4 holds, since in the first row of $\Gamma(GF(q))$ there are $q+1$ incidence signs.

In order to check axiom I_3 it suffices to consider the columns $P_2, P_3, P_{q+2}, P_{q+3}$. We have to find six rows such that each of them intersects two of the four columns in squares containing incidence signs. This requirement is satisfied by the following rows:

$$l_1, l_2, l_{q+2}, l_{q+s+2}, l_{2q+2}, l_{2q+s+2}.$$

In view of the symmetry of the incidence table $\Gamma(GF(q))$, it suffices to verify one of the axioms I_1 and I_2. Therefore, we shall consider two arbitrarily chosen rows and we shall show that there exists one and only one column which intersects each of the two rows in a square containing an incidence sign.

If we look back to Fig. 7 and Fig. 8 *iii* we can readily see, for certain pairs of rows, the unique column satisfying the above requirement.

1° If the two rows are l_r, l_s where $r \neq s$ and $r \leq q+1, s \leq q+1$, then column P_1 is the one in question.

2° If $1 < r \leq q+1$ but $q+1 < s \leq 2q+1$ then the column in question can only be one of the columns passing through the parcel $C^{0,r-2}$, since line l_r only contains incidence signs in these columns; namely, the segment consisting of q signs. Furthermore, the row l_s passes through the parcel $C^{0,r-2}$ and, as the incidence signs form a diagonal pattern, one and only one square of the row can contain a sign • within this parcel. The column required is the one containing this square.

The determination of the column satisfying the above is not so easy when the rows l_r and l_s both pass through $\Gamma\Theta$, i.e. when $r > 2q+1$ and $s > 2q+1$. In this case we first discuss the following question: If the row of index b and the column of index y of the parcel C^{mx} ($0 \leq m, x \leq q-1$) intersect in a square containing a sign •, what kind of relation holds then between the numbers m, b, x, y according to the arithmetic of $GF(x)$? (Another example of the arithmetic in $GF(q)$ is given by the operation tables A and M in Fig. 44; i.e. in the case of the field of $q = 3^2 = 9$ elements.) We shall prove that the relation

$$y = mx - b$$

holds.

In the construction above the parcel C^{mx} corresponds to the square common to the row of index m and the column of index x of table M; this square contains, according to the arithmetic of $GF(q)$, the product $mx = t$. The same t can be found in table A in the squares having row index b and column index y where $b+y=t$ in the addition of the Galois field. Thus in the parcel C^{mx} the square having row index b and column y contains an incidence sign if and only if $b+y=t=mx$, i.e. $y = mx - b$.

Let l_r and l_s be any two rows passing through $\Gamma\Theta$; by using the relation $y = mx - b$ we can determine those squares of l_r and l_s that contain incidence signs. Suppose the coordinates of l_r and l_s are $[m, b]$ and $[m', b']$, respectively. We see that the rows l_r and l_s contain an incidence sign in the same column, (x, y), if and only if

$$y = mx - b \quad \text{and} \quad y = m'x - b'$$

both hold.

The solution of this system of equations over $GF(q)$ can be found by the methods used in common arithmetic (which rely only on the field axioms).

We shall now determine (x, y). Clearly, the following three cases can occur:

a)   If $m \neq m'$ and $b \neq b'$,   then

$$x = (b'-b)(m'-m)^{-1}, \quad y = (mb' - m'b)(m'm)^{-1}.$$

b)   If $m \neq m'$ and $b \neq b'$,   then

$$x = 0, \quad y = -b.$$

c)   If $m = m'$ and $b \neq b'$,   then a solution (x, y) does not exist.

However, this means only that the column intersecting both the rows l_r and l_s, in squares containing an incidence sign, cannot pass through $\Gamma\Theta$. But, because l_r and l_s both pass through the same parcel C^{mx} and every line of this parcel contains an incidence sign $(m+1)$th column. Thus, we have also in case $c)$ a column satisfying the conditions.

We have thus completed the proof of the statement that the incidence table $\Gamma(GF(q))$ fulfils the requirements of axiom I_2.

Therefore we can assign to any field $GF(q)$ by our construction an incidence table $\Gamma(GF(q))$ representing a finite projective plane of order q.

*

We now point out a supplementary result:

The incidence of the line (row) $[m, b]$ *and the point* (column) (x, y) *is equivalent to the equation*

$$y = mx - b$$

defined according to the operations in $GF(q)$.

The finite plane represented by the incidence table $\Gamma(GF(q))$ — briefly $\Gamma(q)$ — and defined over the algebraic structure $GF(q)$ will be denoted by $S_{2,q}$ and will be called a *Galois plane*; $GF(q)$ will be called the *coordinate field* of the plane $S_{2,q}$.

Clearly, the ordered pairs (x, y) of elements (x, y) of the field determine only q^2 points of the plane $S_{2,q}$, namely those which correspond to the columns $P_{q+2}, P_{+3}, \ldots, P_n$ (where $n = q^2 + q + 1$) of $\Gamma(q)$. There are no ordered pairs of elements (x, y) — in other words *pairs of non-homogeneous coordinates* — corresponding to the columns $P_1, P_2, \ldots, P_{q+1}$ of the table; the sign patterns in these columns were not constructed from $GF(q)$, they were prescribed by the standard pattern of the outer and middle Γ-hooks. The $q+1$ points corresponding to these columns are said to be the *ideal points* of the plane $S_{2,q}$.

Similarly, the ordered pairs, $[m, b]$, of elements of the field determine only q^2 lines of the plane, namely the lines corresponding to the rows $l_{q+2}, l_{q+3}, \ldots, l_n$ of the table. No ordered pairs $[m, b]$ of elements correspond to the rows $l_1, l_2, \ldots, l_{q+1}$, these represent the lines of the plane $S_{2,q}$ meeting at the ideal point P_1. The row l_1 of the $\Gamma(q)$ table represents the line containing all the ideal points of the plane $S_{2,q}$; this line is called the *ideal line* of the plane.

We have seen already that the lines defined by the equations $y = mx - b_1$ and $y = mx - b_2$ meet in the point P_{m+1}, that is in an ideal point. In this sense we say that the two lines in question are *parallel*. Similarly, any two of the lines $l_2, l_3, \ldots, l_{q+1}$ are parallel.

In the analytic geometry of the Euclidean plane extended by the ideal elements, the equation

$$y = mx - b$$

is that of a line cutting the y-axis, where m is the slope of the line and $-b$ denotes the distance from the origin (with the appropriate sign) of the point of intersection of the line with the y-axis. Thus a close analogy is revealed between the geometry of the classical projective plane and that of the Galois plane. The equation in question represents a line in the Euclidean plane or in a Galois plane if the coordinate field is taken to be the real number or some $GF(q)$, respectively.

We can also introduce homogeneous coordinates in the plane $S_{2,q}$; these will enable us to include the points $P_1, P_2, \ldots, P_{q+1}$ and the lines $l_1, l_2, \ldots, l_{q+1}$ in the coordinatization. As a matter of fact, the analogy dealt with above enables us to follow the traditional method of introducing homogeneous coordinates. This, however, will not be given here in detail; we mention only the essential statements.

Let the coordinate field be $K = GF(q)$ and let its elements be denoted (with the conventions mentioned before) by $0, 1, 2, \ldots q-1$.

A *homogeneous* coordinate triple (x_1, x_2, x_3) corresponding to the point with (non-homogeneous) coordinates (x, y) is given by: $(x_1, x_2, x_3) = (\lambda x, \lambda y, \lambda)$ where $\lambda \in K$; $\lambda \neq 0$.

The homogeneous coordinates of the points $P_{m+2}(m=0, 1, \ldots, q-1)$ are $(1, m, 0)$, where in these triples the m are considered as elements of the field K.

The coordinate triple corresponding to point P_1 is $(0, 1, 0)$.

Similarly, a homogeneous coordinate triple corresponding to the (non-homogeneous) line with coordinates $[m, b]$ is:

$$[u_1, u_2, u_3] = [\mu m, -\mu, -\mu b] \quad \text{where} \quad \mu \in K, \quad \mu \neq 0.$$

The homogeneous coordinates of the lines $l_{r+2}(r=0, 1, \ldots, q-1)$ are $[1, 0-r]$; where, again, in these coordinate triples the r are considered as elements of the field K.

The coordinate triple corresponding to the line l_1 is $[0, 0, 1]$.

From the above it follows that $(0, 0, 0)$ does not determine a point and $[0, 0, 0]$ does not determine a line; furthermore, point (x_1, x_2, x_3) and line $[u_1, u_2, u_3]$ are incident if and only if

$$u_1 x_1 + u_2 x_2 + u_3 x_3 = 0.$$

The rapid development of Galois geometry is certainly due to the close analogy between the geometry of the Galois plane and the analytic geometry of the classical projective plane, and also to the well developed state of the algebra of finite fields. The theory of Galois planes can today be considered as a closed chapter in the theory of finite geometries.

1.7 Closed subplane of a finite projective plane

In the following discussion the role of axiom I_3 will be emphasized. Indeed, as is shown by the historical development of projective geometry, axioms I_1 and I_2 alone, although expressing the most important properties of the structure of the projective plane, define a much too general structure. During the course of its development attempts were made to restrict the notion of the projective plane, by introducing axioms, so that the theorems of projective plane geometry should not become meaningless because of their excessive generality. Indeed, sometimes the addition of further axioms only complicated the system without leading to a much more general concept than that of the classical projective plane.

We mention two models as examples of planes defined only by the axioms I_1 and I_2.

1° Let the vertices of a triangle in the Euclidean plane be the "points" and its sides be the "lines" of a plane, with the same incidence relation.

2° Let the points $P_1, P_2, \ldots P_k$ ($k>2$) of a line l_0 in the Euclidean plane and a further point P_0 not on l_0 be the "points" and the lines $P_0P_1, P_0P_2, \ldots, P_0P_k, l_0$ be the "lines" of a plane, with the same incidence relation.

In model 1°: theorems (d) and (e) (**1.1**) are still valid, in model 2° they are not. So it is natural to restrict the notion of the projective plane by further axioms.

The role of an axiom equivalent to I_3 — "*the line has at least three points*" — was first investigated by Fano. The notion of the projective plane, if established by axiom system **I**, was so extended to incorporate the finite plane.

We shall now investigate the following simple geometrical model. Consider the regular polygon of 21 sides shown in Fig. 11, i.e. the figure

$$\mathbf{P} = P_1, P_2, \ldots, P_{21} = A_1 B_1 C_1 \ldots A_7 B_7 C_7 = \mathbf{A} \cup \mathbf{B} \cup \mathbf{C}$$

which is decomposed, by the signs ○, ⊙, ● shown in Fig. 7, into the regular heptagons **A**, **B** and **C**. Let us consider the subpentagon

$$P_1 P_4 P_5 P_{10} P_{12} = A_1 A_2 B_2 A_4 C_4 = \varLambda_1$$

of the polygon **P**. The vertices of this pentagon span chords of 10 different lengths, consequently the $\varLambda_1$ pentagon can be considered as a line of the plane of order 4 realized by the polygon **P**. (The other lines can be obtained from this by rotations in the sense indicated, in the figure, by the arrow.)

Let us now take the subtriangle

$$A_1 A_2 A_3 = P_1 P_4 P_{10} = \varLambda_1^A$$

of the heptagon **A**. The side lengths of this triangle are all different, thus the triangle $\varLambda_1^A$ can be considered as a line of the second order plane realized by **A**. The other lines of this plane can be derived from this by rotations, centre to origin, through angles $2k\pi/7$ ($k=1, 2, \ldots, 6$) in the sense indicated by the arrow.

If the figures **P**, **A**, **B**, **C**, $\varLambda_1$ and $\varLambda_1^A$ are considered as point sets and the following statements are valid:

$$\mathbf{P} = \mathbf{A} \cup \mathbf{B} \cup \mathbf{C}, \quad \mathbf{A} \cap \mathbf{B} = \mathbf{B} \cap \mathbf{C} = \mathbf{C} \cap \mathbf{A} = \emptyset;$$
$$\mathbf{A} \cap \varLambda_1 = \varLambda_1^A \subset \varLambda_1, \quad \mathbf{B} \cap \varLambda_1 = B_2, \quad \mathbf{C} \cap \varLambda_1 = C_4.$$

If **A** is rotated through angles of $2\pi/21$ and $4\pi/21$, it is carried into **B** and **C** respectively; moreover the triangles representing second order lines of **A** are carried into triangles representing second order lines of **B** and **C**, respectively.

This example suggests the following definition:

A subplane of a finite projective plane is a finite projective plane every point of which is a point of the original plane but not conversely; and every line of which is a subset of a line of the original plane.

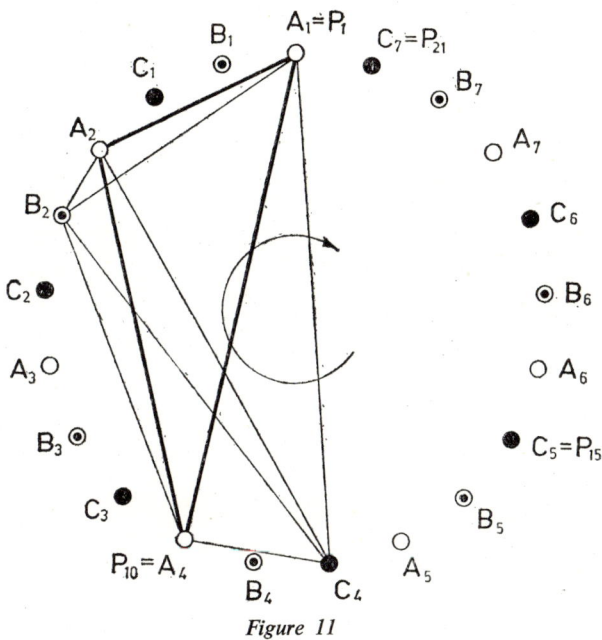

Figure 11

The name subplane is also used in the literature for other subfigures. We shall sometimes distinguish the above definition by introducing the preface "closed" and speak of closed subplanes. However, for the time being, we shall deal only with the above notion of subplane.

Let the plane **S** be a finite plane of order n and let **S*** be a subplane of order m. It follows from our definition that $m < n$. Let $P_1 \in S^*$ and $P_2 \in S^*$. If the line connecting the points P_1 and P_2 on plane **S*** is denoted by l^* then $l^* \subset l$ where l is a line on **S**. The line l is a set consisting of $n+1$ points and the line l^* a set of $m+1$ points, hence the line l contains $n-m$ points which do not belong to the line l^*. We say that the line l^* is *completed* by these $n-m$ points into the line l. Similarly, the plane **S** is derived from the plane **S*** by a completion; this is denoted by saying that **S** is an *extension* of **S***.

In the example in Fig. 11 we have $n=4$, $m=2$. Moreover, as we shall see later, the Galois plane of order 9 has a subplane of order 3, i.e. $n=9$, and $m=3$. So, there exists a case in which $n=m^2$. Also, it is known that there exists a plane — which is not a Galois plane — having a subplane of order 2, so that $n>m^2$.

Theorem: *If a projective plane of order n, has a subplane of order m, then either $n=m^2$, or $n \geq m^2+m$.* (Theorem of Bruck.)

Proof: Let S^* and S be as above. Further, let l^* be a line of S^* and $l \subset S$ be the extended line of l^*. Consider a point $Y \in l$ which is not a point of line l^* and therefore not a point of plane S^*. The point set $S^* - l^* = P^*$ contains $(m^2+m+1) - (m+1) = m^2$ points. By deleting the line l from the lines of the plane S which pass through the point Y, we obtain a set L, of n lines. We shall investigate the incidence relation of sets L and P^*.

Let x^* be the line connecting two points of the set P^* in the plane S^* and let x be the line connecting the same two points in the plane S, thus $x^* \subset x$. The lines x^* and l^* intersect in a point of the plane S^* which does not belong to P^* and since $x^* \cap l^* = x \cap l$, this point of intersection is distinct from Y. Hence, no line of the set L can contain two points of P^*. However, each point of P^* is contained in a line belonging to L (namely, by the line of S connecting the point in question with the point Y), that is m^2 lines of L each "cut" P^* in one point. If L has no other lines, then $n = m^2$.

Suppose now that L contains more than m^2 lines and that y is a line of L having no common point with the set P^*. The m^2+m+1 lines of the plane S which extend the lines of the plane S^* all intersect the line y; but, no two of them can meet in the same point of the line y, since $l \notin L$ and line y has no point in common with P^*. That is, S^* cannot have more lines than y has points, thus $n+1 \geq m^2+m+1$. Hence if $n \neq m^2$, then $n \geq m^2+m$.

We shall see later on, how important the notion of the subplane is in the study of the structure of finite projective planes. Also we shall completely analyse the subplane structure of the plane of order 4, given in Fig. 11.

1.8 The notion of the finite affine plane

The notion of the affine plane is simpler, axiomatically, than the Euclidean plane; although traditionally the Euclidean plane is developed first and then from this the notion of the affine plane is derived. Moreover, there is an alternative approach which develops first the notion of the projective plane from that of the Euclidean plane and then notion of the affine plane is derived from the projective plane by deleting a line (and *a fortiori* all the points on this line).

Thus axiom I₂ loses its validity on the remaining figure, since two lines that originally met in one of the deleted points no longer meet. And the remaining figure has for the property that given any point and a line not through it, there exists one and only one line among those passing through the point which does not cut the given line. Two non-intersecting lines are said to be parallel. In fact by deleting different lines of the original plane different affine planes are obtained, but these planes are isomorphic to each other.*

This approach to affine geometry cannot be followed completely in the case of a plane which does not contain a continuum, or even a countably infinite number, of points on a line.

Before defining the affine plane automatically we shall discuss an example in which the classical approach fails.

Let us consider the finite projective plane determined by Fig. 7. Delete the row (line) l_1 of the table and together with it also the columns (points) P_1, P_2, P_3, P_4 and P_5. Let $\Gamma^{l_1}(4)$ denote the remaining table. The figure so derived will be called an *affine plane (of order 4) associated with a projective plane of order* 4.

The following properties of $\Gamma^{l_1}(4)$ can immediately be derived from the figure.

This affine plane consists of $4^2 = 16$ points and of $4^2 + 4 = 20$ lines; every line has 4 points and 5 lines pass through each point. The totality of lines can be decomposed into five subsets each of four elements so that lines belonging to the same subset are mutually parallel whilst lines belonging to different subsets intersect; any two points determine one and only one connecting line; given any point and any line not containing it then there exists uniquely a line which does not intersect the given line; P_9, P_{10} and P_{11} are not collinear.

In the case of any $\Gamma(q)$ the associate $\Gamma^{l_1}(q)$ has similar properties, only our remarks concerning the number of elements are modified.

However, it is less obvious that $\Gamma^{l_1}(4)$ and the $\Gamma^{l_s}(4)$ are isomorphic. But if we observe that we can regain the cyclic table of Fig. 6 from $\Gamma(4)$ by appropriate column and row permutations and, furthermore, that any row can be carried into the first row by a cyclic transformation without the change of the sign pattern of the table, then it is clear that the two affine planes in question are isomorphic.

The incidence table of a Galois plane transformed into a cyclic form (Theorem of Singer). Hence, in the case of a Galois plane, the deduction of an affine plane from a projective plane follow the traditional treatment in every respect.

However, other kinds of finite projective planes are known in which the deletion of different lines does not lead to isomorphic affine planes.**

---

* Two affine planes S and S' are *isomorphic* if there exists a one-to-one transformation $\alpha: S \to S'$ which takes collinear points into collinear points.

** An example of such a plane is the finite plane of order 9 defined by means of Figs 45 and 46.

Therefore it is significant that in the symbol S^l the superfix indicates which is deleted from the plane S.

We can now turn to the axiom system defining a finite affine plane. We shall understand by the *parallelism* of two lines the fact that they have either all of their points in common or they have no point in common; thus, parallelism is an equivalence relation. Noting the properties of $\Gamma^{l_1}(4)$, the axiom system **A** defining an *affine plane of order q* will be taken to consist of the following four axioms:

- A_1 *Any two distinct points are incident with just one line* (containing as elements the two points).
- A_2 *There is one and only one line incident with a point and parallel to a given line not containing the point.*
- A_3 *There exists a triangle* (three points such that no more than two of them are incident with the same line).
- A_4 *There exists a line consisting of* $q(q>1)$ *points.*

All lines parallel to a given line are said to form a *class*. If we call these classes *ideal points* — following the traditional approach — we arrive at the notion of the projective plane of order q associated with the affine plane. But the affine plane derived from this projective plane by deleting an arbitrary line is not necessarily isomorphic with the original affine plane.

1.9 Different kinds of finite hyperbolic planes

Consider a finite plane; if, for each pair $[l, p]$ consisting of a point and a line not through the point, the number of lines passing through P and not meeting l is 0 respectively 1 then the given finite plane is projective respectively affine.

In view of this the following generalizing assumption will be made: For each non-incident pair $[l, P]$, the lines through P_μ either lie in a set, μ, of lines which intersect l or in a set, ν, of lines which do not intersect l. Let the number of the elements of the sets μ, ν be denoted by $|\mu|$, $|\nu|$, respectively. If, for every $[l, P]$-pair, $|\nu|>1$ then we shall speak of a *hyperbolic plane*. It can occur that for every $[l, P]$-pair $|\nu|=M$ and $|\mu|=m$ (where M and m are two nonnegative integers) in this case we shall speak of a *regular hyperbolic plane*.

In the last decade several papers appeared dealing with the investigation of finite hyperbolic planes and we know now that there exist nonregular hyperbolic finite planes. In what follows, we shall restrict our attention to finite regular hyperbolic planes.

Several authors define a hyperbolic plane, called an $\langle m, n \rangle$ plane, by the following axiom system **H***:

H_1^* *Given any two distinct points there exists just one line incident with both points.*
H_2^* *Given any line, there exists a point not incident with the line.*
H_3^* *For every [l, P]-pair* $|v|=n(>1)$.
H_4^* *There exists a line consisting of* $m(>0)$ *points.*

From this axiom system the following theorems may be derived easily by the reader:

1° The number of the points in the plane is $(m+n)(m-1)+1$ and the number of lines is $\left(\dfrac{m+n}{n}\right)\{(m+n)(n-1)+1\}$

2° The number m is a divisor of the number $(n-1)n$.

3° If $m>2$, we have no $\langle m, 2\rangle$-plane.

Unfortunately, the axiom system above admits, as we shall see, some trivial and irrelevant figures as hyperbolic planes. We shall determine firstly the figure with the minimal number of elements satisfying **H*** and construct a Euclidean model of it.

According to **H*** and theorems 1°, 2°, and 3° the minimal $\langle m, n\rangle$-plane must be the $\langle 2.2\rangle$-plane, if it exists. In fact, this plane does exist, as we shall see below, and consists of 5 points and 10 lines.

Consider *five* points of 3-dimensional Euclidean space, P_1, P_2, P_3, P_4 and P_5, chosen so that no more than two lie on the same line and no more than three lie in the same plane. Consider the *ten* lines $l_{rs}(r\neq s; r, s=1, \ldots, 5)$ where l_{rs} is the line joining the points P_r and P_s. Let these be the "points" and the "lines" of

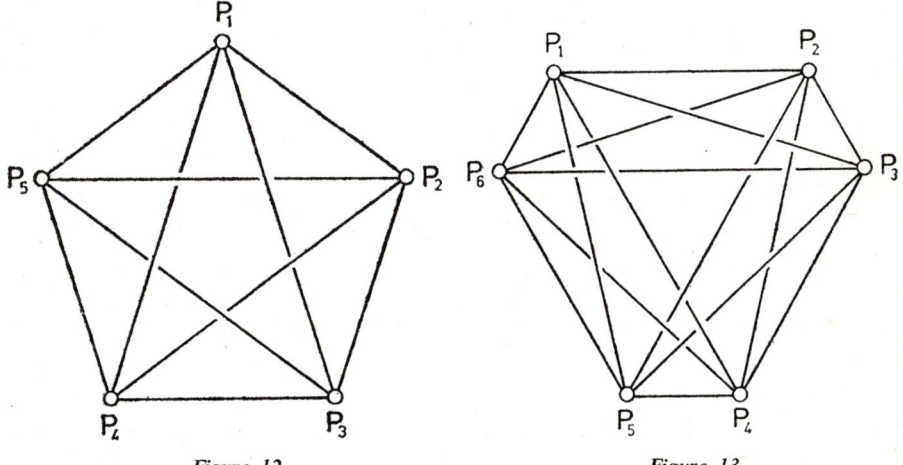

Figure 12 *Figure 13*

the $\langle 2.2 \rangle$-plane. Then this model satisfies the requirements of the axiom system $\mathbf{H}^*$ (Fig. 12).

For the time being let us continue to consider Euclidean space and let us call the figure generated by the 5 points in question $\triangledown(5)$. We can make a generalization as follows: Let $\triangledown(k)$ denote the figure generated by the points $P_1, P_2, \ldots, P_k$; where no more than two of these points lie on the same line and no more than three lie on the same plane. The number of the lines $P_r P_s = l_{rs} (r \neq s; r, s = 1, 2, \ldots, k)$ in the figure is $(k-1)k/2$. It is readily seen that the model $\triangledown(k)$ corresponds to the definition of the $\langle 2, k-3 \rangle$-plane. In Fig. 13 the model $\triangledown(G)$, i.e. the $\langle 2,3 \rangle$-plane, is depicted.

Thus, if we are working with the axiom system $\mathbf{H}^*$, every element of the infinite series of the trivial figures $\triangledown(5)$, $\triangledown(6)$, $\triangledown(7)$ is a model of a finite regular hyperbolic plane. Let us try, therefore, to amend $\mathbf{H}^*$ so that the $\triangledown$-figures will be excluded.

In fact, it suffices to replace in $\mathbf{H}_4^*$ the requirement $m > 0$ by the somewhat stricter condition $m > 2$, which implies, because of Theorem 3, that $n = 2$ cannot hold, i.e. $n > 1$ could be replaced by $n > 2$ in $\mathbf{H}_3^*$. Of course, the modified axiom system will still allow certain trivial figures. In view of these problems we shall follow a procedure due to Crowe for defining hyperbolic planes.

We introduce a concept, the importance of which is shown by a remarkable property of ordinary space. Consider a subset **T** of a plane which is not a line but still possesses the following properties of a line: Every line of the plane determined by two arbitrary points of **T** is a subset of **T**, assuming further that **T** contains at least two points.

Now, for a plane in ordinary space no proper subset **T** can be found. Similarly, in a finite projective plane, such a subset **T** would contain two points, say A and B. The line AB contains altogether $q+1$ points of the plane and these would all belong to **T** as well. But **T** must then contain a point C which does not lie on the line AB. By connecting C with each of the points of the line AB, every point of the lines so obtained would belong to **T** and so all the points of the plane would belong to **T**. However, if we consider the figures $\triangledown(k)$, which are hyperbolic planes according to the axiom system $\mathbf{H}^*$, then proper subsets **T**, with the properties above can be found. In fact, any subtriangle of $\triangledown(k)$ will be a suitable subset. If T a plane contains a proper subset **T** then **T** is said to be a *tract* of the plane. Thus neither the Euclidean plane nor the classical and the finite projective planes contain a tract.

Crowe defined the regular hyperbolic plane by an axiom system which ensures the tractlessness of the plane. Consider the following axiom system $\mathbf{H}$:

H_1 Given any two points there exists just one line incident with both of them.

FINITE HYPERBOLIC PLANES

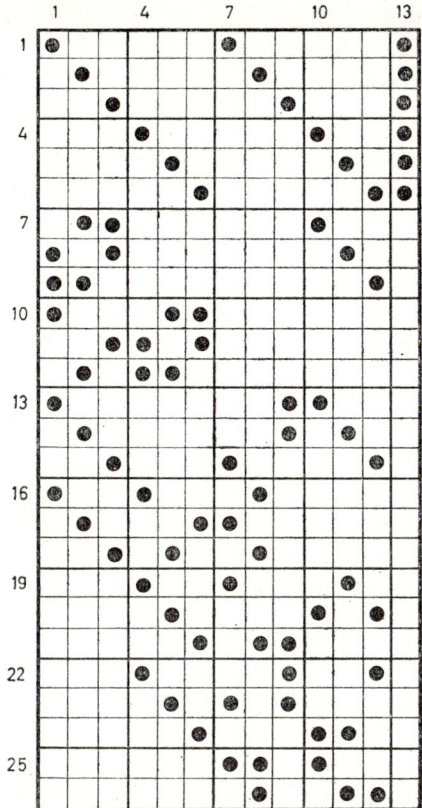

Figure 14

H_2 Given any point P and any line l not containing P then there exist m lines passing through P and intersecting l and n lines passing through P and not intersecting l.

H_3 There exist three points such that no more then two of them lie on the same line.

H_4 $(m-1)^2 > n > 2$.

A figure satisfying the system **H** is said to be an $\langle m, n \rangle$-plane.

It can be easily derived that the number of points and the number of lines of an $\langle m, n \rangle$-plane are the same as those obtained in Theorem 1 following **H***.

Theorem of Crowe: *A plane defined by the axiom system* **H** *has no tract.*

Proof. Suppose that a plane S satisfying **H** contains a tract T. T contains three noncollinear points A, B and C, say. Consider any point $P \in S \setminus T$, the number of lines incident with P is $\lambda = m+n$ and, by $\mathbf{H_4}$, $\lambda \leq m+(m-1)^2 = m(m-1)+1$. Let C be connected with the points of the line AB, thus we obtain, m lines which together contain $m(m-1)+1$ of the points of the plane, and all of these points belong to $\mathbf{T}$. But, by connecting P with each of these points we obtain $m(m-1)+1$ through P, as P cannot lie on a line connecting two points of $\mathbf{T}$. Thus we have a contradiction and the theorem is proved.

Theorem: *There exists a figure satisfying the axiom system* $\mathbf{H}$.

Proof. By $\mathbf{H_4}$ the plane with minimal number of elements, satisfying $\mathbf{H}$, would be the $\langle 3, 3 \rangle$-plane, if it exists. According to Theorem 1, following from $\mathbf{H}^*$, the $\langle 3, 3 \rangle$-plane would consist of 13 points and of 26 lines. In fact, such a plane does exist; and its incidence table is given in Fig. 14.

The point set of this $\langle 3, 3 \rangle$-plane can be decomposed in the following two ways:

$$l_5 \cup l_9 \cup l_{11} \cup l_{25} \cup P_9 \quad \text{and} \quad l_1 \cup l_{18} \cup l_{22} \cup l_{24} \cup P_2$$

where every point of the plane, other than P_2 and P_9, is uniquely determined as the intersection of a line of the first decomposition with a line of the second.

1.10 Galois planes and the theorem of Desargues

When projective geometry was first introduced, as an extension of Euclidean geometry, Desargues' theorem concerning perspective triangles appeared only as a consequence of other, more simple, theorems. Later, when an independent foundation for projective geometry was sought, Desargues' theorem played a role of fundamental importance. In what follows, Desargues' theorem will be sometimes abbreviated to the D-theorem or simply D.

Hilbert, when investigating the possibility of introducing a coordinate system on the projective plane independent of one arising from the Euclidean metric, observed that this can be achieved by using Desargues' theorem.

We fix three points — say O, U, V — of a line. On the set of the points of the line distinct from point V, Hilbert has given two constructions (assuming the validity of the axioms of incidence and of the D-theorem) both of which assign a point of the set to any two given points of the set. The result of these geometrical operations depends only on the triple of points O, U, V. He called one of the operations addition and the other multiplication and he proved that all the axioms of a field are satisfied in this case, except the commutativity of the product. The product, however, is commutative if and only if Pascal's theorem (abbreviated: P-theorem), concerning six points lying three by three on each of two lines, is also valid. This became the basis of the metric-independent introduction of a coordinate system.

Thus, in attempting to put the projective plane on an axiomatic base, Hilbert's attention was directed towards the D-theorem and the P-theorem. He proved that Desargues' theorem cannot be derived from the planar axioms of incidence. So let us add as a fourth axiom to the three axioms of incidence defining the projective plane the validity of Desargues' theorem. Hilbert proved further that even this extended axiom system is not enough for the derivation of Pascal's theorem. On the contrary, as was proved by Hessenberg, by adding Pascal's theorem to three axioms of incidence, the axiom system so obtained — denoted by **IP** — has as a consequence Desargues' theorem.

Thus, if the projective plane is defined by the axiom system **IP** then the analytic geometry of this plane — as was shown by Hessenberg — is very similar to the analytic geometry of the classical projective plane differing only in as much as the coordinate field need not be the real numbers.

If we proceed as sketched above, we discover infinitely many projective planes including the Galois planes mentioned earlier.

We complete now our sketch of the historical development by establishing the Galois planes and discussing some theorems of Galois plane geometry.

1° Consider $\mathbf{K}=GF(q)$, where $q=p^r$ (p is a prime, r a positive integer). Consider the set of all the sequences (x_1, x_2, x_3) consisting of elements of $\mathbf{K}$, with the exception of the sequence consisting of three zero elements. Decompose this set into classes in the following manner: two sequences (x_1, x_2, x_3), and (y_1, y_2, y_3), belong to the same class if and only if there exists $\lambda \in \mathbf{K}$, $\lambda \neq 0$ for which

$$y_1 = \lambda x_1, \quad y_2 = \lambda x_2, \quad y_3 = \lambda x_3.$$

Let these classes be called *points*. Any class is uniquely determined by one of its elements (x_1, x_2, x_3). A sequence is said to be a *homogeneous coordinate sequence* of the point to which it belongs.

Clearly, as $GF(q)$ has q elements and the sequence $(0, 0, 0)$ is excluded, it follows that the number of elements in the set of the sequences is q^3-1. Also, as there are $(q-1)$ choices for λ, every class in the decomposition consists of $q-1$ elements, consequently the number of the points (classes) is q^2+q+1.

A subset of the set of all points defined by an equation

$$a_1 x_1 + a_2 x_2 + a_3 x_3 = 0$$

where $a_1, a_2, a_3 \in K$ and are not all zero, will be called a *line*. Clearly, if

$$b_1 = \lambda a_1, \quad b_2 = \lambda a_2, \quad b_3 = \lambda a_3 \quad (0 \neq \lambda \in \mathbf{K})$$

then the equation

$$b_1 x_1 + b_2 x_2 + b_3 x_3 = 0$$

defines the same set as the first equation. — Conversely, as we shall prove shortly, if two equations of the form above define the same point set then their coefficients satisfy the relations above.

For every field of **K** the following theorems are valid:

A) If the sequences (x_1, x_2, x_3) and (y_1, y_2, y_3) determine two distinct points and if

$$\left.\begin{array}{l} a_1 x_1 + a_2 x_2 + a_3 x_3 = 0 \\ a_1 y_1 + a_2 y_2 + a_3 y_3 = 0 \end{array}\right\} \text{ but } [a_1, a_2, a_3] = [0, 0, 0],$$

then the sequence of coefficients $[a_1, a_2, a_3]$ is determined up to a factor λ.

B) If neither of the equations

$$a_1 x_1 + a_2 x_2 + a_3 x_3 = 0,$$

$$b_1 x_1 + b_2 x_2 + b_3 x_3 = 0$$

is a multiple of the other by any $\lambda \in \mathbf{K}$, then the sequence $(x_1, x_2, x_3) \neq (0, 0, 0)$ is determined up to a factor $\mu \in \mathbf{K}$.

But these two theorems mean precisely that the line and the point so defined satisfy the axioms l_1 and l_2, respectively.

We can see now from Theorem *A)*, that the lines determined by the sequences of coefficients $[a_1, a_2, a_3]$ and $[b_1, b_2, b_3]$ are the same if and only if there exists a $\lambda \neq 0$ for which $b_j = \lambda a_j (j=1, 2, 3)$. Therefore, a sequence $[a_1, a_2, a_3] \neq [0, 0, 0]$ is said to be a *coordinate sequence* of a line and the elements of this sequence are said to be *homogeneous line coordinates*.

Axiom l_3 is satisfied by the following four points:

$$(1, 0, 0), \quad (0, 1, 0), \quad (0, 0, 1), \quad (1, 1, 1).$$

Finally, let us determine the number of the points on a line. In order to do this let us choose — say — the line $[a_1, a_2, a_3] = [1, 0, 0]$, i.e. the line given by the equation

$$1 \cdot x_1 + 0 \cdot x_2 + 0 \cdot x_3 = 0.$$

Obviously, a sequence satisfies this equation if and only if it is of the form $(0, x_2, x_3)$. There are q^2 pairs of elements of the field **K**, but the case $x_2 = x_3 = 0$ must be excluded since no point corresponds to the $(0, 0, 0)$; i.e. the number of allowable sequences $(0, x_2, x_3)$ is $q^2 - 1$ classes. Therefore, this line, and consequently every line, contains $q+1$ points; the plane is therefore of order q.

The plane determined above will be called a *Galois plane over the coordinate field* GF(q) and will be denoted by the symbol $S_{2,q}$.

2° The duality between points and lines is reflected the interchangeability of the signs (...) and [...] in the definitions above.

We must be careful to avoid the pitfalls of applying the language and concepts of vector spaces to projective spaces. For instance, we can speak of a linear combination

$$\lambda(x_1, x_2, x_3) + \mu(y_1, y_2, y_3) = (z_1, z_2, z_3)$$

of the sequences (x_1, x_2, x_3) and (y_1, y_2, y_3) consisting of elements of the field **K**, where this equation means that

$$z_1 = \lambda x_1 + \mu y_1, \quad z_2 = \lambda x_2 + \mu y_2, \quad z_3 = \lambda x_3 + \mu y_3 \quad \text{and} \quad \lambda, \mu \in \mathbf{K}.$$

The sequences (x_1, x_2, x_3), (y_1, y_2, y_3) and (z_1, z_2, z_3) determine, say, the points X, Y and Z respectively; but now the notation $\lambda X + \mu Y = Z$ as well as the statement, "linear combination of the points", are inadmissible because of the lack of uniqueness.

X and Y each stand for a class of $(q-1)$ sequences; clearly by choosing different representatives of these classes, different sequences (z_1, z_2, z_3) will be obtained in the linear combination above. Moreover, if $X \neq Y$, then the sequences (z_1, z_2, z_3) will not all belong to the same class.

We shall discuss this further in connection with the following example.

Let **K** be the class of residues modulo 3, i.e. **K** = GF(3). The two operation tables of this field are:

+	0	1	2		×	0	1	2
0	0	1	2		0	0	0	0
1	1	2	0		1	0	1	2
2	2	0	1		2	0	2	1

A   **M**

We compile the coordinate sequences decomposed into classes:

$P_1: \begin{cases} (1, 0, 0), \\ (2, 0, 0); \end{cases}$ $P_6: \begin{cases} (0, 1, 1), \\ (0, 2, 2); \end{cases}$ $P_{11}: \begin{cases} (2, 1, 1), \\ (1, 2, 2); \end{cases}$

$P_2: \begin{cases} (0, 1, 0), \\ (0, 2, 0); \end{cases}$ $P_7: \begin{cases} (1, 1, 1), \\ (2, 2, 2); \end{cases}$ $P_{12}: \begin{cases} (1, 2, 1), \\ (2, 1, 2); \end{cases}$

$P_3: \begin{cases} (0, 0, 1), \\ (0, 0, 2); \end{cases}$ $P_8: \begin{cases} (2, 1, 0), \\ (1, 2, 0); \end{cases}$ $P_{13}: \begin{cases} (2, 2, 1), \\ (1, 1, 2). \end{cases}$

$$P_4: \begin{cases} (1, 1, 0), \\ (2, 2, 0); \end{cases} \quad P_9: \begin{cases} (2, 0, 1), \\ (1, 0, 2); \end{cases}$$

$$P_5: \begin{cases} (1, 0, 1), \\ (2, 0, 2); \end{cases} \quad P_{10}: \begin{cases} (0, 2, 1), \\ (0, 1, 2); \end{cases}$$

Every class consists of two sequences. In each class we could select as a representative that *sequence having for its last non-zero element the unit element of the field* **K**. This choice of representatives will be useful later.

As an illustration of the problem encountered above, consider what the "linear combination of points" $P_8 + 2P_{13}$ might mean. We have the following choices of representatives for the pair P_8, P_{13}:

(2, 1, 0), (2, 2, 1); (2, 1, 0), (1, 1, 2); (1, 2, 0), (2, 2, 1); (1, 2, 0), (1, 1, 2).

The combination $P_8 + 2P_{13}$ means in turn the sequences (0, 2, 2), (1, 0, 1), (2, 0, 2), (0, 1, 1), i.e. the points P_6, P_5, P_5, P_6 thus we do not obtain a unique point as in the case of the vector space. It is easy to check that the two points, P_5 and P_6, just derived lie on the line connecting the points P_8 and P_{13}; furthermore, these four points are the only points on this line, whose equation is given by:

$$x_1 + x_2 + 2x_3 = 0.$$

In general, if we let $\lambda, \mu \in \mathbf{K} = GF(q)$ and $P_j, P_k \in S_{z,q}$, we would again find that $\lambda P_j + \mu P_k$ is not unique but that the point $q-1$ can be written in this form by varying the coordinate sequences representing the points P_j and P_k. In this case, too, the original points and the $q-1$ derived points exhaust all the points of the connecting line $P_j P_k$.

However, we can remedy the non-uniqueness of linear combinations of points by introducing a normalization of the homogeneous coordinate sequences of a point. We have already pointed at a form of normalization; namely, we chose from the $(q-1)$ equivalent coordinate sequences determining a point the one having for its last non-zero element the unit element; this will be the *normed* coordinate sequence of the point.

We now say that the linear combination $\lambda A + \mu B$ of the points A, B with coefficients λ, μ is the point C determined in the following manner: let the normed coordinate sequences of the points A and B be $(a_1, a_2, a_3,)$ and $(b_1, b_2, b_3,)$, respectively, then the point C is determined by the homogeneous coordinate sequence

$$(\lambda a_1 + \mu b_1, \lambda a_2 + \mu b_2, \lambda a_3 + \mu b_3).$$

Of course the sequence so obtained may not be normed.

Obviously, if $\lambda=\mu=0$, then the linear combination $\lambda A+\mu B$ is not a point, since the sequence $(0, 0, 0)$ does not determine a point. Further, it is clear that if $\lambda\neq 0$ but $\mu=0$, then the point C coincides with the point A; and, similarly, if $\lambda=0$ but $\mu\neq 0$, then C coincides with B. Obviously if $\lambda'=\varrho\lambda$ and $\mu'=\varrho\mu$, where $\varrho\neq 0$, then the linear combinations $\lambda A+\mu B$ and $\lambda' A+\mu' B$ either represent the same point or are both equal to $(0, 0, 0)$. Thus if $\lambda\mu\neq 0$, then the sequence defined by $\lambda A+\mu B$ (be it a point or not) can also be obtained by the pair of coefficients $v, 1$ where $v=\lambda\mu^{-1}$. Hence, if $A\neq B$, to find the number of points obtained by linear combinations $\lambda A+\mu B$, it is enough to consider linear combinations $vA+B$ where v runs through the field $\mathbf{K}$; and we must not forget A itself, which cannot be expressed in this form.

Clearly if A and B are two distinct assumptions then $\lambda A+\mu B=(0, 0, 0)$ if and only if $(\lambda, \mu)=(0, 0)$.

Furthermore, if $v\neq v'$, then the point $vA+B$ is different from the point $v'A+B$, since otherwise we would have

$$(v'-\varrho v)A+(1-\varrho)B = (0, 0, 0)$$

and then, by the statement above, $v'-\varrho v=0$ and $1-\varrho=0$, i.e. $v=v'$ which would contradict our assumption.

Thus we obtain q distinct points corresponding to the q values of v, and so the number of the points $\lambda A+\mu B$ is equal to $q+1$.

Let the equation of the line connecting the points A, B be $u_1 x_1+u_2 x_2+u_3 x_3=0$. Since

$$u_1 a_1+u_2 a_2+u_3 a_3 = 0, \quad u_1 b_1+u_2 b_2+u_3 b_3 = 0,$$

then

$$u_1(\lambda a_1+\mu b_1)+u_2(\lambda a_2+\mu b_2)+u_3(\lambda a_3+\mu b_3) = 0$$

that is, every point represented as a linear combination of the original points belongs to the line; moreover, these points exhaust the set of all points of the line, since a line in a plane of order q contains precisely $q+1$ points.

3° Now we shall use the result above to prove that *Desargues' theorem is valid on every Galois plane.*

Theorem: *Let six distinct points be the vertices of two proper triangles $A_1 A_2 A_3$ and $B_1 B_2 B_3$. If the lines $A_1 B_1$, $A_2 B_2$, $A_3 B_3$ have a common point D, then the three points*

$$C_1 = A_2 A_3 \cap B_2 B_3, \quad C_2 = A_3 A_1 \cap B_3 B_1, \quad C_3 = A_1 A_2 \cap B_1 B_2$$

are distinct and lie on the same line d.

Proof. We discuss firstly a special case when D coincides with one of the six points. Suppose, for example, that $A_1 = D$. Then, A_1, A_2 and B_2 will be collinear and so will A_1, A_3 and B_3. Therefore, $C_3 = A_1A_2 \cap B_1B_2 = B_2$ and $C_2 = A_3A_1 \cap B_3B_1 = B_3$. Further the point $A_2A_3 \cap B_2B_3 = C_1$ cannot coincide with either B_2 or B_3 (since otherwise either $A_2 = B_2$ or $A_3 = B_3$, in contradiction to the hypothesis). The triple $C_1C_2C_3 = C_1B_3B_2$ is now indeed a set of three points which lie on the same line $d = B_2B_3$.

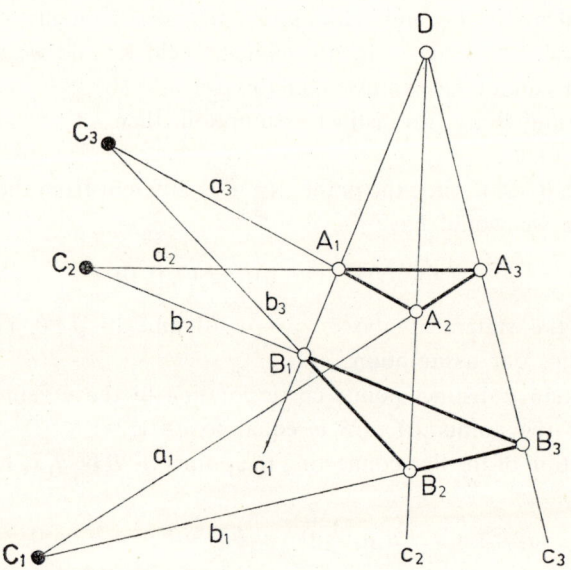

Figure 15

We consider now the general case (Fig. 15). The points B_1, B_2 and B_3 can be obtained (not necessarily in normed form) as linear combinations of the normed forms of the other four points:

$$B_1 = \lambda A_1 + D, \quad B_2 = \mu A_2 + D, \quad B_3 = \nu A_3 + D \quad (\lambda \neq 0, \quad \mu \neq 0, \quad \nu \neq 0).$$

From this we have

$$B_1 - B_2 = \lambda A_1 - \mu A_2, \quad B_2 - B_3 = \mu A_2 - \nu A_3, \quad B_3 - B_1 = \nu A_3 - \lambda A_1.$$

$\lambda A_1 - \mu A_2$ is a point on the line A_1A_2; let this point be denoted by C_3. If the normed forms of B_1 and B_2 are B_1^* and B_2^*, respectively, then there exist $\beta_1 \neq 0$ and $\beta_2 \neq 0$ such that $B_1 - B_2 = \beta_1 B_1^* - \beta_2 B_2^*$, and this linear combination gives a point of the line B_1B_2. Consequently, $C_3 = A_1A_2 \cap B_1B_2$. A similar consequence follows from the two other equations. Thus the points of intersection of corresponding pairs of sides are

$$C_1 = \lambda A_1 - \mu A_2, \quad C_2 = \mu A_2 - \nu A_3, \quad C_3 = \nu A_3 - \lambda A_1.$$

These points are distinct. Since if, for example, the first coincided, there would exist a $\varrho \neq 0$ such that $\lambda A_1 - \mu A_2 = \varrho(\lambda A_2 - \nu A_3)$. From this it would follow that

$$\lambda A_1 + \varrho \nu A_3 = (\varrho + 1)\mu A_2; \quad \lambda \mu \nu \neq 0$$

i.e. A_1, A_2 and A_3 would be collinear, contrary to the hypothesis.

We deduce from the linear combinations above, giving the points C_1, C_2, C_3 that

$$C_1 + C_2 = -1 \cdot C_3.$$

Thus, the points C_1, C_2, C_3 are collinear and the theorem is proved.

Theorem: *The projective plane of order q defined by the axioms* I_1, I_2, I_3, **D** *is isomorphic with the Galois plane* $S_{2,q}$.

A detailed study of Galois (plane) geometry, including a proof of this theorem, will be given in later chapters.

1.11 A non-Desarguesian plane

The first examples (models) proving that Desargues' theorem cannot be derived from the axioms I_1, I_2, I_3 were of planes in which lines are continuous i.e. lines containing a continuum of points. (Examples due to Hilbert and to Moulton.) Subsequently, finite planes were discovered on which Desargues' theorem is not valid. (Examples due to Veblen and Wedderburn.) Four and a half decades later a highly instructive example was given by Hall. Hall's plane contains only countably many points on each line. Though Hall's model is not finite it will be discussed here because we can derive finite non-Desarguesian planes from it.

Consider in Fig. 16 the sequence $T^1, T^2, T^3, \ldots$ of incidence tables. The elements of the sequence are said to be *plane phases* (the columns: points, the rows: lines, etc.). The first phase represents a proper quadrangle for which axiom I_3 is valid. As each phase is a *subphase* of the following one, I_3 is valid in every phase. The sequence is so constructed that I_1 is valid in the phases $T^1, T^3, T^5, \ldots$ and I_2 is valid in the phases $T^2, T^4, T^6, \ldots$ Consider the following sequence:

$$T_1 = T^1, \quad T_2 = T^2 - T^1, \quad T_3 = T^3 - T^2, \ldots.$$

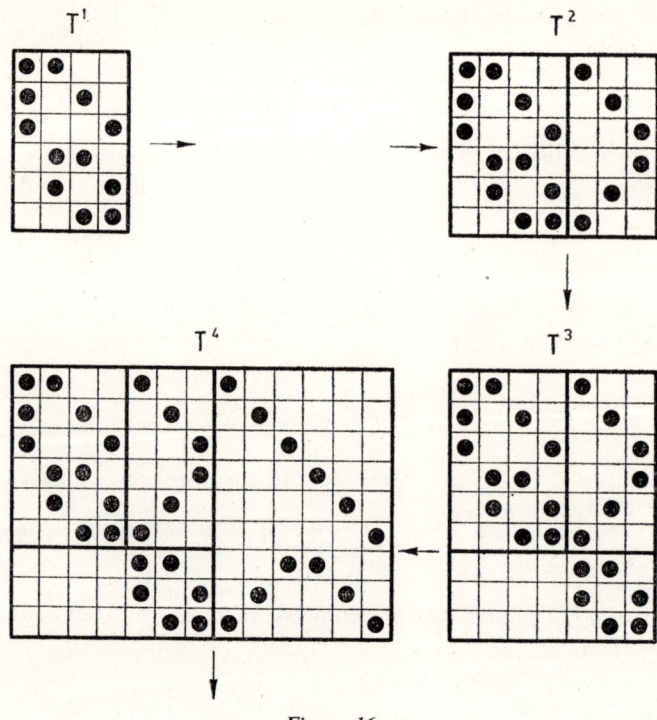

Figure 16

These alternatively extend the table by "points" and "lines"; namely they add to a plane phase either the "points of intersection" or the "connecting lines" missing from the phase in question.

For every pair of rows (columns) which are not already intersected by a column (row) in two squares filled with the sign • we add a column (row) with signs ○ in the two appropriate squares. Of course, at each stage, the number of new columns (rows) is finite, since the phase being extended has only a finite number of pairs of rows (pairs of columns).

This process of alternatively extending by rows and columns continues indefinitely. For, if at some stage a phase could not be extended by a row or by a column, then both of the axioms I_1 and I_2 would be valid (and also I_3 because it is valid for T^1); therefore we would have constructed a finite plane. Now the incidence table of a finite plane has at least three incidence signs in each of its rows and columns. But, for the plane above, each row or column of the final extension could contain only two incidence signs, and we would have a contradiction.

It is obvious from the construction that the figure

$$H = T_1 \cup T_2 \cup T_3 \cup \ldots$$

satisfies the axioms I_1, I_2, I_3, thus it is *a projective plane* although not finite.

Figure 17 shows the part T^5 of H. This figure suffices to prove that H is *non-Desarguesian*. Consider the table and the graphical model next to it. Triangles

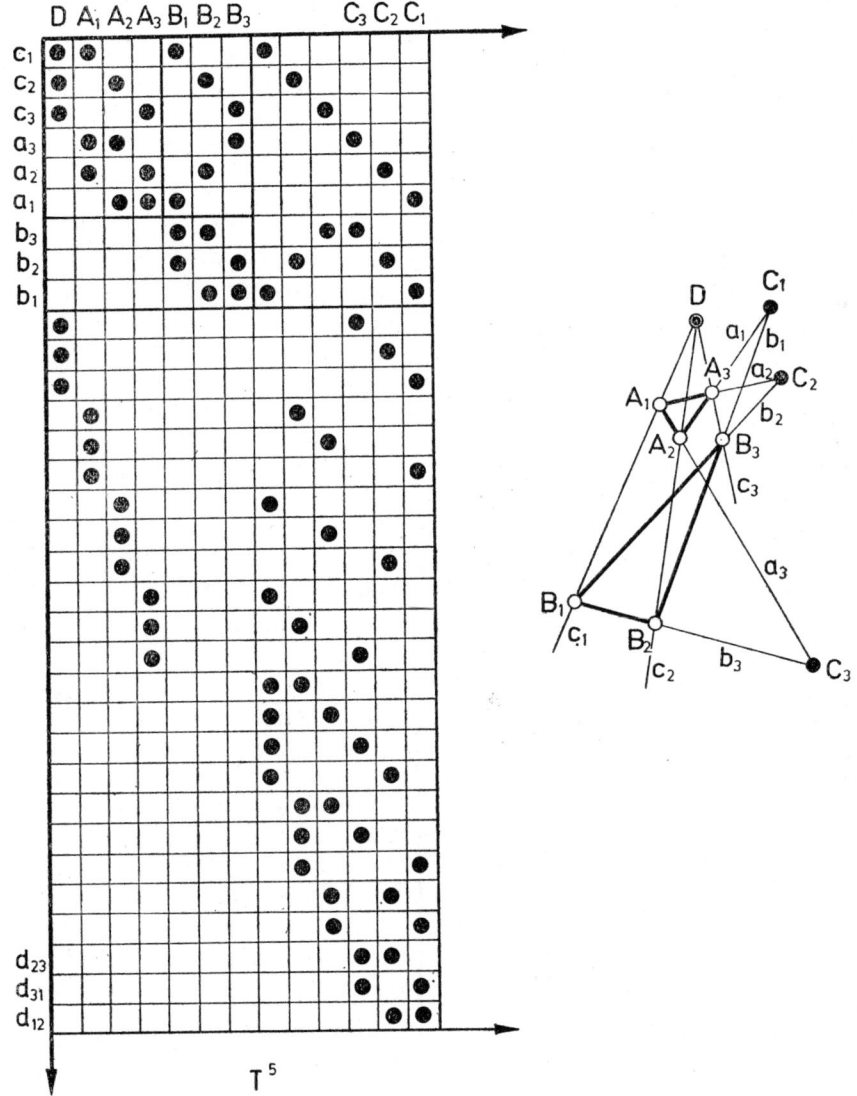

Figure 17

$A_1A_2A_3$ and $B_1B_2B_3$ form a perspective pair with respect to the centre D. This perspectivity, however, is not axial: the points C_1, C_2, C_3 determine pairwise the lines d_{12}, d_{23}, d_{31}, so they are not collinear.

Let us consider the sequence of plane phases and let us note, at each stage, the numbers of points and lines. We obtain the following sequence of pairs of numbers, where the first entry is the number of points and the second is the number of lines:

$$(4, 6), (7, 6), (7, 9), (13, 9), (13, 33) \ldots .$$

Similarly, let us write down the sequence of the pairs of numbers for finite planes, ascending by order. We obtain the following:

$$(7, 7), (13, 13), (21, 21), (31, 31), (57, 57) \ldots .$$

If we compare the two sequences, the following intriguing question arises: Can the tables $\mathbf{T}^2$ and $\mathbf{T}^4$ be extended into the incidence tables of the finite planes with 7 and with 13 lines, respectively; by the addition of rows (one row resp. four rows) of an appropriate sign pattern?

The answer is in the affirmative, as is shown by Figs 18 and 19. By superimposing the sign patterns of the three rows l_7, l_8 and l_9 which extend $\mathbf{T}^2$ into $\mathbf{T}^3$, we obtain the row l^7, say. By completing $\mathbf{T}^2$ wih the line $l^7 (= l_7 \cup l_8 \cup l_9)$ we obtain Fig. 18. Some of the row which extend $\mathbf{T}^4$ into $\mathbf{T}^5$ are superimposed three by three

$$l^{10} = l_{10} \cup l_{11} \cup l_{12}; \quad l^{11} = l_{13} \cup l_{14} \cup l_{15};$$
$$l^{12} = l_{16} \cup l_{17} \cup l_{18}; \quad l^{13} = l_{19} \cup l_{20} \cup l_{21}.$$

By adding the rows $l^{10}, l^{11}, l^{12}, l^{13}$ to $\mathbf{T}^4$ we obtain Fig. 19. It is easy to check that these figures realize planes of order 2 and 3, respectively.

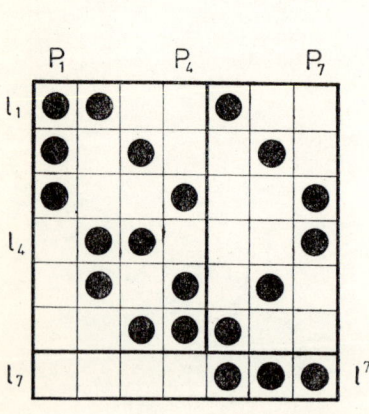

Figure 18

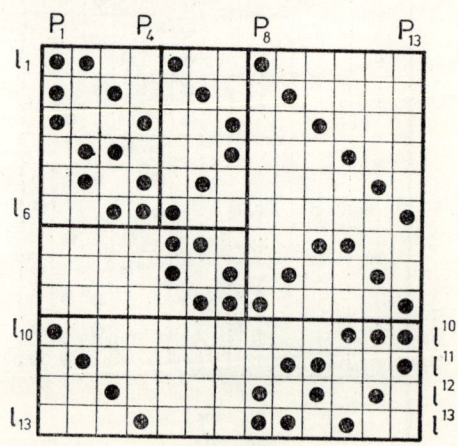

Figure 19

1.12 Collineations and groups of collineations of finite planes

As the reader may already suspect from the title of this section, we are going to introduce certain well-known notions of classical projective geometry into finite geometry. Everything which is not a consequence of the axioms l_1, l_2, l_3 will be discarded and the remainder will be specialized by requiring the validity of axiom l_4. This is the guiding line for the building up of a finite projective geometry.

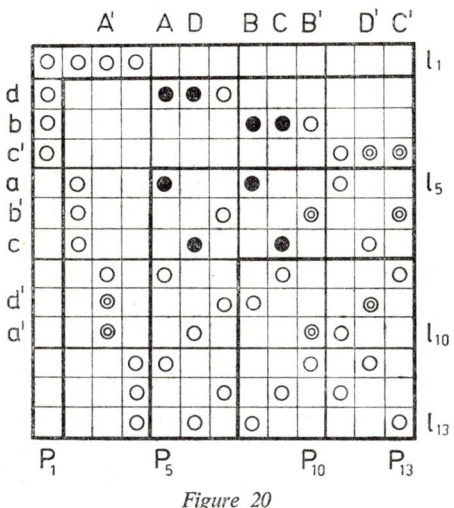

Figure 20

First we shall deal with the *one-to-one line preserving mapping of the plane* — considered as a point set — onto itself. A mapping of this kind is characterized by the following properties:

(1) *We assign to each point of the plane a unique point, called its image point (image)* .
(2) *Given any point of the plane there exists uniquely a point having the given point as image.* (The point so defined is said to be the *primitive point* of the given point.)
(3) *The images of three points are collinear if and only if the points are themselves collinear.*

We shall discuss this in the context of Figs 20 and 21. Both figures represent a plane of order 3 and the common labelling establishes a one-to-one correspondence between them. For the time we shall not be concerned with the three kinds of incidence signs that occur in Fig. 20. In Fig. 21 there are three points each of which corresponds to an ideal point, and three of the lines have no straight lines as images; moreover, the "line" $\{P_1, P_2, P_3, P_4\} = l_1$ is not represented by a straight line.

In the case of finite planes the mappings defined by the properties (1), (2) are described by permutations (both of the points and of the lines). We denote

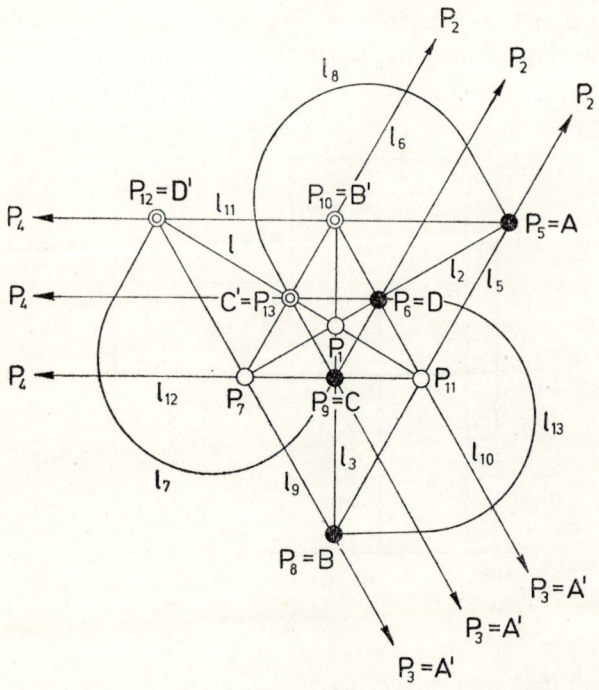

Figure 21

permutation of points and lines by the brackets ( ) and [ ], respectively. Consider a mapping of points ω and a mapping of lines σ in our example, where

$$\omega = \begin{pmatrix} 1 & 2 & 3 & 4 & 5 & 6 & 7 & 8 & 9 & 10 & 11 & 12 & 13 \\ 12 & 8 & 5 & 11 & 10 & 1 & 2 & 3 & 13 & 9 & 4 & 6 & 7 \end{pmatrix},$$

and

$$\sigma = \begin{bmatrix} 1 & 2 & 3 & 4 & 5 & 6 & 7 & 8 & 9 & 10 & 11 & 12 & 13 \\ 4 & 6 & 9 & 2 & 5 & 1 & 3 & 11 & 8 & 7 & 12 & 13 & 10 \end{bmatrix}.$$

Here $\left(\ldots \begin{matrix} j \\ k \end{matrix} \ldots \right)$: stands for $\left(\ldots \begin{matrix} P_j \\ P_k \end{matrix} \ldots \right)$ and means that the image of point P_j is P_k (the primitive point of $\begin{pmatrix} P_k \\ l_r \end{pmatrix}$ is P_j') and similarly for lines.

The dual statements of (1), (2), (3) will be denoted by (1*), (2*), (3*), respectively. We can now see that ω has the properties (1), (2) and σ has the properties (1*), (2*). Clearly any of the 13! permutations of points (lines) will satisfy (1), (2), (resp. (1*), (2*)). But, by no means every point permutation satisfies (3), nor does every line permutation satisfy (3*).

For instance ω transforms the collinear points $P_2 P_3 P_4$ into $P_8 P_5 P_{11}$ which are also collinear, it transforms $P_7 P_8 P_9$ into $P_2 P_3 P_{13}$ and both of these triples are noncollinear; it also transforms a collinear triple $P_1 P_2 P_3$ into a triangle $P_{12} P_8 P_5$, and a triangle $P_5 P_6 P_{10}$ into a collinear triple $P_{10} P_1 P_9$. So ω does not satisfy (3). Similarly σ does not satisfy (3*).

In fact of the 13! point permutations only $2^4 \cdot 3^3 \cdot 13 = 5616$ satisfy (3); we shall not prove this here. It is clear that if a point mapping (permutation) X satisfies (3), then a point set is a line if and only if its image, under π, is a line. This is why a mapping satisfying (1), (2) and (3) is said to be a *collineation*. For instance, in the above example the permutation

$$(1.12.1) \qquad \pi = \begin{pmatrix} 1 & 2 & 3 & 4 & 5 & 6 & 7 & 8 & 9 & 10 & 11 & 12 & 13 \\ 7 & 11 & 9 & 4 & 3 & 12 & 8 & 10 & 13 & 2 & 6 & 1 & 5 \end{pmatrix}$$

is a collineation (Fig. 20). By means of the table and its graphical picture (Fig. 21) we can easily compile the line permutation λ induced by π,

$$(1.12.2) \qquad \lambda = \begin{bmatrix} 1 & 2 & 3 & 4 & 5 & 6 & 7 & 8 & 9 & 10 & 11 & 12 & 13 \\ 12 & 9 & 6 & 2 & 10 & 5 & 4 & 8 & 3 & 7 & 1 & 13 & 11 \end{bmatrix}.$$

From the properties (1), (2) and (3) of the mapping π we see that λ possesses the properties (1*), (2*) and (3*). Similarly, by starting with a λ having the properties (1*), (2*) and (3*), a dualization of the former reasoning leads to the π induced by λ and the validity of (1), (2) and (3) for the mapping π. Thus, although it would be proper to speak of a collineation $\langle \pi, \lambda \rangle$ or $\langle \lambda, \pi \rangle$ as π uniquely determines λ and vice versa, we need only speak of one or the other.

In our example the point P_4 and the line l_8 coincide with their images, i.e.

$$P_4 \xrightarrow{\pi} P_4 \quad \text{and} \quad l_8 \xrightarrow{\lambda} l_8.$$

Elements displaying this property are said to be *fixed points (invariant points)* and *fixed lines (invariant lines)* of the *collineation*.

However, it does not follow from the fact that l_8 is a fixed line, that any of its points are fixed. This can immediately be seen from the permutation

$$\begin{pmatrix} 3 & 5 & 9 & 13 \\ 9 & 3 & 13 & 5 \end{pmatrix}$$

which is the restriction of π to the set

$$l_8 = \{P_3, P_5, P_9, P_{13}\}.$$

Similarly, the lines passing through the fixed point P_4 are not necessarily fixed lines, as can immediately be seen from permutation

$$\begin{bmatrix} 1 & 11 & 12 & 13 \\ 12 & 1 & 13 & 11 \end{bmatrix}$$

which is the restriction of λ to the pencil of lines $\{l_1, l_{11}, l_{12}, l_{13}\}$ through P_4. In general, we can say only that a fixed line and a pencil with a fixed point for its centre are mapped onto themselves by a collineation.

There is, however, a collineation, having a fixed point and a fixed line with a remarkable property. An example of this is the point mapping

(1.12.3) $$\pi = \begin{pmatrix} 1 & 2 & 3 & 4 & 5 & 6 & 7 & 8 & 9 & 10 & 11 & 12 & 13 \\ 1 & 3 & 2 & 4 & 12 & 13 & 11 & 8 & 9 & 10 & 7 & 5 & 6 \end{pmatrix}$$

concerning Figs 20 and 21 and the line mapping

(1.12.4) $$\lambda = \begin{bmatrix} 1 & 2 & 3 & 4 & 5 & 6 & 7 & 8 & 9 & 10 & 11 & 12 & 13 \\ 1 & 4 & 3 & 2 & 9 & 10 & 8 & 7 & 5 & 6 & 11 & 12 & 13 \end{bmatrix}$$

induced by it. The fixed point P_4 of this collineation has the property that *each of the lines $l_1, l_{11}, l_{12}, l_{13}$, passing through it is a fixed line*. Such a point is said to be the *centre* of the collineation. Similarly, the fixed line l_3 has a remarkable property, too: *each one of the points P_1, P_8, P_9, P_{10} lying on it is a fixed point;* a line having this property is said to be the *axis* of the collineation. Of course we may ask the following questions:

1° Is it possible that the collineation has a centre but has no axis or that it has an axis but has no centre?
2° Is it possible for a collineation, which is not the identity, to have several centres or several axes?

These questions will be answered later. The *identity* collineation mapping having every point as a fixed point and every line as a fixed line; moreover, every point is a centre and every line is an axis.

In our example $P_4 \notin l_3$ and we pair off each point with its image with the exception of the fixed points $P_1, P_4, P_8, P_9, P_{10}$. Similarly we pair off each line with its image, except the fixed lines $l_1, l_3, l_{11}, l_{12}, l_{13}$.

We obtain the following:

$(P_2, P_3), (P_3, P_2), (P_5, P_{12}), (P_6, P_{13}),$

$(P_7, P_{11}), (P_{11}, P_7), (P_{12}, P_5), (P_{13}, P_6),$

$[l_2, l_4], [l_4, l_2], [l_5, l_9], [l_6, l_{10}],$

$[l_7, l_8], [l_8, l_7], [l_9, l_5], [l_{10}, l_6],$

where in each pair the second element is the image of the first. It is easy to see that two points of a point pair are connected by a fixed line; *the lines determined by the point pairs meet in the centre* P_4. If a collineation has this property we say that it is *perspective with respect to a point* P_4 (from a point). Similarly, the *point of intersection* of each of the line pairs above lies on the axis l_3. A collineation with this property is said to be *perspectivity with respect to a line* (perspective from a line). Our example is a *central-axial collineation*.

We mention now another example of a central-axial collineation, namely the mapping

(1.12.5) $\quad \pi = \begin{pmatrix} 1 & 2 & 3 & 4 & 5 & 6 & 7 & 8 & 9 & 10 & 11 & 12 & 13 \\ 1 & 2 & 3 & 4 & 7 & 5 & 6 & 10 & 8 & 9 & 13 & 11 & 12 \end{pmatrix}$

and the mapping

(1.12.6) $\quad \lambda = \begin{bmatrix} 1 & 2 & 3 & 4 & 5 & 6 & 7 & 8 & 9 & 10 & 11 & 12 & 13 \\ 1 & 2 & 3 & 4 & 6 & 7 & 5 & 9 & 10 & 8 & 12 & 13 & 11 \end{bmatrix}$

induced by π. The fixed point P_1 of the collineation $\langle \pi, \lambda \rangle$ is a centre and the invariant line l_1 is an axis, but now $P_1 \in l_1$.

If the centre lies on the axis, the collineation is said to be an *elation* if the centre and the axis are nonincident, the collineation is said to be an *homology*. Thus the central-axial collineation given in the first axample is an homology whilst the second example is an elation.

It is known from elementary combinatorics that the set of permutations of n elements forms a group, the *complete permutation group of the set*. The collineations of a finite plane form a subgroup of the complete permutation group of the set of points of the plane, since it is obvious that the product of two collineations is again a collineation and the inverse of every collineation is a collineation. This subgroup is called the *group of collineations* of the finite projective plane. — According to the *Erlangen Programme* of Felix Klein, finite projective geometry is the study of the group of collineations. Hence, it is understandable that, for example, a long chapter of M. Hall's book *The theory of groups* is mainly concerned with the geometry of finite projective planes.

An important role in the study of the groups of collineations is played by the homologies and elations. We may obtain *cyclic subgroups* by taking powers of collineations. If we remember the example given in Fig. 2 it is clear that the powers of the collineation

$$\pi = \begin{pmatrix} 1 & 2 & 3 & 4 & 5 & 6 & 7 \\ 2 & 3 & 4 & 5 & 6 & 7 & 1 \end{pmatrix}$$

form a group. The collineation π has no invariant point. If we want to find the cyclic incidence table of a given finite plane (provided that it exists) then we have to look for such a subgroup in the groups of collineations of the plane.

Collineations whose squares are equal to the identity are of great geometrical significance. Such a collineation is said to be an *involution*. The collineation π defined by (1.12.3) is an involution. Since this involution is a central-axial collineation, it may be considered as analogous to the axial reflection of the Euclidean plane. Here, the axis of reflection corresponds to the axis of the collineation, the line l_3, whereas the lines perpendicular to the axis of reflection correspond to the lines passing through the centre P_4.

The *order* of a collineation π is the smallest positive integer r such that π^r is the identity. For sake of completeness, let us note that in our examples the order of the collineation defined by (1.12.1) is 8, and the order of that defined by (1.12.5) is 3.

The relation of the central-axial collineation to Desargues' theorem will be dealt with later; but it can be seen immediately from Fig. 21 that a triangle of noninvariant points and its image under a central axial collineation are in perspective from the centre as well as from the axis of the collineation. — Thus for instance in the case of the collineation π defined by (1.12.3), with centre P_4 and axis l_3, the image of the triangle $P_5 P_6 P_{11}$ is $P_{12} P_{13} P_7$.

As we have seen, the analogy between the concepts of collineation in the finite plane and collineation in the classical projective plane is quite close; and many theorems of finite geometry can be obtained by using it. However, there are certain properties of finite planes which are entirely different from those of the classical projective plane and which cannot, therefore, be deduced by analogy. For example, on the classical projective plane the following theorem holds: Consider the mapping

$$\pi_0 = \begin{pmatrix} A & B & C & D \\ A' & B' & C' & D' \end{pmatrix}$$

of the vertices of a proper quadrangle $ABCD$ onto the vertices of any other proper quadrangle $A'B'C'D'$. Then there exists uniquely a collineation π such that its effect upon the quadruple of points $ABCD$ is precisely the mapping π_0. This theorem is also true for Galois planes but it is *not* true for every finite projective plane. The first half of this statement will only be illustrated by an example, but we shall prove the second half of it.

Property (3) implies that the image of a proper quadrangle, under a collineation, is again a proper quadrangle. In Figs 20 and 21 the proper quadrangle $ABCD = P_5 P_8 P_9 P_6$ is carried into $A'B'C'D' = P_3 P_{10} P_{13} P_{12}$, the correspondence

$$\pi_0 = \begin{pmatrix} 5 & 8 & 9 & 6 \\ 3 & 10 & 13 & 12 \end{pmatrix}$$

which is realized by the collineation π given in (1.12.1).

In order to prove the second half of our statement we have firstly to make some comments about the Fano plane (Fig. 22).

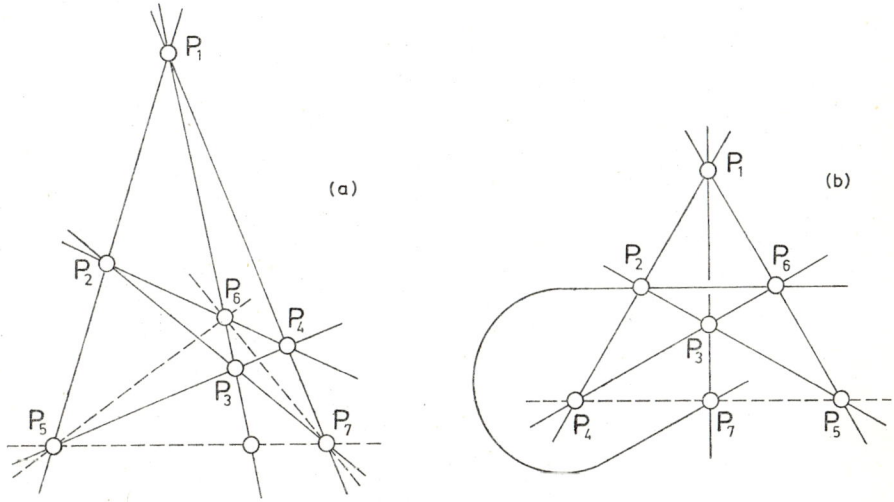

Figure 22

On the classical projective plane the diagonal points of a complete quadrangle form a proper triangle. — This is shown in Part (a) of Fig. 22. The diagonal points are obtained by intersecting the pairs of opposite sides:

$$P_5 = P_1 P_2 \cap P_3 P_4, \quad P_6 = P_1 P_3 \cap P_2 P_4, \quad P_7 = P_1 P_4 \cap P_2 P_3.$$

These form a proper triangle $P_5 P_6 P_7$. — We have already seen an example like this on a finite plane, too. Indeed, let us take the quadrangle $P_1 P_6 P_{13} P_9$ of Fig. 21. Its diagonal points are the points

$$P_5 = P_1 P_6 \cap P_{13} P_9, \quad P_{12} = P_1 P_{13} \cap P_6 P_9, \quad P_8 = P_1 P_9 \cap P_6 P_{13}$$

and the triangle $P_5 P_{12} P_8$ is a proper triangle.

But there exist finite planes in which complete quadrangles may have collinear diagonal points. Let us consider the Fano plane shown in Fig. 22 (b).

In this plane the diagonal points of the quadrangle $P_1P_2P_3P_6$ are:

$$P_4 = P_1P_2 \cap P_3P_6, \quad P_7 = P_1P_3 \cap P_2P_6, \quad P_5 = P_1P_6 \cap P_2P_3.$$

These points lie on the line l_5. There are in all 7 proper quadrangles on the Fano plane. Each of these can be obtained by deleting the points of one line and the remaining points are just the vertices of a proper quadrangle which had the deleted points for its diagonal points.

Thus on the finite plane we have to distinguish between two kinds of proper quadrangles.

1° Quadrangles having a diagonal triangle: *ordinary quadrangles*.
2° Quadrangles having no (proper) diagonal triangle: *Fano quadrangles*. A Fano quadrangle can also be defined by the property that its vertices and diagonal points together form all the points of a finite plane.

We remark without proof that the quadrangles of a Galois plane are either all Fano quadrangles or all ordinary quadrangles. Examples of the first case are the Fano plane and the plane of order 4 shown in Fig. 7. There is a plane of order 9 (a non-Galois plane) which contains a Fano quadrangle, as well as an ordinary quadrangle. We observe that property (3) (see p. 47) ensures that *the image of a Fano quadrangle, under a collineation, is a Fano quadrangle and the image of an ordinary quadrangle is an ordinary quadrangle*.

After this digression let us return to the proof of our statement regarding the correspondence π_0 transforming the quadrangle $ABCD$ into the quadrangle $A'B'C'D'$, on a non-Galois plane. In view of the comment above, assume that one of the two quadrangles is a Fano quadrangle and the other an ordinary quadrangle. Then, clearly, the correspondence π_0 cannot be represented by a collineation.

In what follows, all the planes on which every proper quadrangle is a Fano quadrangle will be called *Fano planes*.

1.13 Line preserving mappings of a finite affine plane and of a finite regular hyperbolic plane

Consider the affine plane of order q defined by the axiom system A. This plane consists of q^2 points and of $q(q+1)$ lines. We select from the complete permutation group of the q^2 points all the permutations possessing the property (3), stated at the beginning of Section 1.12. It is easy to verify that this set of permutations forms a subgroup in the complete permutation group. — This subgroup is said to be the *group of collineations of the affine plane*.

An immediate consequence of the definition is the *invariance of parallelism under affine collineations of the plane*. — In order to prove this we have only to observe that if $a \| b$ but $a \neq b$, then the set $a' \cap b'$ is empty, where a' and b' are the images of a and b respectively under an affine collineation. Because otherwise there would exist point P' such that $P' \in a'$ and $p' \in b'$. But then as P' is the image of some point P, the conditions $P \in a$ and $P \in b$ would be fulfilled, which contradicts the original assumption.

Let S_0 be an affine plane and let its group of collineations be denoted by G_0. Let us add to the plane the ideal points, defined as the common points of parallel lines, and the ideal line l formed by these points. Let the projective plane so obtained be denoted by S and the group of collineations of this plane by G. In fact, G_0 is isomorphic with the subgroup $G_0(l)$ of G, consisting of those permutations for which the line l is an invariant line. We know already that two affine planes obtained from S by deleting different lines are not necessarily isomorphic with each other; this is reflected in the fact that if $l_1, l_2 \in S$, then $G_0(l_1)$ and $G_0(l_2)$ are not necessarily isomorphic groups.

With respect to the ideal line the following central-axial collineations are distinguished, which are useful in the study of the structure of the affine plane:

(a) If the centre of the collineation is not an ideal point but its axis is an ideal line l, then the mapping is said to be *homothetic* and the centre is called the centre of the homothecy.
(b) If the centre of the collineation is an ideal point and its axis is the ideal line l, then the mapping is said to be a *translation*.
Homothecy and translation are analogous to the similarly named transformations in classical geometry.
A collineation of the affine plane known by the descriptive name *affinity*.
(c) If the centre of the collineation is an ideal point but its axis is not the ideal line, then the mapping is called an *axial affinity*.
(d) If the centre of the collineation is an ideal point and lies on the axis but this axis is not the ideal line, then the mapping may be called an *ideal elation*. — It has no generally accepted name, its German name is "Scherung". —

We give the following example of an ideal elation: Consider in Fig. 21 the affine plane of order 3 realized by the points $P_5, P_6, \ldots, P_{13}$ and by the (non-straight) lines $l_2, l_3, \ldots, l_{13}$. Now the point mapping

$$\pi = \begin{pmatrix} 5 & 6 & 7 & 8 & 9 & 10 & 11 & 12 & 13 \\ 7 & 5 & 6 & 8 & 9 & 10 & 12 & 13 & 11 \end{pmatrix}$$

which induces the line mapping

$$\lambda = \begin{bmatrix} 2 & 3 & 4 & 5 & 6 & 7 & 8 & 9 & 10 & 11 & 12 & 13 \\ 2 & 3 & 4 & 9 & 10 & 8 & 12 & 13 & 11 & 6 & 7 & 5 \end{bmatrix}$$

corresponds to the definition *(d)*.

Beside the projective and the affine planes we have already encountered another figure called a finite plane. This was the finite regular hyperbolic plane. Obviously, we could investigate the group formed by the one-to-one line preserving mappings of such a plane onto itself. As we shall not deal with hyperbolic planes again, we examine carefully here the plane shown in Fig. 14. Firstly, we mention the surprising fact that while the order of the group of collineations of the projective plane of order 3 is 6616, that of the group of our hyperbolic plane is only 6, though this plane also contains 13 points. In both of these cases we have to select, from the complete permutation group of 13! elements, those permutations which leave invariant the collinear triples of points. The number of collinear point triples on the projective plane of 13 points is 52, while on our hyperbolic plane of 13 points it is 26. Thus, in the hyperbolic plane there are exactly half as many restrictions on a permutation in order that is satisfy (3) (p. 61) as there are in the projective plane. But nevertheless, the number of suitable permutations is far fewer in the hyperbolic case. The selection of the line preserving mappings seems to be a difficult combinatorial problem; however, it is simplified by knowing certain properties of the following figures in the plane.

Consider two nonintersecting lines a and b of the plane and a point C not incident with either of these lines. Every line of our plane contains 3 points; 6 lines pass through every point of the plane and of the 6 lines passing through a particular point 3 intersect any given line not containing the point in question and 3 do not intersect the given line. Consequently, for the figure Cab, it is not necessarily true that every point of a can be projected from C onto b; moreover, there exist figures Cab such that *no* point of a can be projected onto b and no point of b can be projected onto a either; e.g.

$$C = P_1, \quad a = l_3, \quad b = l_{12}.$$

If every point of a can be projected from C onto b, then the septet of points consisting of the points of the two lines and the point C is called a *perspective septet of points*; e.g. in Fig. 23 Part *i* we have

$$C = P_{11}, \quad a = l_{15} = \{P_3, P_7, P_{12}\}, \quad b = l_{16} = \{P_1, P_4, P_8\}.$$

Here P_{11} is the *centre* of the septet of points in question. We mention here two other examples of perspective septets of points, namely

$$C = P_{13}, \quad a = l_8 = \{P_1, P_3, P_{11}\}, \quad b = l_{23} = \{P_7, P_9, P_5\};$$

and
$$C = P_{13}, \quad a = l_{14} = \{P_2, P_9, P_{11}\}, \quad b = l_{18} = \{P_8, P_3, P_5\}.$$

It will be noted that the point P_{13} is the common centre of these two septets. P_1 is not the centre of any perspective point septet. The points of our hyperbolic plane can be classified according to the above examples: there are *double centres* — the points P_2, P_8, P_9, P_{11} — *simple centres* — the points P_6, P_{12}, P_{13} — and there are *ordinary points* (i.e. points that are not centres) — the points $P_1, P_3, P_4, P_5, P_7, P_{10}$.

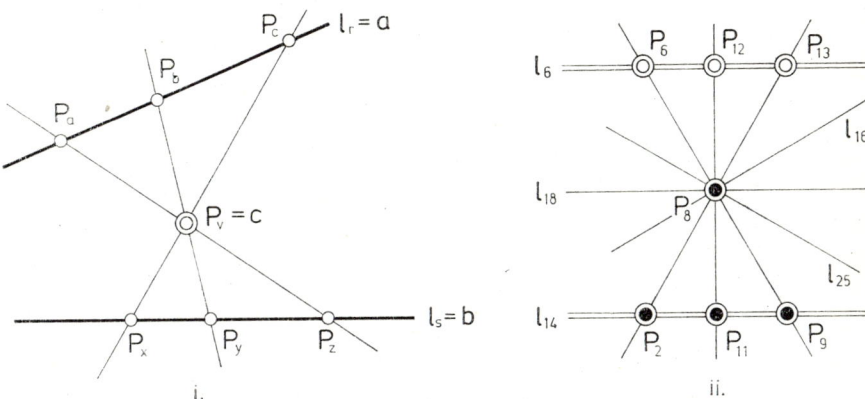

Figure 23

The $4+3=7$ centres (Part *ii* cf. Fig. 23) form a quite remarkable figure. Namely, the simple centres lie on a line: $l_6 = \{P_6, P_2, P_{13}\}$, let this be called a *simple axis*; three of the double centres lie on a line: $l_{14} = \{P_2, P_9, P_{11}\}$ let this be called a *double axis*; the fourth double centre, P_8 is not on either of the two axes but from this point the three simple centres can be projected into the three double centres and each of the three projecting lines will be called a *mixed axis*.

It can be seen from property (3) of the line preserving mappings that the images of perspective point septets and only these are again perspective point septets. It follows that the images of ordinary points, simple centres and double centres are points of the same type. Moreover, the point P_8 must, in every case, be a fixed point, because the other three double centres are collinear. Thus this point, differing in its properties from all other points of the plane, can be called the principal point of the plane. — It follows from our statements concerning points that the images of the simple axes, double axes and mixed axes are axes of the same type.

It is now easy to deduce that only line preserving point transformations of our hyperbolic plane are the following:

	1	2	3	4	5	6	7	8	9	10	11	12	13
π_1:	1	2	3	4	5	6	7	8	9	10	11	12	13
π_2:	7	2	5	10	3	12	1	8	11	4	9	6	13
π_3:	4	9	7	1	10	13	3	8	2	5	11	12	6
π_4:	3	9	10	5	7	12	4	8	11	1	2	13	6
π_5:	10	11	1	7	4	13	5	8	2	3	9	6	12
π_6:	5	11	4	3	1	6	10	8	9	7	2	13	12

The Cayley table of this permutation group is (by writing the index k instead of the element π_k):

×	1	2	3	4	5	6
1	1	2	3	4	5	6
2	2	1	4	3	6	5
3	3	5	1	6	2	4
4	4	6	2	5	1	3
5	5	3	6	1	4	2
6	6	4	5	2	3	1

if we assign to the elements $\pi_1, \pi_2, \pi_3, \pi_4, \pi_5, \pi_6$ the permutations *abc, acb, bac, bca, cab, cba*, respectively, then it turns out that the Cayley table above is just the multiplication table of the complete permutation group of the three elements a, b and c.

The results of this section together with those of Section **1.9** completes our discussion of the geometry of a $\langle 3, 3 \rangle$-plane.

1.14 Finite projective planes and complete orthogonal systems of Latin squares

In this section we shall discuss combinational problems, which are now dealt with in mathematical statistics, but which first occurred in connection with question discussed by Euler in 1782.

The famous problem, the so-called "problem of the 36 officers" is the following: On a military parade 36 officers must be arranged in a 6×6 square satisfying the requirements given below. There are 6 branches of military service and 6 military ranks. Each branch of the service sends one officer of each rank into the parade square. The officers must be arranged in the square so that every branch of service and every rank should be represented in each line and in each column. Euler recognized that the problem had no solution; and a proof was not found until 1900 when Tarry gave a complicated solution.

The nonexistence of the projective plane of order 6 follows from the insolubility of the problem above. This will be proved later.

Before introducing some concepts of combinatorics, we must define a certain algebraic structure. Let **Q** be a finite set containing q elements, and let an operation on **Q** (called multiplication) be defined by the following axioms:

(1) If $a \in \mathbf{Q}$ and $b \in \mathbf{Q}$, then $ab = c \in \mathbf{Q}$.

(2) If $a, b, c \in \mathbf{Q}$ but $a \neq b$, then $ac \neq bc$ and $ca \neq cb$.

The existence of such a structure will be verified by an example. Let Fig. 24 be the operation table of the set $\mathbf{Q} = a, b, c, d$.

We might have chosen as an example the Cayley table of any group satisfying as it would the above axioms, but the structure in question need not be a group. In fact, our example is not a group, since according to the table we have $(bc)d = ad = c$ and $b(cd) = ba = b$, and so the multiplication is not associative.

The finite structure defined by (1), (2), is called a finite *quasigroup*.

In the operation table of the finite quasigroup each of the lines (columns or rows) contains a permutation of the elements of the set; consequently, each of the elements occurs only once in each line, i.e. the squares occupied by a particular element form together a diagonal of the table. Diagonals of this kind are

	a	b	c	d
a	a	b	d	c
b	b	c	a	d
c	c	d	b	a
d	d	a	c	b

Figure 24

0	7	5	8	3	1	4	2	6
1	8	3	6	4	2	5	0	7
2	6	4	7	5	0	3	1	8
3	1	8	5	0	7	2	6	4
4	2	6	3	1	8	0	7	5
5	0	7	4	2	6	1	8	3
6	4	2	1	8	3	7	5	0
7	5	0	2	6	4	8	3	1
8	3	1	0	7	5	6	4	2

Figure 25

said to be the *parallel* diagonals of the table, as they have no squares in common. Examples of these are the two diagonals consisting of the elements 5 and 7 in Fig. 25.

If we can arrange on a $q \times q$ table q^2 elements of q different types so that a permutation of the elements occurs in every line of the table, then the table is called a *Latin square of order* q. Obviously, every Latin square, considered as an operation table, determines a quasigroup. The elements of the ground set **Q** of the Latin square will usually be denoted by the indices $0, 1, 2, ..., q-1$.

Consider now the total class of Latin squares belonging to the ground set $\mathbf{Q} = \{0, 1, 2, ..., q-1\}$. We shall introduce the important concept of the orthogonality of the Latin squares of this class.

Firstly, however, we shall define an operation which assigns a $q \times q$ table, consisting of certain ordered pairs of elements, to every pair $(\mathbf{L}_1, \mathbf{L}_2)$ of Latin squares of order q as follows. The table so obtained will be denoted by $\mathbf{U} = \mathbf{L}_1 \sqcap \mathbf{L}_2$. The first and the second element of a pair of elements standing in a square of **U** is given by the element standing in the homologous square of $\mathbf{L}_1$ and $\mathbf{L}_2$, respectively.*

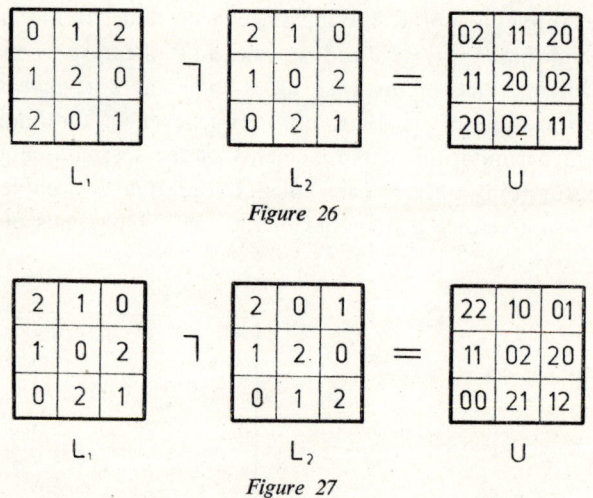

Figure 26

Figure 27

Now there are q^2 ordered pairs of elements of the set **Q**; if each of these occurs in the $q \times q$ table $\mathbf{U} = \mathbf{L}_1 \sqcap \mathbf{L}_2$ then we say that $(\mathbf{L}_1, \mathbf{L}_2)$ is an *orthogonal pair*. Clearly, if $(\mathbf{L}_1, \mathbf{L}_2)$ is an orthogonal pair then so is $(\mathbf{L}_2, \mathbf{L}_1)$. This symmetrical relation between $\mathbf{L}_1$ and $\mathbf{L}_2$ is called *orthogonality*. It is not necessarily true that given

---

* Two squares in different tables of the same size are said to be *homologous* if they are both the k-th square of the j-th row of their tables, for some pair of integers (j, k).

any $q \times q$ table, $\mathbf{V}$, of ordered pairs of elements of $\mathbf{Q}$ in which each pair occurs then there exist Latin squares $\mathbf{L}_1, \mathbf{L}_2$ such that $\mathbf{L}_1 \sqcap \mathbf{L}_2 = \mathbf{V}$. For example, in the case of $q = 3$ the following table:

$$\mathbf{V} = \begin{pmatrix} (0,0) & (0,1) & (0,2) \\ (1,0) & (1,1) & (1,2) \\ (2,0) & (2,1) & (2,2) \end{pmatrix}$$

cannot be derived from the Latin squares by the operation $\sqcap$. An example of a pair of orthogonal Latin squares is shown in Fig. 27, whilst Fig. 25 shows two Latin squares that are not orthogonal. Obviously, no Latin square of order $q > 1$ can be orthogonal to itself.

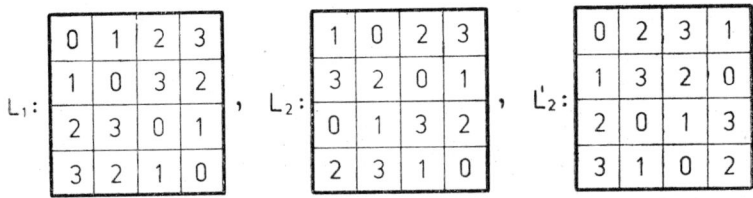

Figure 28

We can now formulate the Euler—Tarry theorem mentioned at the beginning of this section as follows: *An orthogonal pair of Latin squares of order six does not exist.*

It is easy to check that in Fig. 28 $\mathbf{L}_1$ and $\mathbf{L}_2$ form an orthogonal pair, further that $\mathbf{L}_2'$ is obtained from $\mathbf{L}_2$ by permuting the elements as follows:

$$\begin{pmatrix} 0 & 1 & 2 & 3 \\ 2 & 0 & 3 & 1 \end{pmatrix}$$

$\mathbf{L}_1$ and $\mathbf{L}_2'$ are also an orthogonal pair. It is easy to see that a Latin square is transformed by a permutation of the ground set into another Latin square, further that by transforming any one of the Latin squares $\mathbf{L}_1$ and $\mathbf{L}_2$ by a permutation of the ground set, the orthogonality or nonorthogonality of the pair is preserved.

Let us now assume that we have constructed all the possible Latin squares of order q from the set $\mathbf{Q} = \{0, 1, \ldots, (q-1)\}$ and we wish to choose r mutually orthogonal Latin squares from them. It follows from the above remark that to see whether this can be done it suffices to investigate the Latin squares which agree in their first column; moreover, it can be assumed that the first column (when read from the top) is just the permutation $0, 1, 2 \ldots (q-1)$. If we can choose, from the set of the Latin squares, a subset of r mutually orthogonal Latin squares, then this subset will be called an *orthogonal system of r elements of the Latin squares of order* q.

Theorem: *If an orthogonal system of r elements of the Latin squares of order q exists then* $r < q$.

Proof: Let the system in question be $L_1, L_2, \ldots, L_r$ and let the first column of every Latin square be the permutation $0, 1, 2, \ldots, (q-1)$. Assume that $r \geq q$. We have $q-1$ choices for the second element in the first row of each of the squares $L_i (i=1, \ldots, r)$. Since we have at least q Latin squares, there must be two, say L_s and L_t, which contain the same element, k, in the second place of the first row. But then the second element of the first row and the k-th element of the first column of $L_s \sqcap L_t = U$ both contain the pair (k, k), which contradicts the orthogonality of L_s and L_t, and the theorem is proved. For certain values of q, $r = q-1$ can occur, as will be shown by an example. Consider the Latin squares L_1, L_2, L_3 of order 4 in Fig. 29. It is easy to check that these are pairwise orthogonal. The three Latin squares are so ordered that their first columns are identical but as the Euler—Tarry theorem shows, for $q=6, r<2$.

An orthogonal system for which $r=q-1$, when it exists, will be called a *complete* orthogonal system.

It would be useful, particularly in mathematical statistics, to be able to construct, for any given positive integer q, an orthogonal system of Latin squares of order q with the largest possible number of elements. This is a very difficult and, in general, unsolved problem. Indeed, it is only a few years since the existence of an orthogonal pair of Latin squares of order 10 was proved and this was by explicit construction. *If* $q = p^r$ *(where p is prime), then a complete orthogonal system of the Latin squares of order q exists.* This statement will be proved by using some of the properties of the finite projective planes that we have discussed.

We suppose $q = p^r$ and then construct the $\Gamma(q)$ table of the plane $S_{2,q}$. Fig. 30 shows the case $q=2^2$. We separate the part consisting of the $q \times (q-1)$. Consider the rectangle which is q parcels wide and $q-1$ parcels high in the bottom

0	1	2	3
1	0	3	2
2	3	0	1
3	2	1	0

L_1

0	2	3	1
1	3	2	0
2	0	1	3
3	1	0	2

L_2

0	3	1	2
1	2	0	3
2	1	3	0
3	0	2	1

L_3

Figure 29

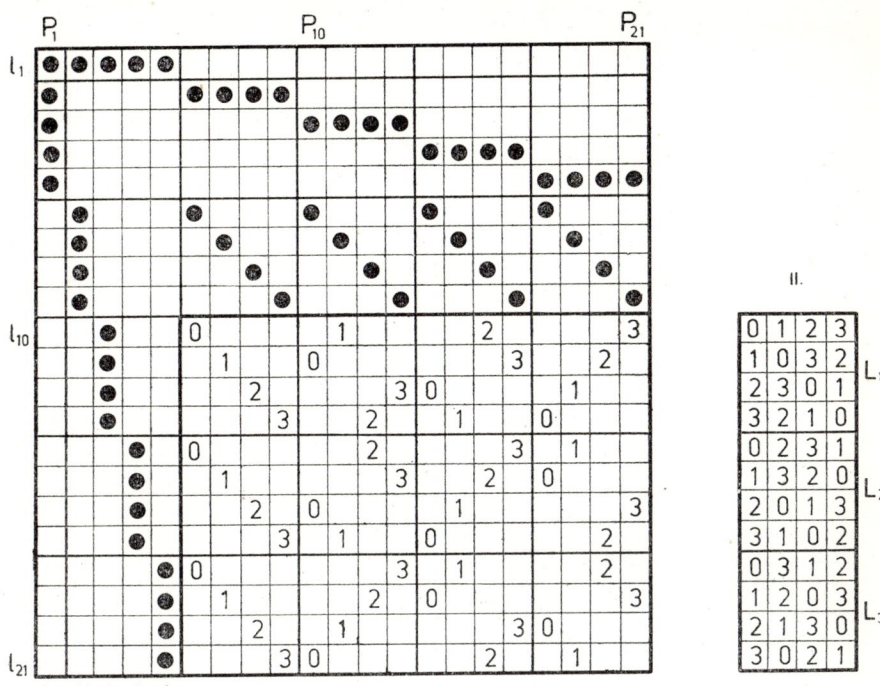

Figure 30

right hand corner of $\Gamma(q)$. In each row of this rectangle replace each incidence sign by the index of the column of the parcel which contains it. Thus, each row of the rectangle contains a permutation of the indices $0, 1, \ldots, q-1$. (Part *i* of Fig. 30.) If we now superimpose the q columns of parcels in the rectangle we obtain a column consisting of $q-1$ parcels (Part *ii* of Fig. 30). Denote these $q-1$ parcels by $L_1, \ldots, L_{q-1}$. The construction just described will be referred as a λ-*procedure* or will be denoted by the symbol $\Gamma(q) \rightarrow (L_1, L_2, \ldots, L_{q-1})$.

In the example shown in Fig. 30 we obtain by $\Gamma(4) \rightarrow (L_1, L_2, L_3)$ a complete orthogonal system of Latin squares, since L_1, L_2, L_3 are precisely the same as in Fig. 29. We claim that a complete orthogonal system of Latin squares of order q is obtained from the $\Gamma(q)$ table of a plane $S_{2,q}$ by the λ-procedure.

A part of our claim, namely that each of the obtained index tables $L_1, L_2, \ldots, L_{q-1}$ is a Latin square, can be easily proved by using the properties D_1, D_2, D_3 given in Section **1.4**. Furthermore, we can prove by means of a geometrical argument that (L_s, L_t) (where $s \neq t$; $s, t = 1, 2, \ldots, q-1$) are orthogonal pairs.

Assume that there exist s and t, $s \neq t$, such that $(\mathbf{L}_s, \mathbf{L}_t)$ do not form an orthogonal pair. This means that we can find two squares in $\mathbf{L}_s$ containing the same index, α say, such that the two homologous squares in $\mathbf{L}_t$ also contain the same index, β say. The pair (α, β) will then occur in two of the squares of $\mathbf{L}_s \sqcap \mathbf{L}_t = \mathbf{U}_{st}$ (e.g. Fig. 31).

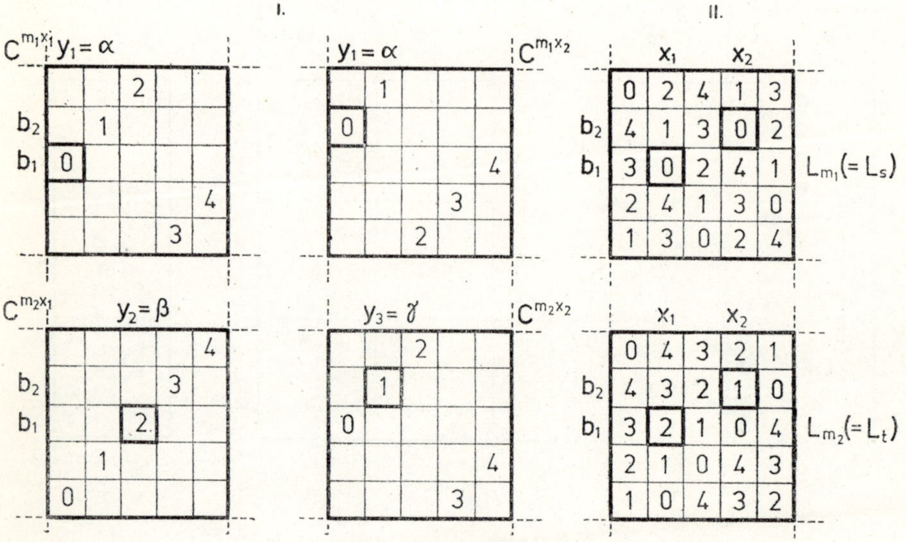

Figure 31

We shall employ the notation introduced in section 1.5. Let $s = m_1$, and $t = m_2$, and let the two chosen squares in $\mathbf{L}_{m_1}$ and their homologous associate in $\mathbf{L}_{m_2}$ be (b_1, x_1) and (b_2, x_2). Let index α occur in each of the squares (b_1, x_1), (b_2, x_2) of $\mathbf{L}_{m_1}$ and let the indices β and γ occur in their homologous associates respectively; we wish to determine the structure of the parcels involved in the λ-procedure for which we obtain $\beta = \gamma$. Let us introduce the notation $\alpha = y_1$, $\beta = y_2$, $\gamma = y_3$. From the prescription of the λ-procedure and the discussions of the Section 1.5 it follows that the generators of the pairs (α, β) and (α, γ) of indices the $\mathbf{U}_{st}$ were parcels $C^{m_1 x_1}$, $C^{m_1 x_2}$, $C^{m_2 x_1}$, $C^{m_2 x_2}$ i.e. in the squares (b_1, y_1), (b_2, y_1), (b_1, y_2), (b_1, y_2) of these parcels occurred the indices α, α, β, γ respectively. This can be formulated by the ternary equation $y = F(x, m, b)$, expressing the incidence of the point (x, y) and the line $[m, b]$, as follows:

$$y_1 = F(x_1, m_1, b_1), \quad y_1 = F(x_2, m_1, b_2),$$
$$y_2 = F(x_1, m_2, b_1), \quad y_3 = F(x_2, m_2, b_2).$$

Thus the equality $\beta=\gamma$ holds, if $F(x_1, m_1, b_1) = F(x_2, m_1, b_2)$ and $F(x_1, m_2, b_1) = F(x_2, m_2, b_2)$. We claim that in the case of our plane these two equations cannot hold. In order to prove this we shall make use of model shown in Fig. 32, where

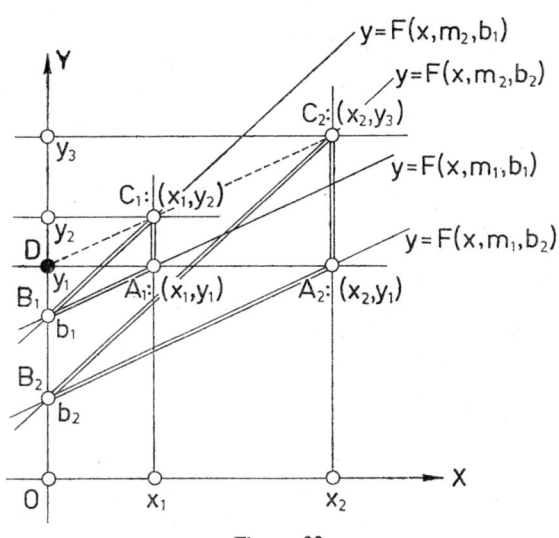

Figure 32

we interpret the ternary equations as we did in Fig. 10 of Section 1.5. In view of the assumptions $m_1 \neq m_2$, $b_1 \neq b_2$ and $x_1 \neq x_2$; we have, in our figure,

$$B_1 \neq B_2, \quad A_1 \neq A_2, \quad C_1 \neq C_2$$

and because

$$y_1 = F(x_1, m_1, b_1) = F(x_2, m_1, b_2)$$

we have

$$A_1 A_2 \| OX.$$

Thus the coordinates of the point $A_1 A_2 \cap OY = D$ are $(0, y_1)$. The pair of triangles $A_1 B_1 C_1$ and $A_2 B_2 C_2$ are in perspective from the ideal line (axis) and $A_1 A_2 \cap B_1 B_2 = D$. As Desargues' theorem on the plane $S_{2,q}$ holds we deduce that $D \in C_1 C_2$. Thus $y_2 = y_3$ (i.e. $\beta = \gamma$) would mean that $D, C_1 C_2$ were points lying on a line parallel to the line OX, which contradicts the other assumptions. This contradiction completes our proof that $(\mathbf{L}_s, \mathbf{L}_t)$ is an orthogonal pair.

We shall now prove that *a projective plane of order q can be derived from a complete orthogonal system of Latin squares of order q when such a system exists.*

We shall not attempt to reverse the λ-procedure as this may not be possible for an arbitrary complete orthogonal system.

Let $\mathbf{L}_1, \ldots, \mathbf{L}_{q-1}$ a given complete orthogonal system of the Latin squares of order q, as usual the elements of the squares are in the set $\mathbf{Q} = \{0, 1, \ldots, (q-1)\}$.

Let $\mathbf{Q}^* = \{1, 2, \ldots, (q-1)\}$. An example of the model we will construct is shown in Fig. 33.

The given complete orthogonal system will be denoted briefly by **LQ** and the individual Latin squares by (ζ) where $\zeta \in \mathbf{Q}^*$. By a square (ξ, η, ζ) of the system **LQ** we shall mean the square of (ζ) at the intersection of the row with the index ξ and the column with the index η, where $\xi, \eta \in \mathbf{Q}$.

(1)					(2)					(3)					(4)				
0	1	2	3	4	0	2	4	1	3	0	3	1	4	2	0	4	3	2	1
1	2	3	4	0	1	3	0	2	4	1	4	2	0	3	1	0	4	3	2
2	3	4	0	1	2	4	1	3	0	2	0	3	1	4	2	1	0	4	3
3	4	0	1	2	3	0	2	4	1	3	1	4	2	0	3	2	1	0	4
4	0	1	2	3	4	1	3	0	2	4	2	0	3	1	4	3	2	1	0

Figure 33

We shall understand by ordinary points the ordered number pairs (x, y) and by ideal points the singletons (z) and (∞), where $x, y, z \in \mathbf{Q}$. Thus the number of points is $q^2 + q + 1$ in all.

We shall understand by the ideal line the set of the points $(0), (1), \ldots (q-1)$ and (∞). We shall also define the following:

A first coordinate line consists of the ideal point (∞) and the ordinary points (x, b) where x runs through the elements of the set $\mathbf{Q}$ and b is a fixed element of this set. The number of the lines of this kind is, therefore, q.

A *second coordinate line* consists of the ideal point (0) and the ordinary points (a, y) where y runs through the elements of $\mathbf{Q}$ and a is a fixed element of this set. The number of these lines is also q.

We know that the squares containing an element $j(\in \mathbf{Q})$ are diagonally arranged in each of the Latin squares. In the example of Fig. 33 we have shown the Latin square (3) and the diagonal formed by its squares containing the element **2**. With respect to the system **LQ** we speak of the diagonal $[j, \zeta]$ as a *diagonal line*; namely, it is the set consisting of the ideal point (ζ) and the ordinary points (x, y) such that the squares (x, y, ζ) all contain the element j. Clearly, the number of diagonal lines $[j, \zeta]$ $j \in \mathbf{Q}$ and $\zeta \in \mathbf{Q}^*$ is equal to $q(q-1)$.

Thus, in all, the number of lines we have obtained is $1 + q + q + q(q-1) = q^2 - q + 1$. Every line consists of $q+1$ points. The figure of the points and lines so derived from the given **LQ** will be called S and correspondence $\mathbf{LQ} \to S$ is called an *ω-procedure*.

Theorem: *The figure S derived from a complete orthogonal system of Latin squares of order q by an ω-procedure is a projective plane of order q.*

Proof. Figure 34 is a graphical model realizing the points of the figure S (corresponding to the example in Fig. 33). The ideal line as well as the first and the second coordinate lines are represented by straight lines. Each diagonal line

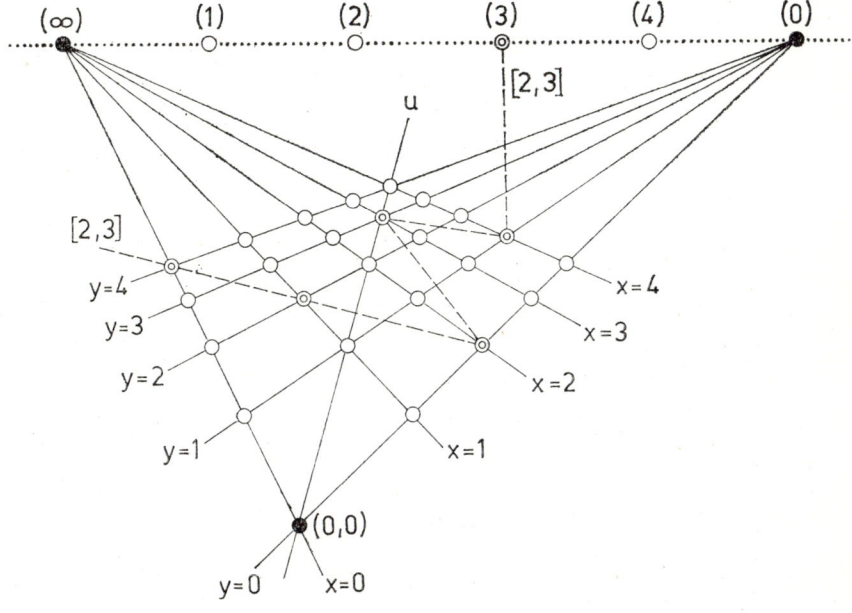

Figure 34

is represented by a broken line consisting of the segments following each other in the order $0, 1, \ldots, q-1$ and the last segment terminates in the ideal point of the line. In Fig. 34 we have only drawn the diagonal line $[2, 3]$.

Firstly we have to prove that any two lines have a unique point of intersection. The only case here which is not immediately obvious from the definition is that of two diagonal lines. Suppose that the two diagonal lines are $[i, r]$ and $[j, s]$, where $r \neq s$. Now $\mathbf{L}_r$ and $\mathbf{L}_s$ are orthogonal therefore the pair (i, j) occurs once and once only in $\mathbf{U}_{rs} = \mathbf{L}_r \sqcap \mathbf{L}_s$, suppose that it occurs in the square (ξ, η) of $\mathbf{U}_{rs}$; this implies that the index i occurs in (ξ, η, r) and the index j occurs in (ξ, η, s). Therefore (ξ, η) is the unique common point of $[i, r]$ and $[j, s]$.

Similarly, the existence and the uniqueness of the line connecting any two points is immediate except when firstly one of the points is and ideal point and the other an ordinary point and secondly when both of them (a, b) and (c, d) say, are ordinary points but $a \neq c$ and $b \neq d$.

In the first case let the ordinary point be (a, b) and let the (ideal) point be (j). If $j = \infty$, then the first coordinate line containing (a, b) is the required line; whilst, if $j = 0$, then the second coordinate line containing (a, b) is the required line; if $j \in \mathbf{Q}^*$, then consider the element occurring in the square (a, b, j), let this element be denoted by i. Then the unique line joining (j) and (a, b) is the diagonal line $[i, j]$.

In the second case consider the $(q+1)$ lines each connecting the point (a, b) with an ideal point. Each pair of these lines have only one point in common and there are $q+1$ points on every line, hence the lines passing through the point (a, b) contain $(q+1)q+1 = q^2+q+1$ points of the figure **S** i.e. all of its points. Thus there exists one and only one of these lines containing the points (c, d). Clearly, the points (∞), (0), $(0, 0)$, $(1, 1)$ are the vertices of a proper quadrangle. The proof of our theorem is now complete.

By means of this simple combinatorial theorem the geometry of the finite projective planes can be considered as a chapter of combinatorics. This chapter is rich in difficult or even as yet unsolved problems.

An interesting question is to find a connection between the λ-procedure and the ω-procedure; this, however, will not be dealt with here.

1.15 The composition of the linear functions and the $D(X, Y)$ plane

In this section we shall establish the connection between finite projective geometry and certain finite algebraic structures.

We shall deal with the linear functions $y = ax + b$ on the field of real numbers **R** by means of their representation on the Euclidean plane. The graphical model of the field **R** is the x-axis. A linear function is a mapping

$$x \to ax + b \quad (a \neq 0)$$

of the field **R** onto itself and is represented in the Euclidean plane by the following geometrical operation (Fig. 35). A coordinate system is given and with it the axes as well as the line l_0 connecting the unit point E with the origin 0. Furthermore, let l be the line whose equation is $y = ax + b$, $(a \neq 0)$. The mapping is now generated by this line. Let x be a point on the x-axis, let the line 1 be drawn through x and parallel to the y-axis. The line 1 cuts the line l at a point P. Let the line 2 be drawn through P and parallel to the x-axis. The line 2 cuts the line l_0 at a point P_0. Let the line 3 be drawn through P_0 and parallel to the y-axis. The line 3 cuts the x-axis in a point x' called the image point of x. If the image point x' is given, its primitive point x can be obtained by reversing the process. Obviously, the mapping can have just one fixed point, if the line l intersects the line l_0, namely the point of the x-axis having the same abscissa as the point of intersection of l with l_0.

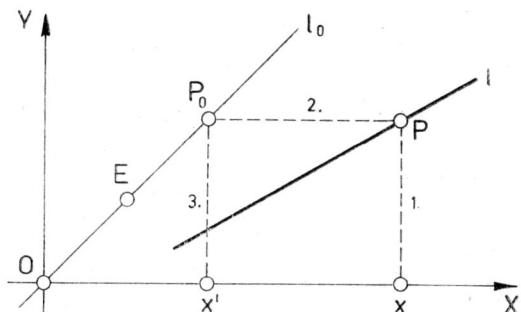

Figure 35

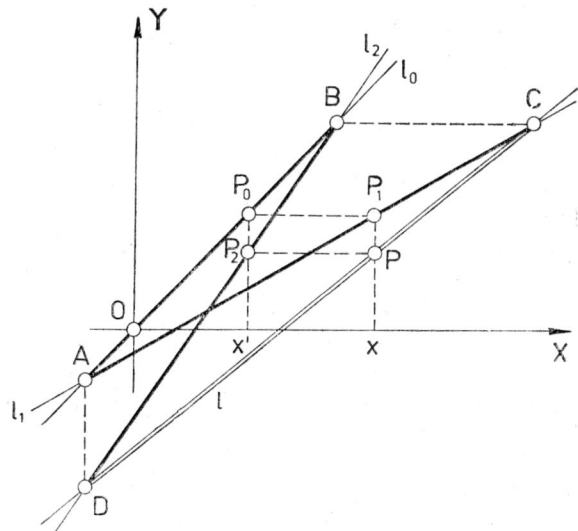

Figure 36

Let the following two linear functions be given (Fig. 36):

$$l_1: x \to ax+b \quad \text{and} \quad l_2: x \to cx+d \quad (a \neq 0, \quad c \neq 0)$$

by the *composition* $l_1 l_2 = l$ of the two functions we shall mean the function

$$l: x \to c(ax+b)+d,$$

this operation of composition is not necessarily commutative even over **R** since $cb+d$ is equal to $ad+b$ if and only if one of the following cases occurs:

1° At least one of the lines l_1, l_2 is identical with the line l_0.
2° The lines l_1 and l_2 are coincident.
3° Both l_1 and l_2 are parallel to l_0.
4° $l_1 \cap l_2 \in l_0$.

These four possibilities can be expressed by saying that *the not necessarily distinct lines* l_0, l_1, l_2 *are the elements of a (flat) pencil; the composition* $l_1 l_2 = l_2 l_1 = l$ *is then, of course, an element of the same pencil.*

The composition is easily seen to be an associative operation.

We shall now interpret geometrically the composition of linear functions (Fig. 36). Let two linear functions be generated by the lines l_1 and l_2, respectively. Given any point x on the x-axis we perform the following construction: Let the line through x and parallel to the y-axis meet l_1 in P_1; let the line through P_1 and parallel to the x-axis meet l_0 in P_0; let the line through P_0 and parallel to the y-axis meet l_2 in P_2 and let the line through P_2 and parallel to the x-axis meet xP_1 in P. Then $P_2 P$ will meet l_0 in a point P' and the line through P' parallel to the y-axis will meet the x-axis in a point x'. As the point x varies on the x-axis the mapping $X \to X'$ generates the required composition of functions. Moreover, as x varies the point P describes the line $l = l_1 l_2$ generating this composition. The quadrangle $P_1 P_0 P_2 P$ is called a *coordinate quadrangle* supported by the lines l_0, l_1, l_2.

Let us assume that $A = l_0 \cap l_1$ and and $B = l_0 \cap l_2$ are distinct and are not ideal points. If x is chosen so that P_0 coincides with the point $A = l_0 \cap l_1$ then P will coincide with the point of intersection, D, of l_2 with the line through A parallel to the y-axis. Similarly, if P_0 coincides with $B = l_0 \cap l_2$ then P will coincide with the point of intersection, C, of l_1 with the line through B parallel to the x-axis. The given lines l_0, l_1 and l_2 and the coordinate system determine uniquely the quadruple of points $ABCD$ and the figure has the appearance of the image of a tetrahedron under orthogonal projection. Furthermore, an arbitrary coordinate quadrangle $P_1 P_0 P_2 P$ can be considered as the image of a plane section of the tetrahedron parallel to the edges AD and BC. The solid figure could also be considered as the union of two triangular prisms: one of these prisms is $AP_1 P_0 \cdot DPP_2$ and the other $BP_0 P_2 \cdot CP_1 P$.

It is possible to introduce into the finite plane a concept analogous to the linear function; namely, the ternary function introduced in section **1.5**. However, the crucial question here is: How do ternary functions behave under composition? Given the two ternary functions $F(-, m_1, b_1)$ and $F(-, m_2, b_2)$, $(m_1, m_2 \neq 0)$; does there exist a ternary function $F(-, m, b)$, $(m \neq 0)$ such that for all x,

$$x' = F(x, m_1, b_1) \,\&\, x'' = F(x', m_2, b_2) \Rightarrow x'' = F(x, m, b)$$

In solving this problem we shall continue to use Fig. 36, but shall now suppose that the construction of the quadrangle $ABCD$ has been done in a finite plane. The question above is then equivalent to the following: For every x, does the vertex P of a coordinate quadrangle supported by the lines l_0, l_1 and l_2 always lie on the line CD?

Consider the triangles CP_1P, BP_0P_2 and pair off the vertices as follows:

$$\begin{pmatrix} C & P_1 & P \\ B & P_0 & P_2 \end{pmatrix}$$

where this means C is paired off with B, etc.; these triangles are in perspective with the ideal point X as a centre. We have the point $A = CP_1 \cap BP_0$ and the ideal point $Y = P_1P \cap P_0P_2$. If Desargues' theorem is valid it suffices to assume a certain restricted validity; namely when a point X is the centre and a pair of corresponding sides intersect in a point Y —, then if the axis will be the line AY and the common point of the lines $BP_2 = l_2$ and CP will also be on the axis. But AY is cut by the line l_2 in the point D, thus the line CP would have to pass through the point D. Obviously, if the point P is not on the line CD, then even in the restricted form Desargues' theorem is not valid on the plane. Clearly in the restricted Desargues' theorem, stated as the necessary and sufficient condition for the incidence of the point P and the line $l = CD$, the role of X and Y can be interchanged, since by considering the pair of triangles

$$\begin{pmatrix} D & P_2 & P \\ A & P_0 & P_1 \end{pmatrix}$$

which are in perspective from the centre Y we can deduce that BY contains the point C. We shall denote the restricted form of Desargues' theorem mentioned above by $D(X, Y)$.

Let us assume for the time being that the theorem $D(X, Y)$ is valid and let us consider Fig. 37. There is a relationship between Figs 36 and 37 which is given by the correspondence

$$\begin{pmatrix} A & B & C & D, & P_1 & P_0 & P_2, & P, & l_0 & l_1 & l_2 & l \\ B_1 & C_3 & A_3 & B_2, & C_1 & A_1 & A_2 & C_2, & c_1 & a_1 & c_2 & a_2 \end{pmatrix}.$$

The latter figure can also be considered as the image of a solid figure. The tetrahedron $B_1C_3A_3B_2$ is cut by a plane containing the line YX, in a quadrangle $C_1A_1A_2C_2$, or more precisely, the sides of the skew quadrangle $A_3B_1C_3B_2$ are cut by the plane in the vertices of the plane quadrangle $C_1 A_1 A_2 C_2$. In this figure the interchangeability of the roles X and the Y is obvious.

Suppose that we are given a plane satisfying the axioms I_1, I_2 and I_3, and two distinct points, X and Y, in this plane. We shall now prove that if the theorem $D(X, Y)$ is valid on this plane then the theorem $D(Y, X)$ is also valid and vice

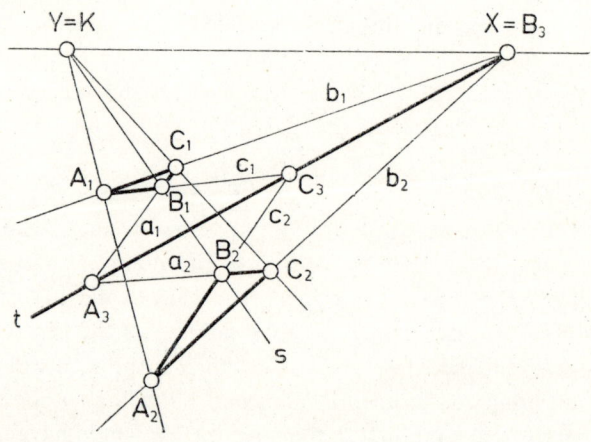

Figure 37

versa. We assume, for the convenience of our proof, that the given plane is of order greater than 2; it is easy to verify that both $D(X, Y)$ and $D(Y, X)$ are valid on a plane of order 2.

Suppose that theorem $D(X, Y)$ is valid and consider two triangles $A_1 B_1 C_1$ and $A_2 B_2 C_2$ such that $YA_1 A_2$, $YB_1 B_2$ and $YC_1 C_2$ are all collinear triples of points, i.e. the two triangles are in perspective from Y. Let $A_1 B_1 \cap A_2 B_2 = C_3$, $B_1 C_1 \cap B_2 C_2 = A_3$ and $C_1 A_1 \cap C_2 A_2 = B_3$. Assume $B_3 = X$ to satisfy the hypothesis of $D(Y, X)$.

Consider the pair of triangles

$$\begin{pmatrix} A_3 & C_1 & C_2 \\ C_3 & A_1 & A_2 \end{pmatrix}.$$

Assume that $A_3 C_3$ does not pass through the point X; suppose C the point of intersection $C(\neq C_3)$ of the lines XA_3 and $A_1 B_1$. Thus we obtain the pair of triangles

$$\begin{pmatrix} A_3 & C_1 & C_2 \\ C & A_1 & A_2 \end{pmatrix}$$

in perspective from the point $X=B_3$. But, by our assumption of the validity of the theorem $D(X, Y)$, it follows that the points B_1, Y and $A_3C_2 \cap A_2C = B$ are collinear. But then the line A_3C_2 would intersect the line YB_1 in both the point B_2 and the point B distinct from B_2, which is impossible. Hence, A_3C_3 passes through X, and the theorem $D(Y, X)$ is valid; similarly, $D(Y, X) \Rightarrow D(X, Y)$.

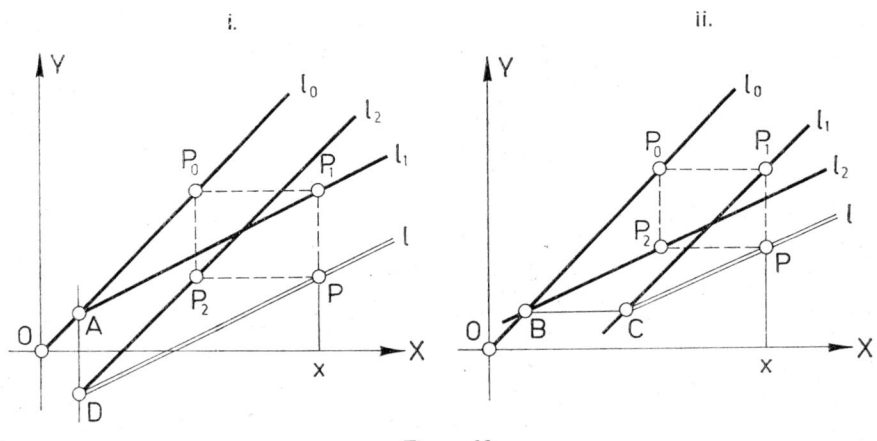

Figure 38

We return now to the discussion in connection with Fig. 36. Here we supposed that A and B were two distinct non-ideal points. We have still to find the locus of the vertex P of the coordinate quadrangle supported by the lines l_0, l_1, l_2, if $A \neq B$, and B is an ideal point; or if $A \neq B$, and A is an ideal point; or if $A = B$ is an ordinary point; or finally, if both A and B are ideal points; assuming $D(X, Y)$ is valid.

To analyse the first two cases we consider parts *i* and *ii* of Fig. 38. If the line l_2 is parallel to the line l_0, then from the pairing

$$\begin{pmatrix} A & P_0 & P_1 \\ D & P_2 & P \end{pmatrix}$$

of triangles perspective from the point Y and from the theorem $D(Y, X)$ it follows that AP_1 and DP are parallel to each other; since, according to our assumption, $AP_0 \| DP_2$ and $P_0P_1 \| P_2P$. Thus the point P always lies on a line drawn through D parallel to l_1. Similarly, in the second case P always lies on the line drawn through C parallel to l_2.

The two other cases are visualized by parts *i* and *ii* of Fig. 39. If $A=B$ is an ordinary point of the line l_0, then take two coordinate quadrangles supported by our lines, giving the point P and other giving the point Q, say.
Consider the pair of triangles

$$\begin{pmatrix} P_0 & Q_0 & Q_2 \\ P_1 & Q_1 & Q \end{pmatrix}$$

which are in perspective from the point X it follows from theorem $\mathsf{D}(X, Y)$, that $P_0 Q_2$ and $P_1 Q$ intersect in an ordinary point S, say, lying on the line AY. Applying theorem $\mathsf{D}(X, Y)$ to the pair of triangles

$$\begin{pmatrix} P_0 & P_2 & Q_2 \\ P_1 & P & Q \end{pmatrix}$$

which are perspective from the point X, it follows that the point S, the point Y and $QP \cap Q_2 P_2$ lie on a line t, say; i.e. the line $l = QP$ passes through the point $A = t \cap l_2$. If we fix the point Q and let x run through every point of the x-axis, then the variable point P always lies on the line AQ. Similar reasoning can be applied in the fourth case; of course, in this case the line l is parallel to the line l_0.

Let us note that, by virtue of the correspondence between ternary functions and oblique lines (i.e. lines not parallel to one of the axes) into a given coordinate system, we may speak of the *product of two oblique lines* instead of the composition of two ternary functions.

After this discussion we can now assert the following

Theorem: *The product of any two oblique lines of a projective plane is an oblique line, if and only if the theorem* $\mathsf{D}(X, Y)$ *is valid for the plane.*

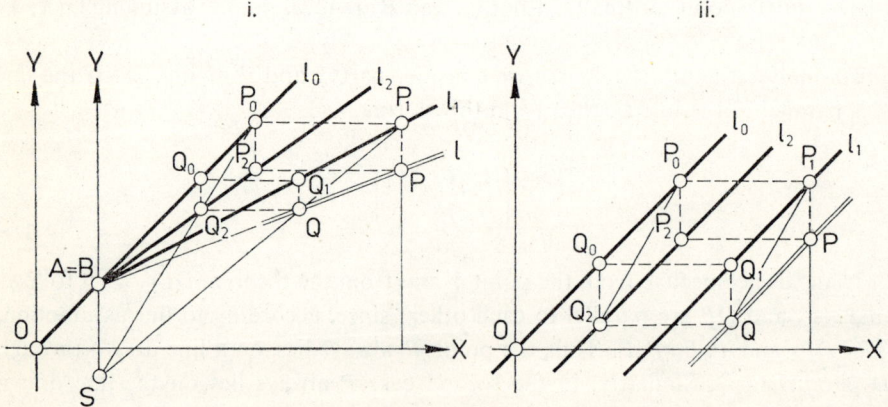

Figure 39

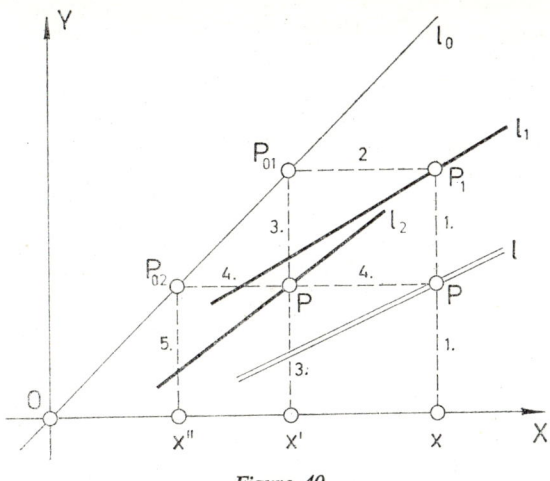

Figure 40

A plane like this will briefly be called a $D(X, Y)$ plane. The mapping $x \to x'$ generated by the oblique line will be called a λ-mapping (Fig. 35). The mapping corresponding to the line l_0 is the identity. If the mappings corresponding to the lines l_1, l_2 and $l_1 l_2 = l$ are, in turn, $\lambda_1 \lambda_2$ and λ, then, obviously, the product of the mappings λ_1 and λ_2 is the mapping λ (Fig. 40). As the composition of mappings is an associative operation, so is the product of oblique lines.

Let L denote the set of oblique lines, together with the operation above. L contains an identity element l_0; i.e. we have $l_0 l = l l_0 = l$ for every $l \in L$. It is easy to see from Fig. 41 that for every $l_1 \in L$ there exists an $l_2 \in L$ such that $l_1 l_2 = l_0$; i.e. each oblique line has an inverse with respect to the product. (In the case of

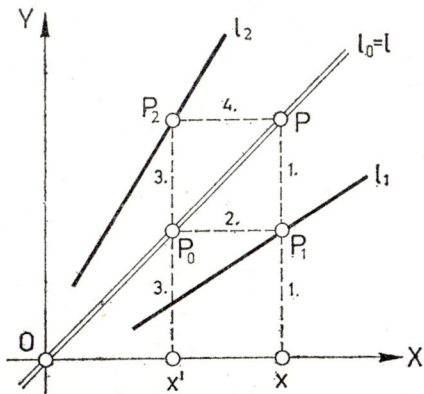

Figure 41

a Euclidean plane the inverse of l_1 can be produced by reflecting l_1 in the line l_0.) Let the set of the λ-mappings be denoted by Λ. Our results can be summarized as follows:

Theorem: *The set* L *is a group under the multiplication of oblique lines and the set* Λ *is a group under the multiplication of* λ*-mappings.*

As is well known, the product of any two lines of a pencil containing l_0 also belongs to the pencil; moreover, the inverse of any line of a pencil containing l_0 also belongs to the pencil. Thus, *a pencil containing* l_0 *together with the multiplication above, forms a subgroup of the group* L. — Of course here the pencil is to be understood not to contain the lines through its vertex parallel to the axes. If the point S is an ordinary or an ideal point of the line l_0, then the set of oblique lines passing through the point S as well as the subgroup established by them will be denoted by S.

To finish this section we shall sketch briefly how the group theoretical constructions we obtained above can be used to transform geometrical problems into questions of the theory of groups. Nowadays, this is a most fruitful method in finite geometry.

Let a group **G**, of order $q(q-1)$ be given. Let **G** have $q+1$ subgroups P such that, one of the subgroups is of order q and the others are of order $(q-1)$ and let every element of **G**, except the identity, belong to just one subgroup P. Let this set of subgroups be denoted by Π. Π is said to be a *covering* of the group G and the subgroups P are said to be the *components* of the covering Π. We shall now establish in this group the following structure:

1. By an *oblique line* we shall mean an element of **G**.
2. By an *ordinary point* we shall mean a coset gP, where the subgroup $P \in \Pi$ is of order $(q-1)$ and $g \in \mathbf{G}$.
3. By an *ideal point* we shall mean a coset gP_0, where $P_0 \in \Pi$ is the subgroup of order q and $g \in \mathbf{G}$.

We shall show now by an example that this construction leads to a finite plane. Let **G** be the complete symmetry group of the regular tetrahedron (in this case $q=4$). We compiled its Cayley table by denoting its elements by positive integers so that the identity element is denoted just by 1 (Fig. 42). The components of the covering are now:

$$A = \{1, 2, 3\}, \quad B = \{1, 5, 12\}, \quad C = \{1, 6, 8\}, \quad D = \{1, 9, 11\},$$
$$U = \{1, 4, 7, 10\}.$$

Indeed, any two of these have only the element 1 in common and we have $A \cup B \cup C \cup D \cup U = \mathbf{G}$.

1	2	3	4	5	6	7	8	9	10	11	12
2	3	1	5	6	4	8	9	7	11	12	10
3	1	2	6	4	5	9	7	8	12	10	11
4	8	12	1	11	9	10	2	6	7	5	3
5	9	10	2	12	7	11	3	4	8	6	1
6	7	11	3	10	8	12	1	5	9	4	2
7	11	6	10	8	3	1	5	12	4	2	9
8	12	4	11	9	1	2	6	10	5	3	7
9	10	5	12	7	2	3	4	7	6	1	8
10	5	9	7	2	12	4	11	3	1	8	6
11	6	7	8	3	10	5	12	1	2	9	4
12	4	8	9	1	11	6	10	2	3	7	5

Figure 42

Let us compile, by means of the Cayley table, the cosets representing the ordinary points:

$P_1 = \{1, 2, 3\}$, $\quad P_2 = \{1, 5, 12\}$, $\quad P_3 = \{1, 6, 8\}$, $\quad P_4 = \{1, 9, 11\}$;

$P_5 = \{4, 8, 12\}$, $\quad P_6 = \{4, 11, 3\}$, $\quad P_7 = \{4, 9, 2\}$, $\quad P_8 = \{4, 6, 5\}$;

$P_9 = \{7, 11, 6\}$, $\quad P_{10} = \{7, 8, 9\}$, $\quad P_{11} = \{7, 3, 5\}$, $\quad P_{12} = \{7, 12, 2\}$;

$P_{13} = \{10, 5, 9\}$, $\quad P_{14} = \{10, 2, 6\}$, $\quad P_{15} = \{10, 12, 11\}$, $\quad P_{16} = \{10, 3, 8\}$.

Now, according to the definition, the lines $1, 2, \ldots, 12$ are the oblique lines, (which, in order to think of these elements as lines, may now be denoted by $l_1, l_2, \ldots, l_{12}$) and $P_1, P_2, \ldots, P_{16}$ are the ordinary points; thus we can interpret certain incidences. For instance $P_{11} = \{7, 3, 5\}$ can be taken to mean that there pass through the point P_{11} the oblique lines l_1, l_3, l_5. Or, for example, we can see immediately from the table, above, that the line $l_3 (=3)$ passes through the points P_1, P_6, P_{11}, P_{16} and we may say that the oblique line l_3 has three ordinary points. In this way we arrive at an incidence table having 12 rows and 16 columns. There are 3 incidence signs in each column and 4 in each row. At a first glance this table does not display any familiar arrangement. If, however, we permutate the rows as follows:

$$\begin{pmatrix} 1 & 2 & 3 & 4 & 5 & 6 & 7 & 8 & 9 & 10 & 11 & 12 \\ 1 & 4 & 7 & 10 & 3 & 12 & 6 & 9 & 2 & 5 & 8 & 11 \end{pmatrix}$$

and leave the columns fixed we obtain the incidence table in Fig. 43.

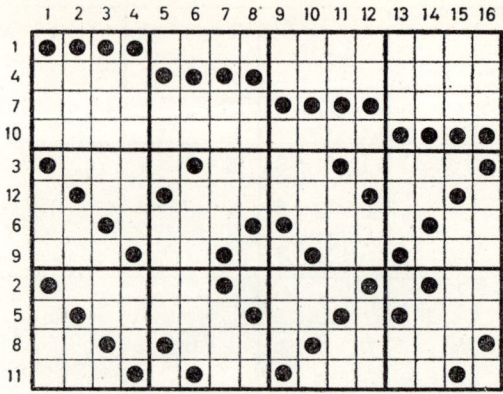

Figure 43

We can see immediately that Fig. 43 can be extended by rows, columns and incidence signs into Fig. 7. The non-oblique lines and the ideal line of the plane will be represented by the new rows and the ideal points by the new columns.

1.16 Problems and exercises to Chapter 1

Problems and exercises, of various degrees of difficulty, will be given at the end of each chapter. These are intended firstly, to complement the material contained in each chapter and, secondly to give the reader practical experience in handling the new ideas and perhaps inspire him to undertake independent investigations.

1. In the Γ-table of a projective plane of order 4 (Fig. 7) the sign pattern in Θ is not determined uniquely. How can we perform row and column transformations so as to obtain a unique sign pattern in Θ, and thereby show that there exists one and only one kind of projective plane of order 4?

2. Starting with the regular 31-gon inscribed in a circle find, by trial and error, its completely irregular subpolygon with the largest number of vertices. How can we simplify this computation?

3. Let a regular polygon of q^2+q+1 sides inscribed in a circle k be given such that a completely irregular subpolygon of $q+1$ exists. The vertices of the latter span convex $(q+1)$-gon, Λ. Prove that if $q=4$, then the centre of the circle k lies in the interior of Λ.

4. Consider the regular polygon with 43 sides inscribed in a circle k. Show, by trial and error, that there are always two equal chords among the 21 spanned by any seven or its vertices.

5. If $q>2$, can we colour each point of the projective plane black or white, so that every line of the plane contains at least one point of each colour? Can this be done, if $q=2$?

6. Prove that the diagonal points of every proper quadrangle of the Galois plane $S_{2,4}$ form a collinear triple of points. (The operation tables of $GF(4)$ were given in Section 1.6.)

7. Check the validity of the following statements by means of the operation tables of $GF(4)$ and with the help of Fig. 7: The coordinates of five points of the plane $S_{2,4}$ satisfy the equation $x_1^2+x_2^2+x_3^2=0$ and, similarly, 5 points satisfy $x_1^2+x_2 x_3=0$. The first set of 5 points lie on a line and there are more than two points in the second set which do not lie on this line.

8. Consider the extension procedure shown in Fig. 16 and 17: how many rows and columns does the table $\mathbf{T}^k$ consist of?

9. The following theorems are valid on the $\langle 3, 3\rangle$-plane defined by the Table in Fig. 14: 1. Parallelism is not a transitive relation. 2. There exists a quadrangle which has no diagonal point. 3. There is a pair of triangles in perspective from a point such that none of the corresponding pairs of sides intersect. How can these be proved by considering the figure?

10. Although the plane of order 4 suffices for the illustration of certain properties, clearly, we cannot hope to illustrate certain other properties, e.g. non-

+	0	1	2	3	4	5	6	7	8
0	0	1	2	3	4	5	6	7	8
1	1	2	0	4	5	3	7	8	6
2	2	0	1	5	3	4	8	6	7
3	3	4	5	6	7	8	0	1	2
4	4	5	3	7	8	6	1	2	0
5	5	3	4	8	6	7	2	0	1
6	6	7	8	0	1	2	3	4	5
7	7	8	6	1	2	0	4	5	3
8	8	6	7	2	0	1	5	3	4

×	0	1	2	3	4	5	6	7	8
0	0	0	0	0	0	0	0	0	0
1	0	1	2	3	4	5	6	7	8
2	0	2	1	6	8	7	3	5	4
3	0	3	6	7	1	4	5	8	2
4	0	4	8	1	5	6	2	3	7
5	0	5	7	4	6	2	8	1	3
6	0	6	3	5	2	8	7	4	1
7	0	7	5	8	3	1	4	2	6
8	0	8	4	2	7	3	1	6	5

Figure 44

Desarguesianism, on this plane. We shall therefore introduce two planes of order q (the minimal order possible for our purposes). We shall not give the complete incidence tables of these planes as they are far too large.

First of all let us consider the two operation tables of the $GF(9)$; these are given in Fig. 44.

11. The construction of the incidence table $\Gamma(9)$. Following the example of Fig. 7 we can enter the incidence sign pattern into the first 10 rows and first 10 columns. The remaining part $\Gamma\Theta$ of the table is divided into 9×9 parcels. Denote by k the diagonal pattern formed by the elements k in the addition table of $GF(9)$. Thus we obtain q diagonal patterns. Let each parcel correspond to the square occupying the same position in the table G of Fig. 45 (which is identical with the multiplication table of Fig. 44).

If a particular parcel corresponds to a square occupied by the element k then enter the sign pattern k in this parcel. This prescription agrees with that which was given in Section 1 for an arbitrary field $GF(q)$ (Fig. 44).

12. Note that Fig. 46 contains altogether 27 sign diagonals, from which we used only the 9 denoted by 0, 1, ..., 8. The construction of the other 18 sign diagonals will not be dealt with here. They are included because we shall use them to construct an incidence table Γ^* of a plane of order 9 which is not a Galois plane.

Let us construct a Γ-lattice of 91×91 squares. The arrangement of the signs in the first $1+4\times 9=37$ rows and in the first 37 column is the same as that of the $\Gamma(9)$ table discussed above. The sign patterns of the remaining 6×6 parcels

G

0	0	0	0	0	0	0	0	0
0	1	2	3	4	5	6	7	8
0	2	1	6	8	7	3	5	4
0	3	6	7	1	4	5	8	2
0	4	8	1	5	6	2	3	7
0	5	7	4	6	2	8	1	3
0	6	3	5	2	8	7	4	1
0	7	5	8	3	1	4	2	6
0	8	4	2	7	3	1	6	5

V

0	0	0	0	0	0	0	0	0
0	1	2	3	4	5	6	7	8
0	2	1	6	8	7	3	5	4
0	3	6	10	11	12	13	14	15
0	4	8	16	17	18	19	20	21
0	5	7	22	23	24	25	26	27
0	6	3	13	15	14	10	12	11
0	7	5	25	27	26	22	24	23
0	8	4	19	21	20	16	18	17

Figure 45

PROBLEMS AND EXERCISES

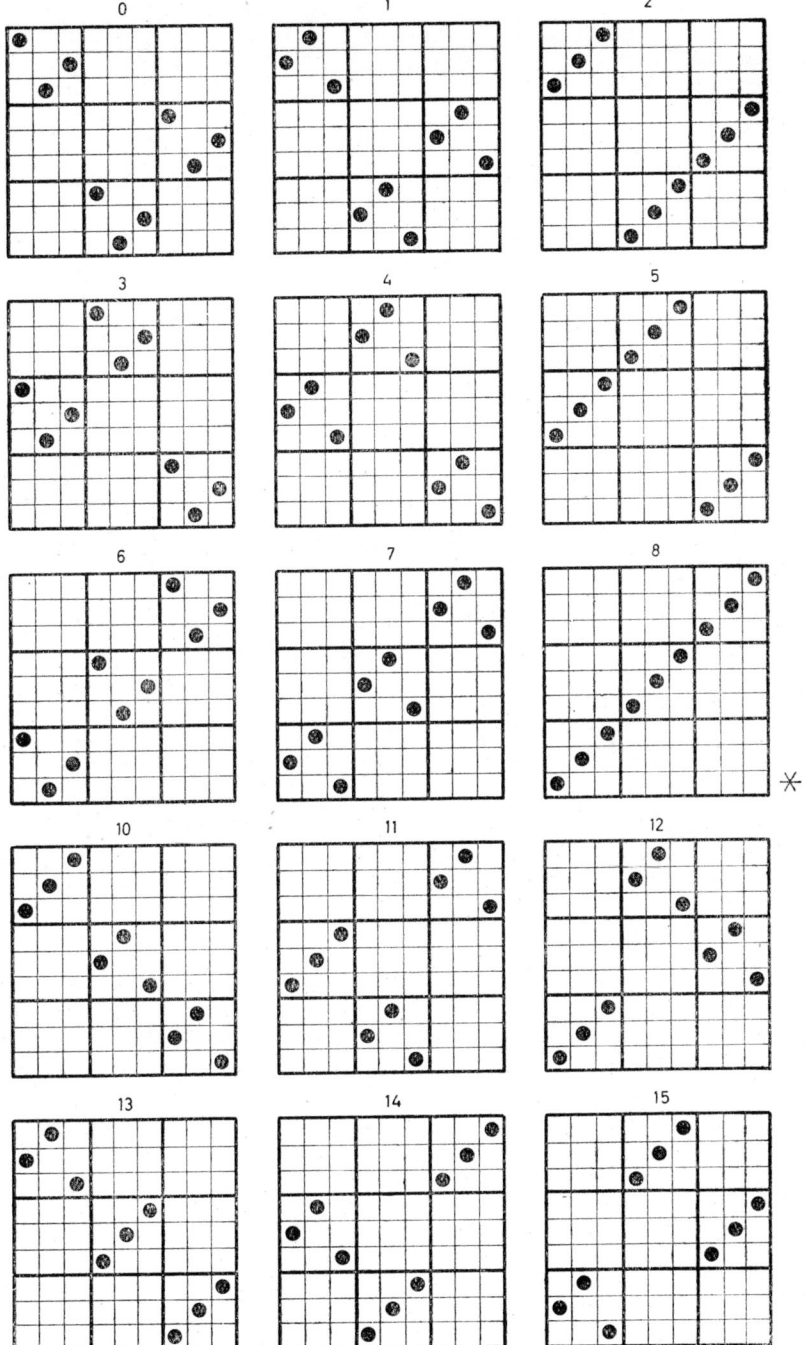

Figure 46a

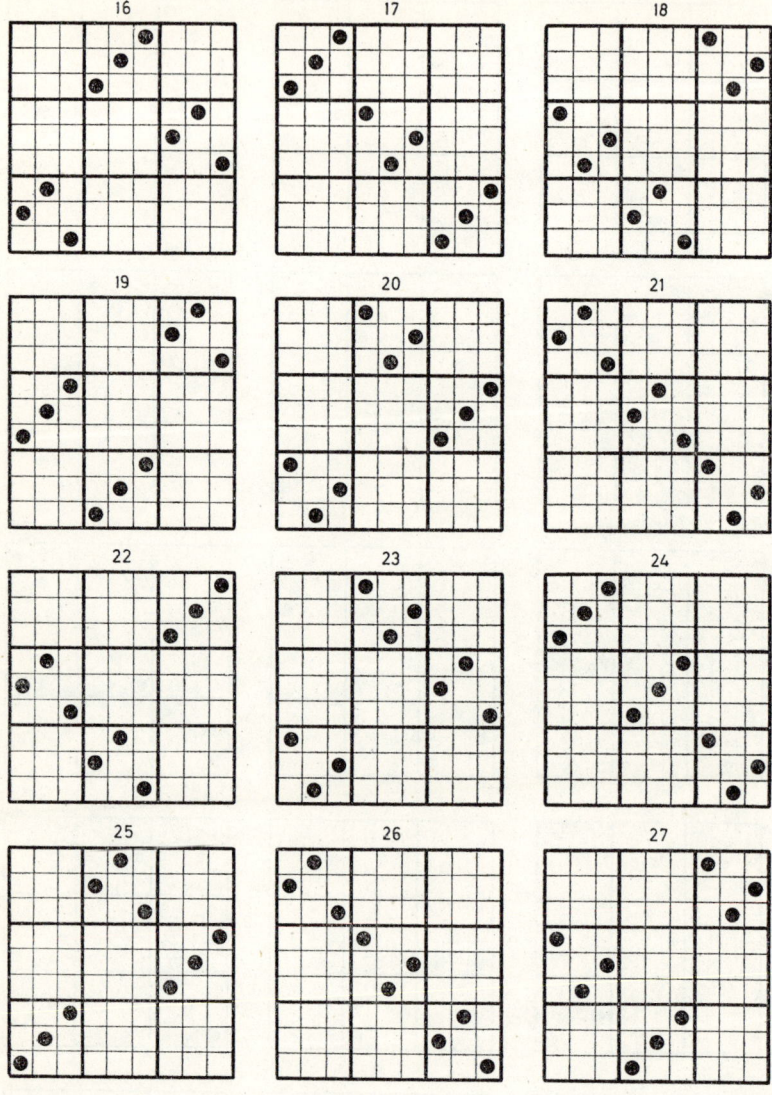

Figure 46b

are shown in Part V of Fig. 45; where, for instance, the index 24 occurring in two squares means that both of the corresponding parcels contain the sign pattern 24.

(The table Γ^* cannot be isomorphic with Γ, since for the former $\delta=27$ and for the latter $\delta=9$.)

13. Prove that on the Galois plane of order 9 there are no Fano quadrangles. Whereas, on the plane of order 9 defined by the Γ^* table there are both Fano quadrangles and ordinary quadrangles; find an example on each on the tables Γ^*.

14. Show that on the plane determined by Γ^* the theorem of Desargues does not hold, i.e. this plane is non-Desarguesian. Notice that this does not mean that it is anti-Desarguesian since — as can be verified by an example — it contains a configuration fulfilling Desargues' theorem as well as one for which this theorem does not hold.

15. Show that the plane defined by the Γ^* has a subplane of order three.

(Because of this we may say that $\Gamma(3)$ can be extended either into $\Gamma(9)$ or into Γ^*, having as it does a Fano quadrangle, that is a subplane of order 2. Thus the planes of order 2 can be extended into both Desarguesian and non-Desarguesian planes (e.g. $\Gamma(2)$ can be extended into $\Gamma(4)$ and also into Γ^*).)

16. Choose one of the Γ tables and determine the composition of two lines (by using the coordinate quadrangles).

*

Finally we give some hints as to the solutions of exercises 13, 14 and 15.

Consider, in the table Γ^* the rows and columns indexed by 2, 6, 11, 17, 56 and 62, respectively. The incidence pattern of the squares forming the intersection of these rows and columns is that of a $\Gamma(2)$ table. Similarly, if we intersect the rows and columns indexed by 1, 2, 3, 4, 11, 12, 13, 20, 21, 22, 29, 30 and 31, respectively, we obtain a table $\Gamma(3)$.]

Also in Γ^*, the pair of triangles

$$\begin{pmatrix} P_{40} & P_{59} & P_{80} \\ P_{54} & P_{65} & P_{86} \end{pmatrix}$$

are in perspective from the point P_5 and the three points of intersection of corresponding sides are P_{48}, P_{21} and P_{34}; these are not collinear because

$$P_{48}P_{21} = l_{13}, \quad P_{21}P_{34} = l_{50}, \quad \text{and} \quad P_{34}P_{48} = l_{42}.$$

CHAPTER 2

GALOIS GEOMETRIES

After the preliminary first chapter, we shall now deal with the most well developed branch of finite geometries, i.e. with the geometry of Galois spaces. The arithmetic generation of this notion of a space and the analytic treatment of its geometry emphasizes strongly the great similarity between the classical projective geometry and Galois geometry. In what follows, we shall deal mainly with plane geometry, although we shall introduce the concept of an n-dimensional Galois space of order q, $S_{n,q}$. The study of the case $n>2$, however, would increase the size of this without really introducing any methods that are not already applied to the study of plane geometry. Because of the analytic method, we shall often appeal to notions and theorems belonging to the theory of the finite algebraic structures, which may not be familiar to the reader. Therefore, all this algebraic background is given concisely in the Appendix, together with some exercises which may be helpful.

2.1 The notion of Galois spaces

The arithmetic characterization of the space $S_{n,q}$ is a generalization of the procedure dealt with in Section 1.10. With respect to the approach, all we have to say here belongs to the theory of algebraic geometry. In fact, as we shall discuss algebraic notions the language of geometry, — since our thinking is led far more by geometric than by algebraic considerations — it would be more appropriate to speak of geometrical algebra instead of algebraic geometry.

Let r be a positive integer, p an arbitrary prime, and $q=p^r$. The field $GF(q)$ will, as before, be denoted by **K**. Assume that $x_j \in \mathbf{K}$, $j=1, 2, \ldots, n$ and consider the set of all sequences $(x_1, x_2, \ldots, x_n)$. The number of such sequences is equal to q^n elements. Decompose the set of sequences into classes as follows. Let two sequences $(x_1, x_2, \ldots, x_n)$ and $(x'_1, x'_2, \ldots, x'_n)$ belong to the same class if and only if there exists $\lambda \in \mathbf{K}$, $\lambda \neq 0$, such that $x'_j = \lambda x_j$ for $j=1, \ldots, n$. Let us now introduce the notation $(x)=(x_1, \ldots, x_n)$, then

$$(x') = (x'_1, x'_2, \ldots, x'_n) = (\lambda x_1, \lambda x_2, \ldots, \lambda x_n) = \lambda(x_1, x_2, \ldots, x_n) = \lambda(x).$$

Let the element $(0, 0, \ldots, 0)$ in the set of sequences be called the *non-point* or *empty space* and let us exclude it from consideration. The remaining $q^n - 1$ sequences of the set are distributed among the classes and every class consists of $q-1$ sequences. Consequently, the number of the classes is

$$(q^n - 1)/(q - 1) = \Theta_{n-1},$$

say.

Let the classes so defined be called *points* and the elements of any sequence $(x_1, x_2, \ldots, x_n)$ of a class be called *homogeneous coordinates* of the point. Thus $q-1$ coordinate sequences belong to every point and any of them determines the point. The point set so defined is called $S_{n-1,q}$; i.e. the $(n-1)$ dimensional Galois space of order q.

For the present we know nothing about this set other than the fact that the number of points in it is

$$\Theta_{n-1} = q^{n-1} + q^{n-2} + \ldots + 1.$$

However, its derivation from the field $\mathbf{K}$ allows us to build a structure on it, reminiscent of classical projective geometry. This was sketched in Section 1.10, for the case of $n=3$. The sketch given there will now be generalized. We introduce the following concepts which play a significant role in the determination of the geometrical structure of $S_{n-1,q}$.

1° Let us consider the coefficients λ_j and the sequences $(a_{j1}, a_{j2}, \ldots, a_{jn})$, where $\lambda_j \in \mathbf{K}$ $(j=1, 2, \ldots, k)$ and $a_{j1} \in \mathbf{K}$ $(j=1, 2, \ldots, k; l=1, 2, \ldots, n)$. We shall understand by a linear combination of the given sequence formed with the given coefficients the sequence $(b_1, b_2, \ldots, b_n)$ where

$$b_l = \lambda_1 a_{1l} + \lambda_2 a_{2l} + \ldots + \lambda_k a_{kl}.$$

We introduce the following notation:

$$(b) = (b_1, b_2, \ldots, b_n) = \lambda_1(a_{11}, a_{12}, \ldots, a_{1n}) +$$
$$+ \lambda_2(a_{21}, a_{22}, \ldots, a_{2n}) + \ldots + \lambda_k(a_{k1}, a_{k2}, \ldots, a_{kn}) =$$
$$= \lambda_1(a_1) + \lambda_2(a_2) + \ldots + \lambda_k(a_k).$$

2° We shall understand by the *fixing* of a coordinate sequence of a point that, at a certain stage in our reasoning we choose one of the $(q-1)$ coordinate sequences determining the point and, henceforth, in speaking about the point we use always this same chosen sequence in order to specify the point.

3° We shall understand by the *norming* of a coordinate sequence a fixing in which the last non-zero element of the sequence is the unit element of the field $\mathbf{K}$. Thus there exists a one-to-one correspondence between the points and their normed coordinate sequences.

4° We shall understand by the *linear combination* $\lambda_1 A_1 + \lambda_2 A_2 + \ldots + \lambda_k A_k$ of the given points A_j formed with the *given* coefficients λ_j ($j=1, 2, \ldots, k$) the element corresponding to the linear combination

$$(b) = \lambda_1(a_1) + \lambda_2(a_2) + \ldots + \lambda_k(a_k)$$

where (a_j) is the normed sequence of the point A_j ($j=1, \ldots, k$). This element is either the point S determined by the (not necessarily normed) sequence $(b) = (b_1, b_2, \ldots, b_n)$ or, if $(b) = (0, 0, \ldots, 0)$ it is the empty space denoted by 0.

We shall say that the points $A_1, A_2, \ldots, A_k$ form a *(linearly) independent point system* if the relation

$$\lambda_1 A_1 + \lambda_2 A_2 + \ldots + \lambda_k A_k = 0$$

implies

$$\lambda_1 = \lambda_2 = \ldots = \lambda_k = 0.$$

Consider the two elements generated by linear combinations

$$\lambda_1 A_1 + \lambda_2 A_2 + \ldots + \lambda_k A_k \quad \text{and} \quad \mu_1 A_1 + \mu_2 A_2 + \ldots + \mu_k A_k$$

of the independent point system

$$\{A_1, A_2, \ldots, A_k\}.$$

Let

$$(b_1, b_2, \ldots, b_n) \quad \text{and} \quad (b'_1, b'_2, \ldots, b'_n)$$

denote the two sequences corresponding to them. Can these two elements be identical? i.e. is it possible that there exists $v \in \mathbf{K}$, $v \neq 0$ such that

$$b'_1 = vb_1, \quad b'_2 = vb_2, \ldots, b'_k = vb_k?$$

Obviously, this can occur if and only if

$$v(\lambda_1 A_1 + \lambda_2 A_2 + \ldots + \lambda_k A_k) - (\mu_1 A_1 + \mu_2 A_2 + \ldots + \mu_k A_k) =$$
$$= (v\lambda_1 - \mu_1) A_1 + (v\lambda_2 - \mu_2) A_2 + \ldots + (v\lambda_k - \mu_k) A_k = 0.$$

According to the definition of the independent point system, we would then have

$$\mu_1 = v\lambda_1, \quad \mu_2 = v\lambda_2, \ldots, \mu_k = v\lambda_k.$$

Thus, for the points derivable by forming linear combinations from an independent point system the sequence of coefficients $(\lambda_1, \lambda_2, \ldots, \lambda_k)$ play a similar role to the homogeneous coordinate sequences in the space $S_{k-1,q}$. Hence we can say that $(\lambda_1, \lambda_2, \ldots, \lambda_k)$ is the *homogeneous coordinate sequence with respect to the independent point system* $\{A_1, A_2, \ldots, A_k\}$ of the point obtained from this system by forming the linear combination with coefficients

$$(\lambda_1, \lambda_2, \ldots, \lambda_k) \neq (0, 0, \ldots, 0).$$

Now let us consider the set of the points $\lambda_1 A_1 + \lambda_2 A_2 + \ldots + \lambda_k A_k$ generated by the independent point system $A_1, A_2, \ldots, A_k$. Obviously, each fundamental point A_j is an element of this set, by taking $\lambda_i = 0$ if $i \neq j$ and $\lambda_j = 1$. It is also clear from the one-to-one correspondence between the points of this point set and the Galois plane $S_{k-1,q}$ that our point set consists of

$$\Theta_{k-1} = q^{k-1} + q^{k-2} + \ldots + 1$$

points. This point set is called the subspace of the Galois space $S_{n-1,q}$ *spanned by the point set* $A_1, A_2, \ldots, A_k$. If $k=2$, $k=3$ and $k=n-1$, then the subspace so defined is also said to be a *line*, a *plane*, and a *hyperplane*, respectively.

Let us now consider the points of $S_{n-1,q}$ defined by the following sequences:

$$A_1: (1, 0, \ldots, 0), \quad A_2: (0, 1, \ldots, 0), \ldots, \quad A_n: (0, 0, \ldots, 1).$$

Obviously, these points form an independent point system. The subspace spanned by the point system $A_1, A_2, \ldots, A_n$ is now the whole space $S_{n-1,q}$; because linear combination $x_1 A_1 + x_2 A_2 \ldots + x_n A_n$ is just the point corresponding to the coordinate sequence $(x_1, x_2, \ldots, x_n)$. Thus $S_{n-1,q}$ is a subspace of itself; we shall call it an *improper* subspace.

The prime p is called the characteristic of the field $GF(p^r)$. Likewise we shall call p the *characteristic* of the space S_{n-1, p^r}. Later we shall give an interesting geometric interpretation to the characteristic of a space, which can also be extended to non-Desarguesian planes.

2.2 The Galois space as a configuration of its subspaces

In this section we shall deal with theorems concerning the incidence properties, connections and intersections of the subspaces of a Galois space.

1° *If the point system* $\{A_1, A_2, \ldots, A_k\}$ *is independent, then any of its subsystems is independent as well.*

Clearly, we can permute the points in an independent order without affecting the subspace. Thus, in order to prove 1° it suffices to prove that $\{A_2, A_3, \ldots, A_k\}$ is an independent point system. If it were not independent, then there would exist a sequence $(\lambda_2, \lambda_3, \ldots, \lambda_k) \neq (0, 0, \ldots, 0)$ for which $\lambda_2 A_2 + \lambda_3 A_3 + \ldots + \lambda_k A_k = 0$, i.e. $\lambda_1 A_1 + \lambda_2 A_2 + \ldots + \lambda_k A_k = 0$ where $\lambda_1 = 0$ and so $(\lambda_1, \lambda_2, \ldots, \lambda_k) \neq (0, 0, \ldots, 0)$ which is, according to our assumption, impossible. By the repeated application of this reasoning the independence of any subsystem of $\{A_1, A_2, \ldots, A_k\}$ can be proved.

2° *If the point system* $\{A_1, A_2, \ldots, A_k\}$ *is not independent, then it has an element which is the linear combination of the other elements, and conversely, if an element of this point system is the linear combination of the other elements, then the point system is not independent.*

If the point system is not independent, there exists a sequence of coefficients not all zero which satisfy the equation $\lambda_1 A_1 + \lambda_2 A_2 + \ldots + \lambda_k A_k = 0$. We may suppose with loss of generality that $\lambda_1 \neq 0$ and then if we define the coefficients $\lambda'_j = \lambda_j \lambda_1^{-1}$ ($j = 2, 3, \ldots, k$) we have

$$A_1 = \lambda'_2 A_2 + \lambda'_3 A_3 + \ldots + \lambda'_k A_k.$$

Conversely, if $A_1 = \lambda_2 A_2 + \lambda_3 A_3 + \ldots + \lambda_k A_k$, then if we define the coefficients $\lambda'_1 = 1$, $\lambda'_j = -\lambda_j$ ($j = 2, 3, \ldots, k$) we have $\lambda'_1 A_1 + \lambda'_2 A_2 + \ldots + \lambda'_k A_k = 0$, where not every coefficient is the zero element, that is the point system $\{A_1, A_2, \ldots, A_k\}$ is not independent.

3° *The set S, consisting, of the points* $\xi_1 A_1 + \xi_2 A_2 + \ldots + \xi_k A_k = P$ *generated by arbitrary points* $A_1, A_2, \ldots, A_k$ *of the space is a subspace.*

If the point system $\{A_1, A_2, \ldots, A_k\}$ is independent, then S is, by the definition, a subspace. If, however, the point system is not independent, then by choosing an appropriate order we have $A_1 = \lambda_2 A_2 + \lambda_3 A_3 + \ldots + \lambda_k A_k$. Let S' be the set of the points $\xi_2 A_2 + \xi_3 A_3 + \ldots + \xi'_k A_k = P$. If $P \in S$, then $P \in S'$, because if $P = \xi_1 A_1 + \xi_2 A_2 + \ldots + \xi_k A_k$, then $P = (\xi_1 \lambda_2 + \xi_2) A_2 + (\xi_1 \lambda_3 + \xi_3) A_3 + \ldots + (\xi_1 \lambda_k + \xi_k) A_k$; obviously if $P' \in S'$, then $P' \in S$; therefore we have shown that $S = S'$.

By repeating this argument we must eventually arrive at an independent point system $\{A_{h+1}, A_{h+2}, \ldots, A_k\}$, $h \leq k-1$, and $S = S' = \ldots = S^{(h)}$ where $S^{(h)}$ denotes the set of points generated by $\{A_{h+1}, \ldots, A_k\}$, but $S^{(h)}$ is a subspace. Thus S is a subspace consisting of

$$\Theta_{h-1} = q^{h-1} + q^{h-2} + \ldots + 1$$

points.

4° *Every element of the point set generated by a subspace S of the space belongs to the subspace S.*

Let S be the subspace generated by the independent point system $\{A_1, A_2, \ldots, A_k\}$. Suppose the vectors

$$B_j = \lambda_{j1} A_1 + \lambda_{j2} A_2 + \ldots + \lambda_{jk} A_k \quad (j = 1, 2, \ldots, l)$$

also span the subspace S, where we assume that the λ_{ji} have been chosen so that the coordinate sequence of every B_j is in normed form.

Consider now a point
$$P = \xi_1 B_1 + \xi_2 B_2 + \ldots + \xi_l B_l.$$
Clearly,
$$P = \eta_1 A_1 + \eta_2 A_2 + \ldots + \eta_k A_k,$$
where $\eta_h = \lambda_{1h}\xi + \lambda_{2h}\xi + \ldots + \lambda_{lh}\xi_h$ ($h=1, 2, \ldots, k$). Thus we have shown that $P \in S$.

5° *In the subspace spanned by the independent points system $\{A_1, A_2, \ldots, A_k\}$ there are not more that k independent points.*

The number of all points of the subspace in question is Θ_{k-1}. By 4° every point of the subspace spanned by the independent point system $\{B_1, B_2, \ldots, B_l\}$ lies in the subspace, therefore $\Theta_{l-1} \leq \Theta_{k-1}$ which implies that $l \leq k$.

We know already that in the space $S_{n-1,q}$ the n points given by the coordinate sequences $(1, 0, 0, \ldots, 0), (0, 1, 0, \ldots, 0) \ldots (0, 0, 0, \ldots, 1)$ form a linearly independent point system, hence in this space there are not more than n independent points. A maximal system of the independent points of the space is said to be a *basis*, the number of the points of the basis minus one is said to be the *dimension* of the space, which agrees with the usage in ordinary geometry. (The line is a one dimensional, the plane a two dimensional space; the line is spanned by two of its points, the plane by three of its points.) Thus in the symbol $S_{n-1,q}$ the first index signifies the dimension. Naturally, a space consisting of more than a single point has several bases; thus for instance each of the two point systems of n elements

$$(1, 0, 0, \ldots, 0), (0, 1, 0, \ldots, 0), (0, 0, 1, \ldots, 0), \ldots, (0, 0, 0, \ldots, 1)$$
and
$$(1, 0, 0, \ldots, 0), (1, 1, 0, \ldots, 0), (1, 1, 1, \ldots, 0), \ldots, (1, 1, 1, \ldots, 1)$$

is a basis of the space $S_{n-1,q}$.

In what follows, the Galois space of dimension n will be denoted by S_n, where we assume throughout that q is fixed. Let S_k, S_l and S_n be given such that $S_k \subset S_n$ and $S_l \subset S_n$ ($k \leq l$). We shall now introduce the notions of the intersection and the join of the spaces S_k and S_l.

6° *If the point set $S_k \cap S_l = S^*$ is non-empty, then the S^* is an m dimensional subspace of S_n where $m \leq k$.*

Let $\{Q_1, Q_2, \ldots, Q_{m+1}\}$ be a maximal independent point system in the set S^*, then clearly $m \leq k$. Every point of the subspace S_m spanned by this point system belongs to the set S^*, belonging as it does — according to 4° both to the subspace S_k and the subspace S_l. If a point Q existed such that $Q \in S^*$ and $Q \notin S_m$, then by 2°, $\{Q_1, Q_2, \ldots, Q_{m+1}, Q\}$ would be an independent point system, in contradiction to our assumption. Thus $S^* = S_m$.

The subspace $S^* = S_m$ is called the *intersection space* of the subspaces S_k and S_l; it is the *maximal subspace* common to the two given subspaces.

7° *If the point set $S_k \cap S_l$ is empty, then $k+l<n$.*

Suppose, conversely, that $k+l \geq n$, then $(k+1)+(l+1) > n+1$. Let $\{A_1, A_2, \ldots, A_{k+1}\}$ and $\{B_1, B_2, \ldots, B_{l+1}\}$ be independent point systems spanning S_k and S_l, respectively. The basis of S_n consists of $n+1$ points, hence the point system

$$\{A_1, \ldots, A_{k+1}, B_1, \ldots, B_{l+1}\}$$

consisting of the $k+l+2$ points is not independent, thus according to 2°, the point B_{l+1} is a linear combination of the other points.

Suppose we have

$$B_{l+1} = \lambda_1 A_1 + \ldots + \lambda_{k+1} A_{k+1} + \mu_1 B_1 + \ldots + \mu_l B_l$$

from which follows

$$\lambda_1 A_1 + \lambda_2 A_2 + \ldots + \lambda_{k+1} A_{k+1} = -\mu_1 B_1 - \ldots - \mu_l B_l + B_{l-1}.$$

At least one of the coefficients on the right hand side is not zero, therefore this combination gives a one point of S_l; but this point is, with respect to the combination on the left hand side, a point of the subspace S_k as well, which contradicts the assumption $k+l \geq n$. Thus statement 7° is proved.

If the point set $S_k \cap S_l$ is empty, then the two subspaces are said to be *skew* or *independent*. The necessary criterion of this is $k+l<n$.

Now let S_k and S_l be two arbitrary subspaces of S_n. The intersection of all the subspaces containing both S_k and S_l is said to be the *join* of the two subspaces. This definition is clearly equivalent to defining the *join* of two subspaces S_k and S_l as the *subspace of the smallest dimension* containing both the subspaces.

8° *If $S_k, S_l \subset S_n$ and $S_k \cap S_l$ is empty, then for the join S_r we have $r = k+l+1$.*

If we assume that the dimension of the join is $r < k+l+1$, then the number of elements of the independent point system spanning this space is $<(k+l+2)$. But then the point system, composed of the point systems $\{A_1, A_2, \ldots, A_{k+1}\}$ and $\{B_1, B_2, \ldots, B_{l+1}\}$ spanning the subspaces S_k and S_l, is not independent. It follows, according to the reasoning given in 7° that the subspaces S_k and S_l have a common point. Thus we have arrived at a contradiction to our assumption and the theorem is proved.

9° *If $S_k, S_l \subset S_n$ and $S_k \cap S_l = S_m$ ($m \geq 0$), then for the dimension of the join S_r of the subspaces S_k and S_l is given by $r+m=k+l$.*

The subspaces S_k and S_l are spanned by $k+1$ and $l+1$ independent points, respectively. These points can be chosen in such a way that the independent point system $\{C_1, C_2, ..., C_{m+1}\}$ spanning S_m is common to both of them. Let the two point systems in question be

$$\{A_1, ..., A_{k-m}, C_1, ..., C_{m+1}\} \quad \text{and} \quad \{C_1, ..., C_{m+1}, B_1, ..., B_{l-m}\}.$$

We claim that the point system $\{A_1, ..., A_{k-m}, C_1, ..., C_{m+1}, B_1, ..., B_{l-m}\}$ is independent, that is if

$$\lambda_1 A_1 + ... + \lambda_{k-m} A_{k-m} + \nu_1 C_1 + ... + \nu_{m+1} C_{m+1} = \mu_1 B_1 + ... + \mu_{l-m} B_{l-m}$$

implies that all of the coefficients are zero.

Indeed, if non-zero coefficients occurred in these equations then at least one of the coefficients μ_j would be non-zero. For if all the μ_i were zero, then, because of the independence of the system composed of the point A and C, each of the coefficients λ, ν in our equation would be zero as well. As the points B are independent the expression on the right hand side of the equation would represent a point; let this point be denoted by P. Since every point B belongs to the subspace S_l, $P \in S$. It follows from the expression of the left hand side of the equation that $P \in S_k$; consequently, $P \in S_m$. Thus the point P could be generated by a linear combination of the points B as well as of the points C:

$$\xi_1 C_1 + \xi_2 C_2 + ... + \xi_{m+1} C_{m+1} = \mu_1 B_1 + \mu_2 B_2 + ... + \mu_{l-m} B_{l-m}.$$

This, however, contradicts the assumption that the union of the points C and B is an independent point system.

Thus we know that we can select $k+l-m+1$ independent points from the subspace S_k and S_l; these points cannot be contained in a subspace of dimension less than $k+l-m$, and the chosen point system spans a space S_r containing S_k, S_l. Hence the proof is complete.

10° *The points $(x_1, x_2, ..., x_{n+1})$ of $S_{n,q}$ satisfying the equation*

$$a_1 x_1 + a_2 x_2 + ... + a_{n+1} x_{n+1} = 0$$

where $a_1, ..., a_{n+1} \in GF(q)$ are not all zero, form a hyperplane. Conversely, all the points of a hyperspace satisfy an equation of the above form.

Let us prove the first part of this theorem. If the sequence $(x_1, x_2, ..., x_{n+1})$ $(\lambda x_1, \lambda x_2, ..., \lambda x_{n+1}) = (x'_1, x'_2, ..., x'_{m+1})$ satisfies the given equation, then the sequence satisfies it as well; we can speak therefore of the points satisfying the equation. But, clearly, not all of the points of the space $S_{n,q}$ satisfying this equation, since we may assume without loss of generality that $a_1 \neq 0$ and then the point $(1, 0, 0, \ldots, 0)$ does not belong to the set S defined by the equation.

Obviously, if every point of an independent point system $\{A_1, \ldots, A_k\}$ belongs to S then every point of the subspace S_{k-1} determined by $\{A_1, \ldots, A_k\}$ belongs to S.

By combining the last two statements we see that S does not contain $n+1$ independent points. However, we shall show that it does contain n independent points. Suppose that $a_1 \neq 0$, by reordering the indices if necessary and let $-a_j a_1^{-1} = c_j$ ($j = 2, 3, \ldots, n+1$). It is easy to see that the n points $(c_2, 1, 0, \ldots, 0)$, $(c_3, 0, 1, \ldots, 0) \ldots (c_{n+1}, 0, 0, \ldots, 1)$ form an independent point system. Obviously, all of these points satisfy the given equation. Let these points be called $C_2, C_3, \ldots, C_{n+1}$, respectively. The points generated by combinations

$$\lambda_2 C_2 + \lambda_3 C_3 + \ldots + \lambda_{n+1} C_{n+1}$$

exhaust S; since, otherwise S would contain more than n independent points, which is impossible. Thus we have proved the first part of 10°.

Let us now consider a subspace S_{n-1}, i.e. a subspace spanned by n independent points.

The points

$$(1, 0, 0, \ldots, 0), \ (0, 1, 0, \ldots, 0), \ \ldots, \ (0, 0, 0, \ldots, 1)$$

forming a basis of the space $S_{n,q}$ cannot all belong to the subspace S_{n-1}; assume that the first one is not in it. Let the lines connecting the first point with each of the others be $l_{12}, l_{13}, \ldots, l_{1,n+1}$, respectively. These each have one point in common with the subspace S_{n+1}; suppose these points are $B_2, B_3, \ldots, B_{n+1}$ and are given by the coordinate sequences

$$(b_2, 1, 0, \ldots, 0), \ (b_3, 0, 1, \ldots, 0), \ \ldots, \ (b_{n+1}, 0, 0, 0, \ldots, 1),$$

respectively. Clearly these form an independent point system and each of them satisfies the equation

$$x_1 - b_2 x_2 - b_3 x_3 - \ldots - b_{n+1} x_{n+1} = 0.$$

The point set defined by this equation contains no points that are not inexpressible in the form

$$\mu_2 B_2 + \mu_3 B_3 + \ldots + \mu_{n+1} B_{n+1}$$

since more than n independent point satisfy the equation. Hence we have proved the second part of the theorem.

We add now some remarks concerning these statements.

First of all let us compare the formulae $r = k + l + 1$ and $r = k + l - m$ occurring in 8° and 9° respectively; i.e. the formulae giving the dimension of the join S_r

of two subspaces S_k and S_l of the space $S_{n,q}$. The first of these formulae refers to the case when $S_k \cap S_l$ is an empty set, the second to the case when the intersection S_m consists at least of one point, i.e. when $m \geq 0$. If we agree that the empty set is -1 dimensional, then the first formula can be incorporated into the second.

Our second remark is of greater importance. Obviously, the equation

$$a_1 x_1 + a_2 x_2 + \ldots + a_{n+1} x_{n+1} = 0$$

in which not all the coefficients are zero, and the equation

$$\lambda a_1 x_1 + \lambda a_2 x_2 + \ldots + \lambda a_{n+1}, x_{n+1} = 0,$$

where $\lambda \neq 0$, define the same point set. Assume that equation

$$b_1 x_1 + b_2 x_2 + \ldots + b_{n+1} x_{n+1} = 0$$

also defines the same point set. What kind of relation exists between the coefficients a and b?

Let us consider again the basis

$$A_1: (1, 0, 0, \ldots, 0),\ A_2: (0, 1, 0, \ldots, 0),\ \ldots,\ A_{n+1}: (0, 0, 0, \ldots, 1)$$

of $S_{n,q}$. By re-indexing, if necessary, we may suppose that the fundamental point A_1 does not satisfy the equation defined by the coefficients a and, consequently, it does not satisfy the equation defined by the coefficient b, either. The subspace S_{n-1} defined by equations intersect each of the lines $A_1 A_j$ ($j=2, 3, \ldots, n+1$) in one point, and the n points so obtained form an independent point system. If we express the coordinate sequences of these points in terms of the coefficient in each equation, then we obtain

$$(-a_1^{-1} a_2, 1, 0, \ldots, 0),\ (-a_1^{-1} a_3, 0, 1, \ldots, 0),\ \ldots,\ (-a_1^{-1}, a_{n+1}, 0, 0, \ldots, 1)$$

and

$$(-b_1^{-1} b_2, 1, 0, \ldots, 0),\ (-b_1^{-1} b_3, 0, 1, \ldots, 0),\ \ldots,\ (-b_1^{-1} b_{n+1}, 0, 0, \ldots, 1).$$

However, according to our assumption, these two sets represent the same n points, consequently

$$b_1 = \lambda a_1,\quad b_2 = \lambda a_2,\quad b_j = \lambda a_j, \ldots, \quad b_{n+1} = \lambda a_{n+1},$$

where $\lambda = b_1 a_1^{-1} \neq 0$.

Thus we can identify the hyperplanes with the classes of sequences of $(n+1)$ elements which are not all zero, as we did with the points. In order to distinguish hyperplanes from points we use the notation $[u_1, u_2, \ldots, u_{n+1}]$ where the elements $u_1, u_2, \ldots, u_{n+1}$ are called the *homogeneous coordinates* of the hyperplane. We say that a point (x_1, x_{n+1}) and a hyperplane $[u_1, \ldots, u_{n+1}]$ are *incident* if and only if $u_1 x_1 + \ldots u_{n+1} x_{n+1} = 0$. Because of the symmetry of the elements u and

x in the condition of incidence, $S_{n,q}$ the *principle of duality* holds in the space $S_{n,q}$. According to this principle, if a theorem is true in the space $S_{n,q}$ then by interchanging the roles of the points and hyperplanes in the statement of the theorem we obtain another theorem which also is valid in $S_{n,q}$.

A concept closely analogous to the cross ratio of the classical projective space can be introduced into the Galois space. Let the normed sequences $(a_1, a_2, \ldots, a_{n+1})$ and $(b_1, b_2, \ldots, b_{n+1})$ — i.e. points A and B — be mutually independent. Consider the points $\lambda A + \mu B = C$ and $\varrho A + \sigma B = D$ where C is distinct both from A and B, and D is distinct from A. We shall understand by the *cross ratio* (ABCD) *of points* A, B, C, D *in the given order* the element

$$(ABCD) = \lambda^{-1}\mu\varrho\sigma^{-1} = \theta$$

of the coordinate field.

We shall see that this concept enables a development of Galois geometry which is similar to Steiner's approach to projective geometry.

Let us note finally that the set of the nontrivial solutions of a system of homogeneous linear equations defined over a Galois field is the same thing as this intersection space of the set of hyperplanes, in a Galois space defined by the equation, in the first case we use the language of algebra and in the second that of geometry.

2.3 The generalization of Pappus' theorem on the Galois plane

In what follows we shall restrict ourselves to the geometry of Galois planes and shall use the results of sections 1.10 and 2.2. First of all we shall deal with the following theorem of Pappus in the classical projective plane:

If concurrent lines u_1, u_2, u_3, u_4 are cut by a line, which passes through their common point, in the points X_1, X_2, X_3, X_4, respectively, then the cross ratio (u_1, u_2, u_3, u_4) of the four lines is equal to the cross ratio $(X_1 X_2 X_3 X_4)$ of the four points.

The cross ratio of four concurrent lines is defined as the dual of the cross ratio of four collinear points. Let $[a_1, a_2, a_3]$ and $[b_1, b_2, b_3]$ be two normed, linearly independent sequences of the line-coordinates, i.e. two distinct lines a and b. Any line c passing through the common point of these two lines has for its (not necessarily normed) coordinates the linear combination

$$[\lambda a_1 + \mu b_1, \lambda a_2 + \mu b_2, \lambda a_3 + \mu b_3] = [c_1, c_2, c_3]$$

which is denoted by $\lambda a + \mu b = c$. Let another similarly derived line be $d = \varrho a + \sigma b$.

We understand by the cross ratio (*abcd*) of these concurrent lines, in the given order, the element

$$(abcd) = \lambda^{-1}\mu\varrho\sigma^{-1} = \theta$$

of the coordinate field.

We shall introduce the following concise notation:

$$[u_1, u_2, u_3] = [u], \quad (x_1, x_2, x_3) = (x), \quad u_1x_1 + u_2x_2 + u_3x_3 = [u](x);$$
$$[\lambda u_{k1}, \lambda u_{k2}, \lambda u_{k3}] = \lambda[u_{k1}, u_{k2}, u_{k3}] = \lambda[u]_k,$$
$$(\lambda x_{k1}, \lambda x_{k2}, \lambda x_{k3}) = \lambda(x_{k1}, x_{k2}, x_{k3}) = \lambda(x)_k.$$

On the Galois plane, by two distinct lines we mean two normed, independent sequences of line-coordinates. Let these be $[u]_1$ and $[u]_2$. Let two lines passing through the point of intersection of $[u]_1$ and $[u]_2$ be $[u]_3 = \lambda[u]_1 + \mu[u]_2$ and $[u]_4 = \varrho[u]_1 + \sigma[u]_2$. Similarly, one can define four points of a line by sequences of coordinates: let these be

$$(x)_1, (x)_2; \quad (x)_3 = \alpha(x)_1 + \beta(x)_2; \quad (x)_4 = \gamma(x)_1 + \delta(x)_2$$

The conditions of the theorem of Pappus are represented on the Galois plane by the following

$$[u]_1(x)_1 = 0, \quad [u]_1(x)_2 = \xi \neq 0, \quad [u]_2(x)_1 = \eta \neq 0,$$
$$[u]_2(x)_2 = 0, \quad [u]_3(x)_3 = 0 \quad [u]_4(x)_4 = 0.$$

From the linear combinations giving $[u]_3$, $[u]_4$, $(x)_3$ and $(x)_4$, the last two equations, above, imply that the pair of elements (ξ, η) satisfy the following:

$$\lambda\beta \cdot \xi + \mu\alpha \cdot \eta = 0,$$
$$\varrho\delta \cdot \xi + \sigma\gamma \cdot \eta = 0.$$

As these equations are satisfied by a pair of elements $(\xi, \eta) \neq (0, 0)$, the determinant of the system is the zero element of the coordinate field

$$\lambda\beta\sigma\gamma - \mu\alpha\varrho\delta = 0.$$

If λ, σ, α and δ are non-zero, then this relation can be expressed in the form

$$\lambda^{-1}\mu\varrho\sigma^{-1} = \alpha^{-1}\beta\gamma\delta^{-1}.$$

But the expression on the right hand side is the cross ratio of the quadruple of points $(x)_1, (x)_2, (x)_3, (x)_4$; whilst that on the left hand side is the cross ratio of the quadruple of lines $[u]_1, [u]_2, [u]_3, [u]_4$. Thus we have stated and proved Pappus' theorem for Galois planes as well.

Consider the distinct points X_1, X_2, X_3 on a line. To each point X_4 distinct from x_1 we assign the cross ratio (X_1, X_2, X_3, X_4). It is easy to see that, in this way, we get a 1—1 correspondence between the elements of the coordinate field K and the points of the line, distinct from X_1. Moreover, in this correspondence, the zero and unit elements of K are assigned to X_3 and X_2, respectively. These results can be dualized for a set of four concurrent lines.

The following properties of cross ratio follow easily from a definition:

(i) $\qquad (X_1 X_2 X_3 X_4) + (X_1 X_3 X_2 X_4) = 1;$

(ii) $\qquad (X_1 X_2 X_3 X_4) - (X_3 X_4 X_1 X_2) = 0;$

(iii) $\qquad$ if $(X_1 X_2 X_3 X_4) \neq 0$ then $(X_1 X_2 X_3 X_4) \cdot (X_1 X_2 X_4 X_3) = 1.$

The proofs are left to the reader.

A glance at the results dealt with in this chapter might suggest that there exists a complete identity between the Galois geometry and the classical projective geometry. However, we shall now discuss certain properties which show that there are significant differences between the two geometries.

For the time being, our approach will not be algebraic and, whenever it is practical, we shall refer to a figure.

We prove now that *cross ratio is invariant under projection*. Consider Fig. 47; according to Pappus' theorem we have $(ABCD)=(abcd)$ and $(A'B'C'D')= =(abcd)$, from which it follows that $(ABCD)=(A'B'C'D')$. We need not have assumed that $B=B'$, as shown in the figure, since we used only the fact that $b \cap u = B$ and $b \cap u' = B'$. Furthermore we need not assume that the lines a, b, c, d are distinct, since the theorem of Pappus is valid even if $d=b$ or, $d=c$. Indeed, Pappus' theorem is valid on $S_{2,2}$ where only three lines pass through any point.

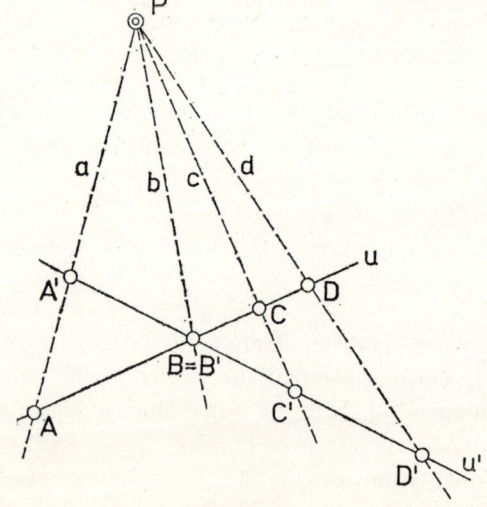

Figure 47

The proof of the statement $(ABXY) = -1$, with respect to Fig. 48, is more difficult. We apply the result, above, first to a projection from the point C and then the projection from the point D. We obtain

$$(ABXY) = (UVZY), \quad \text{and} \quad (UVZY) = (BAXY).$$

From this we have $(ABXY) = (BAXY) = \Theta$, say. If we use the properties of cross ratio, given above, it follows that $\vartheta^2 = 1$. Notice that we have relied upon

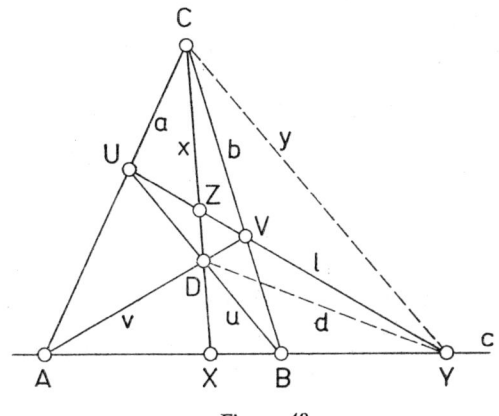

Figure 48

the fact that the points A, B, C, D determine a proper quadrangle, and from this follows the existence of three further points, namely:

$$U = AC \cap BD, \quad V = AD \cap BC \quad \text{and} \quad X = AB \cap CD.$$

We solve the equation $\vartheta^2 = 1$ and we shall need the geometric interpretation of the solution. The solution when $K = GF(2^r)$ is significantly different from that when $GF(p^r)$ $(p \neq 2)$.

If the field of coordinates is the $GF(2^r)$, then the equation $\vartheta^2 = 1$ has one and only one solution, namely $\vartheta = 1$. (Every element in a field of characteristic 2 is its own additive inverse.) Therefore, in this case we have $(ABXY) = 1$, which implies that $X = Y$, i.e. the line $l = UV$ cuts the line AB at the point $X = Y$. This cannot be visualized by Fig. 48; Fig. 22b is more appropriate for the visualization of this property.

We now state a corollary of this result: *Every (proper) quadrangle of a Galois plane of even order is a Fano quadrangle.*

A field of coordinates $GF(p^r)$ contains, in the case of $p \neq 2$, two solutions of the equation $\vartheta^2 = 1$: the one is the unit element, the other the additive inverse of the unit element, and these are not equal. We shall prove that $(ABXY) \neq 1$. Let us consider the normed coordinate triples of the points A, B, C and D. D

can be expressed in the following form:

$$\lambda A + \mu B + \nu C = D, \quad \lambda \cdot \mu \cdot \nu \neq 0.$$

Since $X = AB \cap CD$, the, not necessarily normed, coordinates of this point is given by $\lambda A + \mu B = D - \nu C = X$. Similarly $\lambda A + \nu C = D - \mu B = U$ and $\mu B + \nu C = D - \lambda A = V$. Thus we have $U - V = \lambda A - \mu B = Y$. The expressions

$$\lambda A + \mu B = X \quad \text{and} \quad \lambda A - \mu B = Y$$

hold irrespective of the characteristic of the coordinate field. Thus it follows immediately that $(ABXY) = \lambda^{-1}\mu\lambda(-\mu)^{-1} = -1$.

This unified proof, which avoids the theorem of Pappus, also clears up the question as to when X and Y coincide. The points X and Y coincide if and only if there exist a ϱ and a σ such that $(\varrho, \sigma) \neq (0, 0)$ but $\varrho X + \sigma Y = 0$, i.e. if $\lambda(\varrho + \sigma)A + \mu(\varrho - \sigma)B = 0$. From this, because of the independence of A and B, follows $\varrho + \sigma = \varrho - \sigma = 0$, and therefore $\sigma = -\sigma \neq 0$. Therefore $X = Y$ if and only if $\sigma = -\sigma$, i.e. if and only if the field is of characteristic 2.

As a corollary we obtain the following:

The diagonal points of any (proper) quadrangle of a Galois plane of odd order determine a proper triangle (i.e. they are ordinary quadrangles).

Wee see by the above results that the analogy between planes of even order and the classical projective plane is not as close as that between planes of odd order and the classical projective plane. Sometimes it will be necessary to approach the odd and even cases in quite different ways.

2.4 Coordinates on a Galois plane

Let $\{A_1, A_2, A_3\}$ be any independent point system in $S_{2,q}$, where A_1, A_2 and A_3 are given by the normed coordinate sequences

(2.4.1) $\quad (a_{11}, a_{12}, a_{13}), \quad (a_{21}, a_{22}, a_{23}) \quad \text{and} \quad (a_{31}, a_{32}, a_{33})$,

respectively

Let us consider the point ϱE generated by the linear combination $A_1 + A_2 + A_3$. Here the notation

(2.4.2) $\quad A_1 + A_2 + A_3 = \varrho E$

means that the point E is defined by the sequence (e_1, e_2, e_3) obtained by norming the sequence

$$(a_{11} + a_{21} + a_{31}, a_{12} + a_{22} + a_{32}, a_{13} + a_{23} + a_{33}).$$

Obviously, the point system $\{A_1, A_2, A_3, E\}$ is not independent, but the four point triplets which can be chosen from it all give rise to independent point systems, i.e. we are dealing with the vertices of a proper quadrangle. Let the figure of these four points be called a *system of coordinates* (or *figure of reference*), the first three points are called *fundamental points,* the fourth is called the *unit point.* — The system of axioms defining the finite planes ensure that such a quadrangle exists.

If we take the systems of coordinates as follows:

(2.4.3) $\quad A_1^\circ: (1, 0, 0), \quad A_2^\circ: (0, 1, 0), \quad A_3^\circ: (0, 0, 1),$
$$E^\circ: (1, 1, 1)$$

then the coordinates of the point $P: (x_1, x_2, x_3)$ can be considered as the coefficients of the linear combination

$$x_1 A_1^\circ + x_2 A_2^\circ + x_3 A_3^\circ.$$

When we consider the points of the plane with respect to an arbitrarily chosen system of coordinates, i.e. when we generate a point P by the linear combination

(2.4.4) $\quad x_1' A_1 + x_2' A_2 + x_3' A_3 = \sigma P,$

we assign another sequence of coordinates to the point, namely the sequence (x_1', x_2', x_3').

We speak about the *old* coordinates (x_1, x_2, x_3) of the point with respect to the system of coordinates $\{A_1^\circ, A_2^\circ, A_3^\circ, E^\circ\}$ and the *new* coordinates (x_1', x_2', x_3') of the point with respect to the system of coordinates $\{A_1, A_2, A_3, E\}$. It is easy to find, with the help of (2.4.1) and (2.4.4), the relation between the new and the old coordinates:

(2.4.5) $\quad \begin{cases} \sigma x_1 = a_{11} x_1' + a_{21} x_2' + a_{31} x_3', \\ \sigma x_2 = a_{12} x_1' + a_{22} x_2' + a_{32} x_3', \\ \sigma x_3 = a_{13} x_1' + a_{23} x_2' + a_{33} x_3'. \end{cases}$

The linear independence of the fundamental points is equivalent to the non-vanishing of the determinant

$$\begin{vmatrix} a_{11} & a_{21} & a_{31} \\ a_{12} & a_{22} & a_{32} \\ a_{13} & a_{23} & a_{33} \end{vmatrix} = A.$$

Consequently, the coordinates x_1', x_2', x_3' can be expressed from (2.4.5) — up to a factor $\varrho \neq 0$ as linear functions of the coordinates x_1, x_2, x_3:

(2.4.6) $\quad \begin{cases} \varrho x_1' = c_{11} x_1 + c_{12} x_2 + c_{13} x_3, \\ \varrho x_2' = c_{21} x_1 + c_{22} x_2 + c_{23} x_3, \\ \varrho x_3' = c_{31} x_1 + c_{32} x_2 + c_{33} x_3, \end{cases}$

where the matrix of this system is the inverse of the matrix of the system (2.4.5).

The linear transformation $(x) \to (x)'$ defined by (2.4.6) is said to be a non-singular linear transformation. (Where no confusion can occur we shall call this a "transformation".)

From (2.4.2) we get $A_1+A_2 = \varrho E - A_3$ which gives the point $E_3 = A_1 A_2 \cap A_3 E$. Similarly, the linear combinations $x'_1 A_1 + x'_2 A_2 = \sigma P - x'_3 A_3$ obtained from (2.4.4) give the point $P_3 = A_1 A_2 \cap A_3 P$ (Fig. 49). We shall assume that $x'_1 \cdot x'_2 \cdot x'_3 \neq 0$. Consider the quadruple of points $A_1, A_2, E_3 = A_1 + A_2, P = x'_1 A_1 + x'_2 A_2$. Obviously,

$$(A_1 A_2 E_3 P_3) = x'_1 x'^{-1}_2.$$

Similarly we obtain $E_1 = A_2 A_3 \cap A_1 E$ and $P_1 = A_2 A_3 \cap A_1 P$, where

$$(A_2 A_3 E_1 P_1) = x'_2 x'^{-1}_3;$$

also $E_2 = A_3 A_1 \cap A_2 E$ and $P_2 = A_3 A_1 \cap A_2 P$, where

$$(A_3 A_1 E_2 P_2) = x'_3 x'^{-1}_1.$$

From these it follows that

$$(A_1 A_2 E_3 P_3) \cdot (A_2 A_3 E_1 P_1) \cdot (A_3 A_1 E_2 P_2) = 1.$$

Conversely, if we are given a sequence of non-zero elements (ξ_1, ξ_2, ξ_3) then, the points $A_2 A_3$ and $E_1 = A_2 + A_3$ determine a unique point P_1 of the line $A_2 A_3$ for which

$$(A_2 A_3 E_1 P_1) = \frac{\xi_2}{\xi_3}.$$

Similarly the points A_3, A_1 and $E_2 = A_3 + A_1$ determine a unique point P_2 for which

$$(A_3 A_1 E_2 P_2) = \frac{\xi_3}{\xi_1}.$$

Let the point $P = A_1 P_1 \cap A_2 P_2$ be given by the linear combination $\eta_1 A_1 + \eta_2 A_2 + \eta_3 A_3$, then according to our first statement

$$(A_2 A_3 E_1 P_1) = \frac{\eta_2}{\eta_3} \quad \text{and} \quad (A_3 A_1 E_2 P_2) = \frac{\eta_3}{\eta_1}$$

we have therefore

$$\frac{\xi_2}{\xi_3} = \frac{\eta_2}{\eta_3}, \quad \frac{\xi_3}{\xi_1} = \frac{\eta_3}{\eta_1}, \quad \text{which imply} \quad \frac{\xi_2}{\xi_1} = \frac{\eta_2}{\eta_1},$$

i.e.

$$(\xi_1, \xi_2, \xi_3) = (\varrho \eta_1, \varrho \eta_2, \varrho \eta_3),$$

where $\varrho \neq 0$.

On the basis of the third quality we deduce that the line determined by the cross ratio $(A_1A_2E_3P_3) = \xi_1\xi_2^{-1}$ and connecting point P_3 with A_3 passes through the point $P = A_1P_1 \cap A_2P_2$.

Note that if $\xi_1\xi_2\xi_3 = 0$ then the point P will lie on one of the lines A_1A_2, A_2A_3 and A_3A_1. More precisely, if $\xi_1 = 0$ but $\xi_2\xi_3 \neq 0$ then P is the point of the line A_2A_3 determined by the cross ratio $(A_2A_3E_1P) = \xi_2/\xi_3$, and similarly for points on the other two sides; if $\xi_1 = 0$ and $\xi_2 = 0$ then $P = A_3$, if $\xi_2 = 0$ and $\xi_3 = 0$ then $P = A_1$ and if $\xi_3 = 0$ and $\xi_1 = 0$ then $P = A_2$.

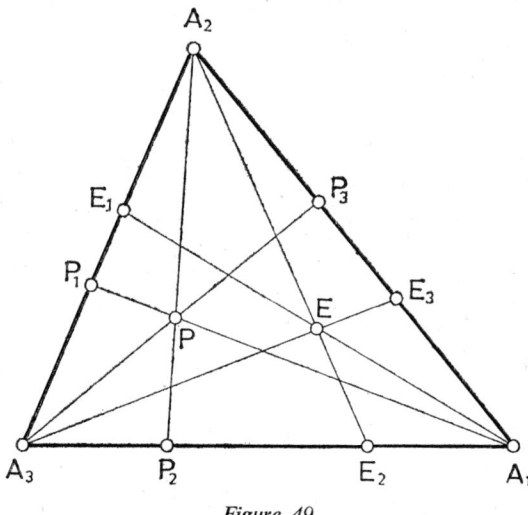

Figure 49

Changes of coordinate systems are sometimes advantageous when dealing with specific problems.

We can now give a geometrical description of linear transformations i.e. changes in the system of coordinates. Suppose A_1, A_2, A_3 and E are the reference points and unit point of a particular coordinate system. Of the $q^2 + q + 1$ points of the plane, the number lying on the sides of the fundamental triangle $A_1A_2A_3$ is $3(q+1) - 3 = 3q$, since every line contains $q+1$ points. We can establish a one-to-one correspondence between the remaining $q^2 - 2q + 1$ points of the plane and the class sequences (x_1, x_2, x_3) with the property $x_1 \cdot x_2 \cdot x_3 \neq 0$, as follows:

Let E_1 (resp. E_2, E_3) be the projection of E from A_1 (resp. A_2, A_3) onto A_2A_3 (resp. A_3A_1, A_1A_2). Given a sequence (x_1, x_2, x_3) where $x_1 \cdot x_2 \cdot x_3 \neq 0$. Consider the three points P_1, P_2 and P_3 determined by the cross ratios

(2.4.7) $\quad (A_1A_2E_3P_3) = \dfrac{x_1}{x_2}, \quad (A_2A_3E_1P_1) = \dfrac{x_2}{x_3} \quad$ and $\quad (A_3A_1E_2P_2) = \dfrac{x_3}{x_1}$.

Then, the lines A_1P_1, A_2P_2 and A_3P_3 meet in a point P. By assigning the sequence (x_1, x_2, x_3) to the point P we get the required correspondence. We have already characterized the points lying on the sides of the triangle $A_1A_2A_3$.

We can readily dualize everything presented in this section by dealing with the line coordinates. The details will be left to the reader, we sketch only the results.

The system of coordinates is formed by a system of lines $\{a_1, a_2, a_3, e\}$ in which no three lines have a point in common. The role of the points E_l, P_l is taken over here by lines e_l and p_l, connecting the point of intersection $a_j \cap a_k$ with the points of intersection $e \cap a_l$ and $p \cap a_l$, respectively; where (jkl) is in turn (123), (231) and (312). In the role of (2.4.7) we have

(2.4.8) $$(a_j a_k e_l p_l) = u_j u_k^{-1}$$

characterizing the line coordinates $[u_1, u_2, u_3]$; and the sequence $[u_1, u_2, u_3]$ determines a line passing through the point $a_j \cap a_k$ or just the line a_j according to whether $u_l = 0$ or $u_l = u_k = 0$.

In terms of the original coordinate system (2.4.3) the coefficients in the equation of a line we have taken as line coordinates, and the incidence relation of a point (x_1, x_2, x_3) and a line $[u_1, u_2, u_3]$ was expressed by

$$u_1 x_1 + u_2 x_2 + u_3 x_3 = 0.$$

If we change the coordinate system of the points, can we always find a coordinate system of lines so that the incidence relation takes the same form, i.e. if the new coordinates of a point and a line are (x_1', x_2', x_3') and $[u_1', u_2', u_3']$, respectively, then the point lies on the line if and only if

$$u_1' x_1' + u_2' x_2' + u_3' x_3' = 0?$$

In order to answer this question we must investigate the relation between the original point and line coordinates. The coordinates (x_1, x_2, x_3) belong to the figure of reference defined by (2.4.3), the line coordinates $[u_1, u_2, u_3]$ belong (in a dual manner) to the figure of reference formed by

(2.4.9) $\quad a_1^\circ: [1, 0, 0], \quad a_2^\circ: [0, 1, 0], \quad a_3^\circ: [0, 0, 1];$

$\quad\quad\quad\; e_1^\circ: [1, 1, 1].$

From the incidence relation $[u](x) = 0$ it follows that the lines $a_1^\circ, a_2^\circ, a_3^\circ$ are the lines $A_2^\circ A_3^\circ$, $A_3^\circ A_1^\circ$ and A_1° and A_2° respectively. That is, the reference triangle is common to both coordinate system, where in the case of point coordinates it is considered as a figure of three points and in the case of line coordinates as a figure of three lines.

We have now to find the relationship between the unit point and unit line, and how these are related to the fundamental triangle. It turns out that here the characteristic of the coordinate field has a significant effect upon the figure. If the coordinate field is $K=GF(p^r)$, we have to discuss the three cases $p=2$, $p=3$ and $p\neq 2, 3$ separately.

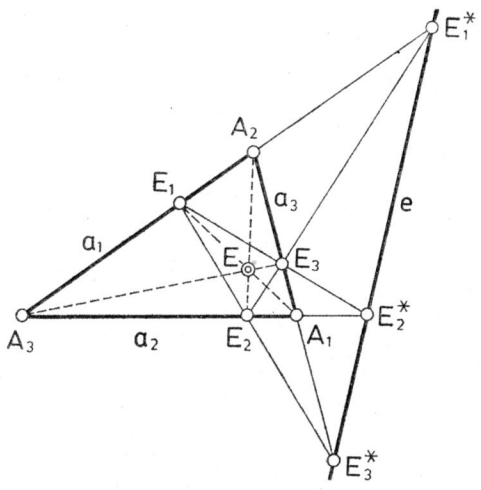

Figure 50

Consider first the case $p\neq 2,3$. (Cf. Fig. 50, where we replace E_1 by E_1°, etc.) We obtain readily that

(2.4.10) $\qquad E_1^\circ : (0, 1, 1), \quad E_2^\circ : (1, 0, 1), \quad E_3^\circ : (1, 1, 0).$

Let
$$e^\circ \cap a_1^\circ = E_1^{*\circ}, \quad e^\circ \cap a_2^\circ = E_2^{*\circ}, \quad e^\circ \cap a_3^\circ = E_3^{*\circ}.$$

The equation of the four lines are:

$$\begin{aligned} e^\circ &: \ x_1 + x_2 + x_3 = 0; \\ a_1^\circ &: \qquad\quad x_1 = 0, \\ a_2^\circ &: \qquad\quad x_2 = 0, \\ a_3^\circ &: \qquad\quad x_3 = 0. \end{aligned}$$

From these we immediately obtain the coordinates of the points of intersection, above, in normed form:

(2.4.11) $\qquad E_1^{*\circ} : (0, -1, 1), \quad E_2^{*\circ} : (-1, 0, 1), \quad E_3^{*\circ} : (-1, 1, 0).$

From (2.4.10) and (2.4.11) it is easy to discover the structure of the complete figure of reference, when $p\neq 2, 3$.

1° *The point E° is not on the line e°.*
2° *The point triple $E_1^\circ E_2^\circ E_3^\circ$ is a proper triangle.*
3° *The point system $\{A_j^\circ, A_k^\circ, E_l^\circ, E_l^{*\circ}\}$, where $(jkl)=(123)$, (231) and (312), each consist of four points forming a harmonic set*, i.e. a set of four collinear points where cross ratio (in some order) is equal to -1. In fact, $A_j^\circ + A_k^\circ = E_l^{*\circ}$ and $A_j^\circ - A_k^\circ = E_l^{*\circ}$, from which we have $(A_j^\circ A_k^\circ E_l^\circ E_l^{*\circ}) = -1$, and $E_l^\circ \neq E_l^\circ$.
4° *The fundamental triangle $A_1^\circ A_2^\circ A_3^\circ$ and the projection triangle $E_1^\circ E_2^\circ E_3^\circ$ are in perspective from the centre E° and from the axis l°.* This follows immediately from $A_j^\circ - A_k^\circ = A_l^{*\circ}$ and $E_j^\circ - E_k^\circ = E_l^{*\circ}$.

Figure 50 demonstrates the four properties in question.

We deal now with the case $p=3$. Here the figure has properties 2°, 3° and 4°; but, in view of the fact that $3 \cdot 1 = 0$, we have $E^\circ \in e^\circ$.

In the case $p=2$, only the first, properties hold; instead of properties 2°, 3° and 4° we have the following: The points $E_1^\circ = E_1^{*\circ}$, $E_2^\circ = E_2^{*\circ}$, $E_3^\circ = E_3^{*\circ}$ are three distinct points on the line e°.

The property that the fundamental triangle and the unit points determine uniquely the unit line holds, in all three cases.

Now we can return to the question of choosing the figures of reference (A_1, A_2, A_3, E) and (a_1, a_2, a_3, e) in the new coordinate system.

Consider the line of the equation $u_1 x_1 + u_2 x_2 + u_3 x_3 = 0$, referred to the original coordinate system. For the new coordinates of the points of the line we obtain, by (2.4.5), the equation

$$\frac{1}{\sigma}(a_{11}u_1 + a_{12}u_2 + a_{13}u_3)x_1' + \frac{1}{\sigma}(a_{21}u_1 + a_{22}u_2 + a_{23}u_3)x_2' +$$

$$+ \frac{1}{\sigma}(a_{31}u_1 + a_{32}u_2 + a_{33}u_3)x_3' = 0.$$

Let the coefficients of this equation be denoted in turn by u_1', u_2', u_3'.

Therefore the equation of the line in the new coordinates is

$$u_1' x_1' + u_2' x_2' + u_3' x_3' = 0$$

and let us call the lines expressed in the new coordinates by the equations $x_1' = 0$, $x_2' = 0$, $x_3' = 0$ the (new) reference lines and the line

$$x_1' + x_2' + x_3' = 0$$

the (new) unit line. Thus the new coefficients are determined by the following transformation:

(2.4.12)
$$\begin{cases} \sigma u_1' = a_{11}u_1 + a_{12}u_2 + a_{13}u_3 \\ \sigma u_2' = a_{21}u_1 + a_{22}u_2 + a_{23}u_3 \\ \sigma u_3' = a_{31}u_1 + a_{32}u_2 + a_{33}u_3. \end{cases}$$

The matrix of this linear transformation is the transpose of (2.4.5). Consequently, this matrix has determinant $A \neq 0$ and thus it is invertible and its inverse is given by

(2.4.13)
$$\begin{cases} \varrho u_1 = c_{11}u_1' + c_{21}u_2' + c_{31}u_3' \\ \varrho u_2 = c_{12}u_1' + c_{22}u_2' + c_{32}u_3' \\ \varrho u_3 = c_{13}u_1' + c_{23}u_2' + c_{33}u_3'. \end{cases}$$

By (2.4.11)—(2.4.13) it is clear that the incidence relation of the figures $\{A_1, A_2, A_3, E\}$ and $\{a_1, a_2, a_3, e\}$ is the same as that of the figures $\{A_1^\circ, A_2^\circ, A_3^\circ, E^\circ\}$ and $\{a_1^\circ, a_2^\circ, a_3^\circ, e^\circ\}$.

Therefore we may regard the coefficients $[u_1', u_2', u_3']$ of a line in the new coordinates as line-coordinates.

Finally we examine how cross ratio behaves under a linear transformation.

Let (x_1, x_2, x_3) and (y_1, y_2, y_3) be normed coordinate sequences defining the points X and Y respectively. Consider the point Z defined by the, not necessarily normed, sequence

$$(z_1, z_2, z_3) = (\lambda x_1 + \mu y_1, \lambda x_2 + \mu y_2, \lambda x_3 + \mu y_3).$$

Under the transformation (2.4.6) the new coordinates of X and Y, in normed form, will be $(\varrho x_1'; \varrho x_2', \varrho x_3')$ and $(\varrho^* y_1, \varrho^* y_2, \varrho^* y_3)$ respectively, where $\varrho, \varrho^* \neq 0$. The point 2 will now be given by the coordinate sequence

(2.4.14)
$$(\lambda \varrho x_1' + \mu \varrho^* y_1', \lambda \varrho x_2' + \mu \varrho^* y_2', \lambda \varrho x_3' + \mu \varrho^* y_3').$$

In the same manner, the new coordinates of a point T, obtained from the points X, Y by a linear combination with coefficients v, ω, will be

(2.4.15)
$$(\nu \varrho x_1' + \omega \varrho^* y_1', \nu \varrho x_2' + \omega \varrho^* y_2', \nu \varrho x_3' + \omega \varrho^* y_3').$$

Now in the original coordinates the cross ratio $(XYZT) = \lambda^{-1}\mu\nu\omega^{-1}$, and in the new coordinates

$$(XYZT) = (\lambda \varrho)^{-1} \cdot (\mu \varrho^*) \cdot (\nu \varrho) \cdot (\omega \varrho^*)^{-1}$$

but

(2.4.16)
$$(\lambda \varrho)^{-1} \cdot (\mu \varrho^*) \cdot (\nu \varrho) \cdot (\omega \varrho^*)^{-1} = \lambda^{-1}\mu\nu\omega^{-1}$$

and so we can state that: *cross ratio is an invariant of the coordinate system.*

2.5 Mappings determined by linear transformations

Let the linear transformation

(2.5.1) $$\begin{cases} x'_1 = c_{11}x_1 + c_{12}x_2 + c_{13}x_3 \\ x'_2 = c_{21}x_1 + c_{22}x_2 + c_{23}x_3 \\ x'_3 = c_{31}x_1 + c_{32}x_2 + c_{33}x_3 \end{cases}$$

be non-singular (i.e. the determinant, C, of the matrix formed by the coefficients is non-zero). We shall interpret this substitution on the plane $S_{2,q}$. Let both $(x_1 x_2 x_3)$ and (x'_1, x'_2, x'_3) be normed sequences and let both be referred to the same coordinate system, say to the original system $\{A_1^\circ, A_2^\circ, A_3^\circ, E^\circ\}$. Clearly one sequence is zero if and only if the other is, since the condition $C \neq 0$. Hence the *non-singular linear transformation* (2.5.1) *expresses a one-to-one mapping of the plane* $S_{2,q}$ onto itself; let this be written by $C(X) = X'$, where C denotes the matrix formed by the coefficients of (2.5.1), X and X' denote the points determined by the coordinate sequences (x_1, x_2, x_3) and (x'_1, x'_2, x'_3) respectively. If we denote by $C^{-1}(X)$ the mapping determined by the inverse matrix C^{-1} of the matrix C then, obviously $C^{-1}(X') = X$; i.e. the mapping $C^{-1}(X)$ is the *inverse* of the mapping $C(X)$.

We shall determine now some properties of the mapping $C(X)$. We know that the equations

(2.5.2) $$u_1 x_1 + u_2 x_2 + u_3 x_3 = 0$$

and

(2.5.3) $$u'_1 x'_1 + u'_2 x'_2 + u'_3 x'_3 = 0$$

are equivalent, if (2.5.1) and the related equations

(2.5.4) $$\begin{cases} \sigma u'_1 = C_{11}u_1 + C_{12}u_2 + C_{13}u_3, \\ \sigma u'_2 = C_{21}u_1 + C_{22}u_2 + C_{23}u_3, \\ \sigma u'_3 = C_{31}u_1 + C_{32}u_2 + C_{33}u_3 \end{cases}$$

are satisfied, where C_{jk} denotes the cofactor of c_{jk} in the matrix C. Thus, the image of the set of points satisfying (2.5.3) is the set of points satisfying (2.5.4) and we have the following result:

1° *The mapping* $C(X)$ *defined by* (2.5.1) *is line preserving*.

The equations (2.5.4) express the fact that the line determined by the sequence $[\sigma u'_1, \sigma u'_2, \sigma u'_3,]$ is the image of the line determined by the normed sequence $[u_1, u_2, u_3]$. — We may also say that the mapping $C(X)$ induces a one-to-one mapping of the set of lines of the plane onto itself. From the observations made at the end of section 2.4 we conclude the following.

2° *The mapping C(X) is cross ratio preserving.*

That is, if the mapping assigns to the points A, B, C, D of a line the images A', B', C', D', then the latter are collinear and we have the relation $(ABCD) = (A'B'C'D')$.

The duals of 1° and 2° can be obtained by considering the transformation (2.5.4):

The image of a set of concurrent lines is a set of concurrent lines and if the lines a', b', c', d', are the images of a, b, c, d, respectively, we have the relation $(a, b, c, d) = (a'b'c'd')$.

For the time being the mapping of points defined a linear substitution and the induced mapping of lines will together be called a *linear mapping*. In Section **1.12** we encountered the notion of collineation, i.e. the one-to-one line preserving mapping of the plane onto itself. Obviously, *every linear mapping is a collineation*. However, it is far more difficult to determine whether every collineation is a linear mapping.

2.6 Linear mapping of a given quadrangle onto another given quadrangle

Consider the proper quadrangle formed by the points $A_1^\circ : (1, 0, 0)$, $A_2^\circ : (0, 1, 0)$, $A_3^\circ : (0, 0, 1)$ and $E^\circ : (1, 1, 1)$ of the plane $S_{2,q}$; also the quadrangle formed by the points

$$A_j : (a_{j1}, a_{j2}, a_{j3}), \quad (j = 1, 2, 3, 4)$$

of which each set of three points is linearly independent, and the coordinate sequences are assumed to be normed.

The linear transformation:

(2.6.1)
$$\varrho x_1' = \lambda_1 a_{11} x_1 + \lambda_2 a_{21} x_2 + \lambda_3 a_{31} x_3,$$
$$\varrho x_2' = \lambda_1 a_{12} x_1 + \lambda_2 a_{22} x_2 + \lambda_3 a_{32} x_3,$$
$$\varrho x_3' = \lambda_1 a_{13} x_1 + \lambda_2 a_{23} x_2 + \lambda_3 a_{33} x_3,$$

referred to the coordinate system $\{A_1^\circ, A_2^\circ, A_3^\circ, E^\circ\}$ provided that $\varrho, \lambda_1, \lambda_2, \lambda_3$ are elements distinct from zero — assign points A_1, A_2 and A_3 to the points

A_1°, A_2° and A_3°, respectively. Clearly this transformation is non-singular, because its determinant

$$(2.6.2) \qquad A^* = \lambda_1 \lambda_2 \lambda_3 A, \quad \text{where} \quad A = \begin{vmatrix} a_{11} & a_{21} & a_{31} \\ a_{12} & a_{22} & a_{32} \\ a_{13} & a_{23} & a_{33} \end{vmatrix}$$

and as A_1, A_2 and A_3 are independent we have $A \neq 0$.

By a suitable choice of the coefficients λ_1, λ_2, λ_3 we can satisfy the equations

$$(2.6.3) \qquad \begin{aligned} a_{41} &= \lambda_1 a_{11} + \lambda_2 a_{21} + \lambda_3 a_{31}, \\ a_{42} &= \lambda_1 a_{12} + \lambda_2 a_{22} + \lambda_3 a_{32}, \\ a_{43} &= \lambda_1 a_{13} + \lambda_2 a_{23} + \lambda_3 a_{33}, \end{aligned}$$

i.e. the point A_4 is the image of point E°. In fact, because of the independence of the points A_j every minor of the third order of the determinant

$$\begin{vmatrix} a_{11} & a_{21} & a_{31} & a_{41} \\ a_{12} & a_{22} & a_{32} & a_{42} \\ a_{13} & a_{23} & a_{33} & a_{43} \end{vmatrix}$$

is distinct from zero. Thus the equations (2.6.3) have a unique solution for λ_1, λ_2, and λ_3 and all of these will be non-zero.

Summarizing this, we can say that *there exists a linear mapping which carries the vertices of the proper quadrangle $A_1^\circ A_2^\circ A_3^\circ E^\circ$ into the vertices of an arbitrarily chosen proper quadrangle $A_1 A_2 A_3 A_4$*.

Let this mapping be denoted by φ and the inverse of this mapping by φ^{-1}. Let the inverse transformation be denoted simply by

$$(2.6.4) \qquad \varphi^{-1}: \begin{cases} \varrho x_1' = a_{11} x_1 + a_{12} x_2 + a_{13} x_3 \\ \varrho x_2' = a_{21} x_1 + a_{22} x_2 + a_{23} x_3 \\ \varrho x_3' = a_{31} x_1 + a_{32} x_2 + a_{33} x_3 \end{cases} \quad (\det(a_{jk}) \neq 0).$$

Let $B_1 B_2 B_3 B_4$ be an arbitrary proper quadrangle. Let the linear mapping carrying the points A_1°, A_2°, A_3° and E° into the points B_1, B_2, B_3 and B_4 respectively be denoted by and let this be given by

$$(2.6.5) \qquad \psi: \begin{cases} \sigma x_1' = b_{11} x_1 + b_{12} x_2 + b_{13} x_3 \\ \sigma x_2' = b_{21} x_1 + b_{22} x_2 + b_{23} x_3 \\ \sigma x_3' = b_{31} x_1 + b_{32} x_2 + b_{33} x_3 \end{cases} \quad (\det(b_{jk}) \neq 0).$$

If the mapping φ^{-1} assigns to the point X defined by the normed coordinate sequence (x_1, x_2, x_3) the image X' defined by the normed coordinate sequence (x_1', x_2', x_3') and then the mapping ψ assigns to the point X' the image X'' defined by the normed coordinate sequence (x_1'', x_2'', x_3''), then the mapping $X \to X''$ is said to be the *composition* (or *product*) of the (first) mapping φ^{-1} and the (second) mapping ψ and will be denoted by $\varphi^{-1}\psi$.

We claim that the mapping $\varphi^{-1}\psi$ is *also a linear mapping*.

Namely, consider the linear transformation

(2.6.6)
$$\varrho\sigma x_1'' = c_{11}x_1 + c_{12}x_2 + c_{13}x_3,$$
$$\varrho\sigma x_2'' = c_{21}x_1 + c_{22}x_2 + c_{23}x_3,$$
$$\varrho\sigma x_3'' = c_{31}x_1 + c_{32}x_2 + c_{33}x_3,$$

where the coefficients are

(2.6.7) $\qquad c_{jk} = b_{j1}a_{1k} + b_{j2}a_{2k} + b_{j3}a_{3k} \quad (j, k = 1, 2, 3)$

We can see immediately, by a direct calculation according to (2.6.4) and (2.6.5), that the mapping $\varphi^{-1}\psi = \tau$ is given by the linear transformation of (2.6.5). The prescription (2.6.6), means precisely that the matrices $\mathbf{A}, \mathbf{B}, \mathbf{C}$ determining the mappings φ^{-1}, ψ, τ are related by $\mathbf{A} \cdot \mathbf{B} = \mathbf{C}$. Hence we have also for the determinants $A \cdot B = C$ and therefore the conditions $A \neq 0$, $B \neq 0$ imply that $C \neq 0$.

Summarizing, we can say that *given any two proper quadrangles $A_1A_2A_3A_4$ and $B_1B_2B_3B_4$ of the plane there exists a linear mapping which assigns the points B_1, B_2, B_3 and B_4 to the points A_1, A_2, A_3 and A_4, respectively*.

The uniqueness of the linear mapping satisfying the requirements $\tau(A_j) = B_j$ ($j = 1, 2, 3, 4$) would also appear from a detailed analysis of the reasoning leading to (2.6.5) but we shall show this in a more instructive way. Suppose that the requirements in question are satisfied by the distinct linear mappings τ_1 and τ_2. Because $\tau_1 \neq \tau_2$ there exists at least one point P such that $\tau_1(P) = P'$, $\tau_2(P) = P''$ and $P' \neq P''$, and, clearly the point P must be distinct from the points A_j. Also, P must be distinct from the points $A_1A_2 \cap A_3A_4$, $A_2A_3 \cap A_1A_4$, $A_3A_1 \cap A_2A_4$, since a linear mapping preserves points of intersection. Further it is not possible for P to lie on any of the lines A_jA_k. For if it could lie, say, on the line A_3A_4 and $P \neq U = A_1A_2 \cap A_3A_4$, then the distinct points P' and P'' would be incident with the line B_3B_4 but distinct from the point $U^* = B_1B_2 \cap B_3B_4$. A linear mapping is crosss ratio preserving thus $(A_3A_4 UP) = (B_3B_4 U^*P')$ and $(A_3A_4 UP) = (B_3B_4 U^*P'')$. But by an earlier result $(B_3B_4U^*P') = (B_3B_4U^*P'')$ implies $P' = P''$ which is in contradiction with our assumption. Thus P cannot be a point on any of the lines A_jA_k. Consequently, neither P' nor P'' lies on any of the lines B_jB_k. The line $P'P''$ can pass, at most, through one of the points B_j. Thus we may assume that it does not pass through the point B_1, say. Hence the lines l' and l'' connecting point B_1 with P' and P'', respectively, i.e. the images of the line $A_1P = l$ under τ_1 and τ_2, respectively, do not coincide and clearly neither of them coincides with any of the lines $B_j B_k$. If we use the notations $A_1 A_k = l_{1k}$ and $B_1 B_k = l'_{1k}$, the cross ratio preserving property of τ_1 and τ_2 implies that

$$(l_{12} l_{13} l_{14} l) = (l'_{12} l'_{13} l'_{14} l')$$

and

$$(l_{12} l_{13} l_{14} l) = (l'_{12} l'_{13} l'_{14} l'');$$

and from these $(l'_{12} l'_{13} l'_{14} l') = (l'_{12} l'_{13} l'_{14} l'')$ from which follows $l' = l''$. And this contradiction completes our proof of the uniqueness of τ.

2.7 The concept of an oval on a finite plane

Once again we shall investigate a certain analogy with classical geometry. But here odd planes and even planes have to be given distinct treatments.

The ellipse is a special kind of the closed curve called *oval* in Euclidean geometry. The number of common points of an oval and a line is 2 or 1 or 0. Corresponding to these cases the line is said to be a *secant*, a *tangent* or an *exterior line* with respect to the oval. We shall try to find figures characterized by this property on a finite plane (not necessarily a Galois plane). The line of the finite plane is a subset of the plane consisting of $q+1$ points, hence it seems to be natural that an oval should also be a subset of the same number of points.

We shall understand by an *oval* of $q+1$ points Ω on a finite plane of order q which has no more than two points lying on the same line. A line is said to be a *secant*, a *tangent*, or an *exterior line* with respect to the oval, if the number of common points of the line with the oval is 2, 1, of 0, respectively.

It is easy to find an example of an oval on a Galois plane.

Consider, on the $S_{2,q}$, the point set Ω_0 which is defined by the equation

(2.7.1) $$x_1 x_2 - x_3^2 = 0.$$

Clearly, the points of Ω_0 are: the point $(1, 0, 0)$ and q other points determined by the sequences $(\omega^2, 1, \omega)$, where ω runs through the elements of the coordinate field. Thus, Ω_0 consists of $q+1$ points.

We have still to verify that any three (distinct) points of the point set Ω_0 are linearly independent, but this follows immediately from the identities

$$\begin{vmatrix} 1 & 0 & 0 \\ \omega_1^2 & 1 & \omega_1 \\ \omega_2^2 & 1 & \omega_2 \end{vmatrix} = \omega_2 - \omega_1 \quad \text{and} \quad \begin{vmatrix} \omega_1^2 & 1 & \omega_1 \\ \omega_2^2 & 1 & \omega_2 \\ \omega_3^2 & 1 & \omega_3 \end{vmatrix} =$$

$$= (\omega_1 - \omega_2)(\omega_2 - \omega_3)(\omega_3 - \omega_1).$$

It is worth-while noting how many of the secants of Ω_0 pass through the fundamental point $(0, 0, 1)$. If the points $(0, 0, 1)$, $(\omega_1^2, 1, \omega_1)$, $(\omega_2^2, 1, \omega_2)$ are collinear, then the expansion of the determinant formed by their coordinates leads to the relation $\omega_1^2 - \omega_2^2 = 0$. From this, because of the condition $\omega_1 \neq \omega_2$, it follows that $\omega_2 = -\omega_1$. If q is even, then $\omega_2 = -\omega_1 = \omega_1$ which we have excluded; therefore no secant passes through the fundamental point $(0, 0, 1)$. However, the number of the lines passing through this point is $q+1$, and each of them contains one and only one point of Ω_0, i.e. every line through $(0, 0, 1)$ is a tangent of the oval. This property is not exhibited by the ovals of classical geometry. If q is odd, then the fundamental lines $[1, 0, 0]$ and $[0, 1, 0]$ are the only tangent lines through $(0, 0, 1)$: they pass through the point $(0, 1, 0)$ and the point $(1, 0, 0)$, respectively. Through the fundamental point $(0, 0, 1)$ there pass as many secants as there are unordered pairs consisting of an element and its additive inverse, discounting $(0, 0)$ i.e. there are $(q-1)/2$ secant lines through $(0, 0, 1)$. The remaining $(q-1)/2$ lines through $(0, 0, 1)$ are clearly exterior lines. — Thus in the case of an odd q the fundamental point $(0, 0, 1)$ has properties which are similar to those of an exterior point of an oval in the classical plane: two tangents pass through the point as well as lines which meet the oval and other lines which do not.

The $q+1$ points of an oval Ω_0 in a Galois plane of even order determine pairwise $(q+1)/2$ secants, q secants passing through any of its points. The line connecting a point of the oval with the fundamental point $(0, 0, 1)$ is a tangent and this tangent and the q secants through the point in question exhaust all lines of the plane incident with the point in question. As we have shown, all the tangents of the oval Ω_0 meet at the fundamental point $0 : (0, 0, 1)$. This special point with respect to the oval — which, however, does not lie on it — is said to be the *nucleus* of the oval. The $q+1$ points of the oval and its nucleus form a set of $q+2$ points such that no more than two of them lie on the same line. A point set of $q+2$ elements with this property is said to be a hyperoval. — We shall see later that hyperovals exist only on planes of even order. —

In what follows a plane of order q does not mean necessarily a Galois plane of order q. The problems to be dealt with here require only the most elementary combinatorial reasoning for their solutions.

1° We shall study first the ovals of the *planes of odd order*.

An oval Ω of the plane consists of $q+1$ points, i.e. of an even number of points. Let these points be denoted by $A_0, A_1, \ldots, A_q$. It follows from the definition of an oval that if we connect a point A_k with each of the other points in turn, we obtain q distinct lines and the $(q+1)$-th line passing through the point is necessarily a tangent. Thus through every point of an oval passes a tangent having this point as its *point of contact*.

Let us take an arbitrary point A_0, say, of Ω and the tangent a_0 touching the oval at this point. Let the other points of this line be: $P_1, P_2, \ldots, P_q$. Each of the other q tangents to Ω must pass through one of the points $P_1, \ldots, P_q$, since otherwise there would exist a point P through which only the tangent a_0 would pass and all other lines through P would be secants and exterior lines. This, however, is impossible, since the points $A_1, \ldots, A_q$ would then lie on secants through P, and q would then have to be even. Therefore, two tangents pass through every point of a tangent, except the point of contact; such points are said to be *exterior* points with respect to the oval. From this it follows that *the number of exterior points is* $(q+1)q/2$. If we delete, from the q^2+q+1 points of the plane, the $q+1$ points of contact and the $(q+1)q/2$ exterior points, there remain $q(q-1)/2$ points, none of which lie on a tangent. These points are said to be *interior* points with respect to the oval.

Like the lines, the points also are separated into three classes by the oval: exterior points, the points of the oval and interior points. One might imagine that this separation into classes is completely analogous to the situation in the classical plane. However, in the finite plane of order q, an *exterior* (or in other word a *skew*) *line consists of* $(q+1)/2$ *exterior points and of the same number of interior points*.

Our statement follows from the fact that each tangent meets an exterior line in an exterior point but two tangents pass through every exterior point.

It follows from a similar consideration that any secant has $(q-1)/2$ exterior points and as many interior points.

2° We shall deal now with the ovals of a finite planes of *even order*.

The number of the points of an oval, i.e. $q+1$, is now an odd number and also in this case one tangent only passes through each point of the oval, all other lines passing through a point on the oval are secants.

Consider now the secant l connecting the points A_j and A_k of the oval.

A tangent to the oval passes through each of the remaining $q-1$ points of l. Since, if the line l contained a point P through which no tangent passed, then the connecting P with any of the $(q-1)$ points of the oval which do not belong to l, we would obtain a secant, which is impossible, since $q-1$ is an odd number. As the number of the tangents is equal to the number of the points of l, *there passes just each one tangent through every point of* l.

Let M be the common point of two tangents to an oval. Let these two tangents be the lines a_0 and a_1, say. Let us connect M with each of the points of the oval. None of the lines so obtained is a secant, according to the satement proved above, since otherwise two tangents, a_0 and a_1 would pass through a point of a secant, namely M. Consequently, every line through M must be a tangent to the oval. The point M is said to be the *nucleus* of the oval. The set of $q+2$ points formed by the oval and its nucleus is said to be a *hyperoval*.

3° A set of k points of a finite plane, no more than two of which are collinear, will be called a *k-arc*. If a k-arc is contained in no $(k+1)$-arc then it is said to be *complete*. An oval in a plane of even order is a non-complete $(q+1)$-arc, since it contained in a hyperoval, which is a $(q+2)$-arc.

Theorem of Bose[*]: *If there exists a k-arc on a finite plane of order q, then we have* $k \leq q+2$ *or* $k \leq q+1$ *if q is even or odd, respectively.*

This theorem can easily be proved by using 1° and 2°. Assume that q is even. Let A_0 be an arbitrary point of a k-arc. By connecting the other points of the k-arc with A_0 we obtain $k-1$ distinct lines, but the total number of the lines passing through A_0 is $q+1$, hence $k-1 \leq q+1$, i.e. $k \leq q+2$.

Assume that q is odd and there exists a k-arc for which $k > q+1$. By taking $q+1$ points of the k-arc, we obtain an oval Ω. Let A_0 be a point of the k-arc which does not belong to the oval Ω. Let the point A_0 be connected with the points of Ω by $q+1$ distinct lines, these are secants of the k-arc but are tangents to the oval Ω, hence $q+1 \leq 2$, but for a finite plane $q+1 \geq 3$. This contradiction verifies the second statement of the theorem.

2.8 Conics on a Galois plane

For the time being we introduce the notion of conics by considering the point set Ω_0 defined by the equation (2.7.1) and we shall study its properties, which are similar to those of conics in classical projective plane.

The quadratic polynomial occurring in (2.7.1) is irreducible over every finite field. However, we shall encounter quadratic polynomials which are reducible over certain finite fields.

The following examples hold for any coordinate field $GF(q)$.

1° The equation $x_1 x_2 = 0$ defines a set consisting of $2q+1$ points: the set of the points formed by the lines $x_1 = 0$ and $x_2 = 0$. This point set is clearly not an oval.

---

[*] Since about 1940 much work has been done in the theory of k-arcs, especially in the case of Galois planes. One reason for this is that the results obtained are often of importance in other subjects, e.g. in mathematical statistics.

2° The point set defined by the equation $x_1^2 - x_2^2 = 0$ is the join of the point sets defined by the equations $x_1 - x_2 = 0$ and $x_1 + x_2 = 0$ and thus it is not an oval.

The following example refers to the case when the coordinate field is of characteristic 2:

3° The equations $x_1^2 + x_2^2 + x_3^2 = 0$ and $x_1 + x_2 + x_3 = 0$ define the same point set, i.e. we are dealing with $q+1$ points forming a line, consequently it is not an oval.

+	0	1	2	3	4	5	6	7
0	0	1	2	3	4	5	6	7
1	1	0	3	2	5	4	7	6
2	2	3	0	1	6	7	4	5
3	3	2	1	0	7	6	5	4
4	4	5	6	7	0	1	2	3
5	5	4	7	6	1	0	3	2
6	6	7	4	5	2	3	0	1
7	7	6	5	4	3	2	1	0

×	0	1	2	3	4	5	6	7
0	0	0	0	0	0	0	0	0
1	0	1	2	3	4	5	6	7
2	0	2	4	6	3	1	7	5
3	0	3	6	5	7	4	1	2
4	0	4	3	7	6	2	5	1
5	0	5	1	4	2	7	3	6
6	0	6	7	1	5	3	2	4
7	0	7	5	2	1	6	4	3

Figure 51

Our following two examples refer also to a coordinate field $GF(2^r)$ but the first one is investigated in the case $r=2$ and the second in the case $r=3$. We give the operation tables of $GF(8)$, as they will be useful later (Fig. 51).

In the case of a coordinate field $GF(2^r)$ the set of the points defined by the equation $x_1^2 + x_1 x_2 + x_2^2 = 0$ contains the fundamental point $(0, 0, 1)$; any other point in this set will have both x_1 and x_2 non-zero. Suppose $x_2 \neq 0$, then by the substitution of $x_1 x_2^{-1} = z$ we obtain the equation $z^2 + z + 1 = 0$. From the assumption $q = 2^r$ it follows that $-1 = 1$, hence our equation can be written in the form $z(z+1) = 1$.

4° Consider the solutions of the equation over $GF(8)$. We can immediately see from the operation tables that in case of $z = 0, 2, 3, 4, 5, 6, 7$ the product $z(z+1)$ is in turn equal to $0, 0, 6, 6, 2, 2, 4, 4$. In other words the equation $z(z+1) = 1$ has no solution over $GF(8)$ consequently the point set defined by the equation $x_1^2 + x_1 x_2 + x_2^2 = 0$ consists of a single point namely, the fundamental point $(0, 0, 1)$.

5° Consider now the solutions of the equation $z(z+1) = 1$ over $GF(4)$. The operation tables were given in **1.6**. Now if $z = 0, 1, 2, 3$, the product $z(z+1)$

is in turn 0, 0, 1, 1. As $z=2$ and $z=3$ are solutions of $z(z+1)=1$ we see that the equation $x_1^2+x_1x_2+x_2^2=0$ can also be expressed in the form $(x_1+2x_2)(x_1+3x_2)=0$ over $GF(4)$ thus it defines the union of the points of two lines, which is not an oval.

Upon the basis of the above examples we may ask the following question: Is there a criterion for the coefficients in a quadratic equation so that the point set defined by the equation will be an oval? Similarly, we may put a further question: Is it possible to make a change of coordinates so that the quadratic equation defining the oval becomes $x_1x_2-x_3^2=0$? These questions will be answered later. Here we investigate merely the geometrical properties of the special oval Ω_0 defined by the quadratic equation (2.7.1). In general, we shall deal with an arbitrary coordinate field $GF(q)=K$, but in particular cases we may specialize the values of q. In what follows we shall refer to Fig. 52.

Consider the set of (q^2-1) pairs $(\lambda, \mu) \neq (0, 0)$, where $\lambda, \mu \in K$. Decompose the set into classes by saying that (λ, μ) and (λ', μ') belong to the same class if and only if there exists a non-zero $\varrho \in K$ such that $\varrho\lambda=\lambda'$ and $\varrho\mu=\mu'$. Obviously, we obtain in this way $(q^2-1)/(q-1)=q+1$ classes, i.e. as many classes as there are points on the oval Ω_0. It is easy to see that we can establish a one-to-one correspondence between the set of classes and the point set Ω_0 by the prescription

(2.8.1) $$x_1 = \lambda^2, \quad x_2 = \mu^2, \quad x_3 = \lambda\mu.$$

Thus denoting the point $(\lambda^2, \mu^2, \lambda\mu)$ by (λ, μ) we have parametrized the point of Ω_0. This parametrization will greatly simplify our investigations.

Now we shall derive a significant geometrical property of the oval Ω_0. The points:

(2.8.2) $\quad\quad A_1: (1, 0, 0), \quad A_2: (0, 1, 0) \quad$ and $\quad E: (1, 1, 1)$

of Ω_0 are parametrized by the pairs $(1, 0)$, $(0, 1)$ and $(1, 1)$, respectively. Let P be a point of Ω_0 distinct from A_1, A_2 and E, parametrized by (λ, μ), say. Let us call the line connecting an arbitrary point of Ω_0 with A_1, the line *projecting* the point from A_1; also let us call the tangent to Ω_0 at A_1; the line projecting the point A_1 from itself. Further, we shall denote by $A_1(A_1A_2EP)$ the cross ratio of the lines projecting the points A_1, A_2, E, P from the point A_1. Similarly, $A_2(A_1A_2EP)$, denotes the cross ratio of the lines projecting the same points from point A_2. Thus (Fig. 52) according to the theorem of Pappus:

$$A_1(A_1A_2EP) = (a_2a_3e_1p_1) = (A_3A_2E_1P_1) = x_3x_2^{-1},$$

and

$$A_2(A_1A_2EP) = (a_3a_1e_2p_2) = (A_1A_3E_2P_2) = x_1x_3^{-1},$$

where $(x_1, x_2, x_3)=(\lambda^2, \mu^2, \lambda\mu)$ and consequently

$$x_3x_2^{-1} = x_1x_3^{-1} = \lambda\mu^{-1} = \vartheta$$

9 Introduction

hence

(2.8.3) $$A_1(A_1A_2EP) = A_2(A_1A_2EP) = \vartheta.$$

If the point P runs through the points of $\Omega_0 - A_1$ then P_1 runs through the points $a_1 - A_1$, consequently $(A_3A_2E_1P_1) = \lambda\mu^{-1} = \vartheta$ runs through the elements of $\mathbf{K}$. The values $\Theta = 0$ and $\Theta = 1$ correspond to the positions $P = A_2$ and $P = E$, respectively.

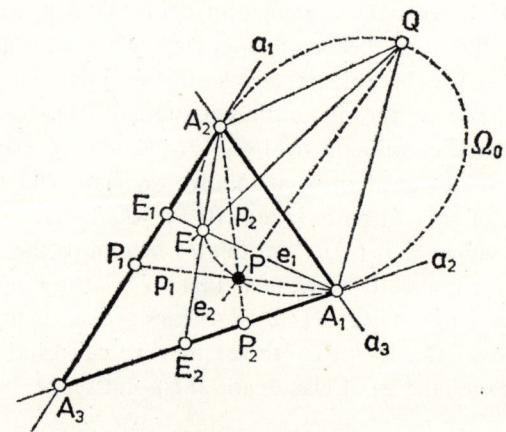

Figure 52

Conversely, for each value of Θ we obtain a point P_2 of the set $a_2 - A_3$ and a point P_1 of the set $a_1 - A_1$. Thus each value of ϑ determines one point of intersection $A_2P_2 \cap A_1P_1 = P$ on Ω_0.

Thus we arrive at the following lemma:

If we establish a correspondence between the lines projecting the points A_1, A_2, E, P of the oval Ω_0 from the point A_1 and the lines projecting the same points from A_2, then we have the relation (2.8.3); we can establish, by means of the inhomogeneous parameter ϑ, a one-to-one correspondence between the point set $\Omega_0 - A_1$ and the coordinate field $\mathbf{K}$.

If $\vartheta \in \mathbf{K}$, then denote the point of Ω corresponding to ϑ by P_ϑ.

Let $Q: (\xi^2, \eta^2, \xi\eta)$ be a point of the oval distinct from $P: (\lambda^2, \mu^2, \lambda\mu)$. The coordinates of the point $A_1A_3 \cap OP = P_3^*$ are easily seen to be

(2.8.4) $$(\lambda\eta + \mu\xi, 0, \mu\eta), \quad (\mu, \eta) \neq (0, 0).$$

Clearly, we can regard P_3^* as the projection of P onto the line $A_1A_3 = a_2$ from the point Q. If we replace P in turn by A_1, A_2, E then we obtain the points

$$A_1^* = A_1: (1, 0, 0), \quad A_2^*: (\xi, 0, \eta), \quad E_3^*: (\xi + \eta, 0, \eta),$$

therefore

$$E_3^* = \eta A_1^* + A_2^* \quad \text{and} \quad P_3^* = \lambda\eta A_1^* + \mu A_2^*,$$

and hence

(2.8.5) $\quad (A_1^* A_2^* E_3^* P_3^*) = \eta^{-1} \cdot 1 \cdot \lambda\eta \cdot \mu^{-1} = \lambda\mu^{-1} = (A_1 A_2 E_3 P_3).$

By applying the theorem of Pappus to the projecting quadruple we obtain the following lemma:

All quadruples of lines projecting the fixed points A_1, A_2, E, P *of the oval* Ω_0 *from any other point of the oval have the same cross ratio.*

We can speak therefore of the cross ratio of this quadruple of points.

By means of this lemma and the theorem of Pappus we can extend the notion of cross ratio to any four points of the oval. In order to do this we apply a third lemma which establishes a one-to-one correspondence between two projective pencils under certain conditions.

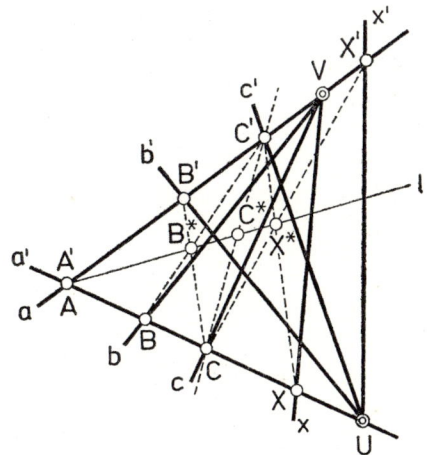

Figure 53

The trivial case of $q=2$ and the uninteresting case of $q=3$ will be omitted in the following discussion. Let U and V be two distinct points of the plane, let a, b and c be three distinct lines through U and let a', b' and c' be three distinct lines through V (Fig. 53). Let A, B and C denote the points of intersection of the lines a, b, c with the line a' respectively, similarly, let A', B', C' be, respectively, the points of intersection of the lines a', b', c' with a. Clearly $A = A' = = A^*$, say, and let $B^* = CB' \cap C'B$ and $C^* = A^*B^* \cap CC'$; denote the line A^*B^*

by l. Consider any two lines x and x' passing through the points U and V, respectively, and satisfying the condition

(2.8.6) $\quad (abcx) = (a'b'c'x')$, let $X = x \cap a'$ and $X' = x' \cap a$.

Assume that $C'X$ cuts the line l in a point X_1 and CX' cuts it in a point X_2. By applying the theorem of Pappus it follows that

$$(abcx) = (ABCX)$$
$$= C'(ABCX)$$
$$= (A^*B^*C^*X_1).$$

Similarly
$$(a'b'c'x') = (A^*B^*C^*X_2)$$

therefore by (2.8.6) we have

$$(A^*B^*C^*X_1) = (A^*B^*C^*X_2).$$

From this we have $X_1 = X_2 = X^*$, say, and the lines $C'X$ and CX' intersect in a point which lies on l.

For each choice of the line x among the $q-2$ lines, distinct from a, b and c, passing through U we obtain, by (2.8.6), a corresponding line x' through V and distinct from a', b', and c'. It is easy to see that in this way we get a one-to-one correspondence

(2.8.7) $\qquad \varphi : \begin{pmatrix} a & b & c & x & \ldots \\ \downarrow & \downarrow & \downarrow & \downarrow & \\ a' & b' & c' & x' & \ldots \end{pmatrix}.$

Moreover, for each choice of x, we have shown that in the construction above the lines $C'X$ and CX' intersect on the line l. Thus it is clear that for any four lines x_1, x_2, x_3 and x_4 through U and the corresponding lines (under φ) x'_1, x'_2, x'_3 and x'_4, respectively we must have

$$(x_1, x_2, x_3, x_4) = (x'_1, x'_2, x'_3, x'_4).$$

Our statements are summarized in the third lemma:

The correspondence φ is a one-to-one correspondence and is cross ratio preserving, in other words it is a projectivity.

The theorem below follows immediately from our lemmas. It is the analogue of a theorem of Steiner concerning a well-known property of conics in the classical projective plane;

The relation between two pencils formed by the lines projecting the points of the oval Ω_0 from any two points of itself is a projectivity, if the corresponding lines of the two pencils are those which project the same point of the oval.

The close relationship between the point set Ω_0 of the Galois plane and the conic of the classical projective plane justifies us in calling this point set a conic. For the time being this will be the only point set we shall call a conic on a Galois plane.

2.9 Point configurations of order 2 on a Galois plane of even order

Throughout this section the coordinate field **K** will be assumed to be of characteristic 2.

Consider the most general form of quadratic equation with coefficients in **K**:

(2.9.1) $f(x) = a_1 x_1^2 + a_2 x_2^2 + a_3 x_3^2 + b_1 x_2 x_3 + b_2 x_3 x_1 + b_3 x_1 x_2 = 0.$

We shall establish a criterion for the irreducibility of a quadratic equation over **K**.

If $b_1 = b_2 = b_3 = 0$ then the equation (2.9.1) can be written in the form

$$f(x) = 0: \quad c_1^2 x_1^2 + c_2^2 x_2^2 + c_3^2 x_3^2 = (c_1 x_1 + c_2 x_2 + c_3 x_3)^2 = 0,^*$$

where $c_l^2 = a_l$, $l = 1, 2, 3$.

If $(b) = (b_1, b_2, b_3) \neq (0, 0, 0)$ it is still possible for $f(x)$ to be factorized into two non-equivalent linear forms, i.e. $f(x) = \alpha(x) \cdot \beta(x)$ where

$$\alpha(x) \equiv \alpha_1 x_1 + \alpha_2 x_2 + \alpha_3 x_3$$

and

$$\beta(x) \equiv \beta_1 x_1 + \beta_2 x_2 + \beta_3 x_3.$$

Then the reducible quadratic equation will be of the form:

$$\alpha_1 \beta_1 x_1^2 + \alpha_2 \beta_2 x_2^2 + \alpha_3 \beta_3 x_3^2 + (\alpha_2 \beta_3 + \alpha_3 \beta_2) x_2 x_3 + $$
$$+ (\alpha_2 \beta_1 + \alpha_1 \beta_3) x_3 x_1 + (\alpha_1 \beta_2 + \alpha_2 \beta_1) x_1 x_2 = 0.$$

Now the non-trivial solutions of the system of equations

$$\begin{cases} \alpha_1 x_1 + \alpha_2 x_2 + \alpha_3 x_3 = 0, \\ \beta_1 x_1 + \beta_2 x_2 + \beta_3 x_3 = 0 \end{cases}$$

are

$$\varrho x_1 = \alpha_2 \beta_3 + \alpha_3 \beta_2, \quad \varrho x_2 = \alpha_3 \beta_1 - \alpha_1 \beta_3, \quad \varrho x_3 = \alpha_1 \beta_2 - \alpha_2 \beta_1.$$

However, as our coordinate field is of characteristic 2 we can rewrite these as follows:

$$\varrho x_1 = \alpha_2 \beta_3 + \alpha_3 \beta_2, \quad \varrho x_2 = \alpha_3 \beta_1 + \alpha_1 \beta_3, \quad \varrho x_3 = \alpha_1 \beta_2 + \alpha_2 \beta_1.$$

---

* In the field $GF(2^r) = \mathbf{K}$ we have $ab + ab = 0$, consequently $(a+b)^2 = a^2 + ab + ab + b^2 = a^2 + b^2$, by a repeated application of this we obtain that $(a+b+c)^2 = a^2 + b^2 + c^2$.

Thus we can make the following statement:
If the quadratic form $f(x)$ is the product of two linear forms, then $f(b)=0$. (1)

We shall show now that *every point of the plane satisfies $f(x)=0$ only if every coefficient of $f(x)$ is equal to zero* (2).

If every point satisfies $f(x)=0$, then substituting the coordinates of the fundamental points $(1, 0, 0)$, $(0, 1, 0)$, $(0, 0, 1)$ we obtain $a_1=0, a_2=0, a_3=0$. Hence

$$f(x) = 0: \ b_1 x_2 x_3 + b_2 x_3 x_1 + b_3 x_1 x_2 = 0$$

and if this is satisfied also by the points $(0, 1, 1)$, $(1, 0, 1)$, $(1, 1, 0)$, then we obtain that $b_1=0, b_2=0, b_3=0$ which proves statement (2).

*

Let X: (x_1, x_2, x_3) and Y: (x_1, y_2, y_3) be given. Suppose that a point Z: $(\lambda x_1 + \mu y_1, \lambda x_2 + \mu y_2, \lambda x_3 + \mu y_3) = (z_1, z_2, z_3)$ of the line XY satisfies the equation in question, i.e. $f(z)=0$. This can be written more explicitly as follows:

(2.9.2) $$f(x) \cdot \lambda^2 + \varphi(x, y) \cdot \lambda \mu + f(y) \cdot \mu^2 = 0,$$

where, in characteristic 2 only,

(2.9.3) $\quad \varphi(x, y) = (b_3 x_2 + b_2 x_3) y_1 + (b_3 x_1 + b_1 x_3) y_2 + (b_2 x_1 + b_1 x_2) y_3.$

It can be seen directly that $\varphi(x, y) = \varphi(y, x)$.

We can define now a symmetrical relation between a pair of points X and Y. We say that the two points are conjugate, with respect to the point set determined by the vanishing of f if $\varphi(x, y) = 0$.

Suppose that two conjugate points X and Y, with respect to the figure, exist; and neither of them belongs to the figure, i.e. $f(x) = \alpha \ne 0$, $f(y) = \gamma \ne 0$ and $\varphi(x, y) = 0$.

Clearly a point $Z = \lambda x + \mu x + \mu y$, $\lambda, \mu \ne 0$ will be a point of the figure if and only if

$$\alpha \lambda^2 + \gamma \mu^2 = 0.$$

Suppose $\alpha \lambda^2 + \gamma \mu^2$ then $\sigma^2 = \left(\dfrac{\lambda}{\mu}\right)^2 = \dfrac{-\alpha}{\gamma}$ has a unique solution for σ, in a field of characteristic 2. Thus we can say that the *line connecting the conjugate points X, Y, not on the figure $f(x)=0$, has a single point Z in common with the figure;* i.e. XY is a tangent to the figure and Z is its *point of contact* (3).

Let us consider now the point B: (b_1, b_2, b_3) and let us look for all the points conjugate to B. Clearly if $(x_1, x_2, x_3) = (b_1, b_2, b_3)$ then the coefficients in (2.9.3) will all be zero, hence

(2.9.4) $$\varphi(b, y) = 0.$$

Consequently, *every point of the plane is conjugate to the point* B (4).

Also by putting $(x_1, x_2, x_3)=(y_1, y_2, y_3)$ in (2.9.3) we see that $\varphi(x,y)=0$. Consequently *every point of the plane is conjugate to itself* (5).

By assuming that $f(b)=\alpha\neq 0$, so that $f(x)$ is irreducible, and by using the fact that B is conjugate to every point of the plane, we can easily find the number of points of the figure determined by the vanishing of f. Given any point Y of the plane we know, by (3), that the line BY contains a single point Z of the figure, i.e. it touches the figure at the point Z. The number of lines passing through B is $q+1$, consequently our figure consists of $q+1$ points. B is said to be the *nucleus* of the figure. Thus we can also say that *a figure with a nucleus consists of $q+1$ points* (6).

Suppose now that B belongs to the figure, if a point $Y: (y)\neq (b)$ belongs to the figure then *every point of the line BY belongs to the figure* (7). Since if $f(b)==f(y)=\varphi(b,y)=0$ then (2.9.2) will be satisfied for any pair (λ,μ).

Assume that $X, Y, Z=\lambda x+\mu Y$ are three distinct points of the figure belonging to the same line, that is $f(x)=f(y)=0$ and therefore

$$f(z) = \varphi(x,y)\cdot\lambda\mu = 0,$$

but by assumption $\lambda\mu\neq 0$ therefore $\varphi(x,y)=0$. This, however, implies that every point of the line XY belongs to the figure. Thus we can say that *if three points of a line belong to the figure, then every point of the line belongs to the figure.* (8)

If B and $Y(\neq B)$ belong to the figure then according to (7) every point of BY belongs to the figure. If the figure has a point Y^* which does not belong to the line BY, then the points of the line BY belong to the figure as well. It follows from (8) that the figure can contain no further point. For if a point Y_0^* of the figure did not belong to either of the lines BY and BY^*, then a line through Y_0 would meet BY in a point Y_1, say, and would meet BY^* in a point Y_1^*, say. But the points $Y_0^* Y_1 Y_1^*$ would all belong to the figure and so, by (8) every point of the line would belong to the figure. The same would hold for every line l passing through the point Y_0^*. Therefore every point of the plane would belong to the figure, i.e. according to (2) every coefficient of $f(x)$ is equal to zero, which we have excluded. Thus we can say that *if $f(b)=0$, then the figure is not an oval* (9).

We can now summarize the statements above in the following

Theorem: *The point figure corresponding to the quadratic equation $f(x)=0$ is an oval if and only if $f(b)\neq 0$.*

An oval defined in this manner — i.e. by a quadratic equation — is said to be a *curve of the second order* and will be denoted by $\mathscr{C}$.*

---

* This deviates from the usual notation. Every point figure corresponding to an equation of the second order is usually called a curve of the second order.

It is known that on a *Galois plane of order* 2^r, $r>2$, *there exists an oval which is not a curve of the second order*. This will not be dealt with here.

By analogy with the classical projective plane, let us consider the *polarity* determined by a curve of the second order. The set of the points Y conjugate to a fixed point X with respect to the curve $\mathscr{C}$ is said to be the *polar of point* X. From the equation (2.9.3) a Y: (y_1, y_2, y) will lie on the polar of X if and only if it satisfies the equation (expressed in normed line coordinates):

(2.9.5) $$u_1 y_1 + u_2 y_2 + u_3 y_3 = 0,$$

where

(2.9.6) $$\begin{cases} \varrho u_1 = 0 + b_3 x_2 + b_2 x_3, \\ \varrho u_2 = b_3 x_1 + 0 + b_1 x_3, \\ \varrho u_3 = b_2 x_1 + b_1 x_2 + 0. \end{cases}$$

As X is conjugate to itself and also to B, the polar of X is clearly the line BX. Hence points on the same line through B have the same polars. The polar of the nucleus B is the whole plane since by putting $(x_1, x_2, x_3) = (b_1, b_2, b_3)$ in (2.9.6) we obtain $[u_1, u_2, u_3] = [0, 0, 0]$.

Hence, we have the following

Theorem: *If the nucleus of a curve $\mathscr{C}$ is the point* B, *then the set of the polars is identical with the set of the lines passing through this point,* i.e. with the set of tangents to the curve.

This theorem shows that the concept of polarity in a plane of even order is quite unlike polarity in the classical projective plane.

Finally we may ask what kind of relationship can be found between an arbitrary curve $\mathscr{C}$ and the special oval Ω_0. This question will be dealt with in the following section.

2.10 The canonical equation of curves of the second order on the Galois planes of even order

We continue to assume that **K** is of characteristic 2.
Consider the non-singular linear transformation

$$\varphi: \begin{cases} \sigma x_1 = c_{11} y_1 + c_{12} y_2 + c_{13} y_3, \\ \sigma x_2 = c_{21} y_1 + c_{22} y_2 + c_{23} y_3, \\ \sigma x_3 = c_{31} y_1 + c_{32} y_2 + c_{33} y_3, \end{cases}$$

and denote the matrix of φ by **C**. If we take the triangle of reference for the coordinate system (y_1, y_2, y_3) as in (2.4.3), namely $A_1^\circ = (1, 0, 0)$, $A_2^\circ = (0, 1, 0)$ and

$A_3^\circ = (0, 0, 1)$; then the point A_i° corresponds, under φ; to the point A_i^* given by the coordinates $(c_{1i}, c_{2i}, c_{3i}) = (c_i)$ ($i = 1, 2, 3$). Also the side a_i of the triangle of reference, above, corresponds, under φ, to the line

$$C_{1i}x_1 + C_{2i}x_2 + C_{3i}x_3 = 0 \quad (i = 1, 2, 3),$$

where C_{jk} is the cofactor of c_{jk} in the matrix $\mathbf{C}$ ($j = 1, 2, 3$; $k = 1, 2, 3$).

Consider the quadratic form occurring in (2.9.1) and let $c_{1i}b_1 + c_{2i}b_2 + c_{3i}b_3 = c_i$ ($i = 1, 2, 3$). After some calculation we find that φ transforms $f(x)$ into a quadratic form $g(y)$ where

(2.10.1)
$$g(y) \equiv f(c_1) \cdot y_1^2 + f(c_2) \cdot y_2^2 + f(c_3) \cdot y_3^2 +$$
$$+ C_1 \cdot y_2 y_3 + C_2 \cdot y_3 y_1 + C_3 \cdot y_1 y_2.$$

If we now choose the coefficients in φ so that A_2^* and A_3^* lie on the curve $\mathscr{C}$, defined by $f(x) = 0$, and so that $A_1^* = B$ is the nucleus of the curve; then $f(c_1) = f(b) \neq 0$ and $f(c_2) = f(c_3) = 0$. Also $C_1 \neq 0$ but $C_2 = C_3 = 0$ since $B (= A_1^*)$ lies on the lines $A_1^* A_2^*$ and $A_1^* A_3^*$, ($A_1^* A_2^*$ and $A_1^* A_3^*$ touch the curve at $A_1^* A_3^*$, respectively, and pass through the nucleus.)

So the equation $g(y) = 0$ reduces to

(2.10.2)
$$g(y) \equiv f(c_1) \cdot y_1^2 + C_1 \cdot y_2 y_3 = 0.$$

Let α denote the element of $\mathbf{K}$ such that $\alpha^2 = \dfrac{1}{f(c_1)}$ and let β and γ be two elements of $\mathbf{K}$ such that $\beta \gamma = \dfrac{1}{C_1}$.

Consider the non-singular linear transformation

$$\psi : \begin{cases} \varrho y_1 = \alpha z_3 \\ \varrho y_2 = \beta z_2 \\ \varrho y_3 = \gamma z_1 \end{cases}$$

This transformation carries the equation (2.10.2) into $z_1 z_2 + z_3^2 = 0$, which, in view of the fact that $\mathbf{K}$ is of characteristic 2, is the same as

(2.10.3)
$$z_1 z_2 - z_3^2 = 0.$$

The effect of the transformation $\varphi \psi = \omega$ is a change of coordinates so that the reference points A_2° and A_3° and the unit point E° now lie on $\mathscr{C}$ and the reference point A_1° is the nucleus of $\mathscr{C}$. That is, the relationship of the new coordinate system with $\mathscr{C}$ is the same as the relationship of the old coordinate system with Ω_0. We say that the equation of the curve $\mathscr{C}$ has been put into a *canonical* form (2.10.3). Thus we have the following

Theorem: *Any two curves of the second order of a Galois plane of even order are equivalent* (with respect to a linear mapping).*

In view of this theorem we can determine properties of curves of the second order by considering the particular case $\mathscr{C}_0$: $x_1 x_2 - x_3^2 = 0$. For example, let us consider the analogue of the following theorem of classical projective geometry: "*The polar of any of the diagonal points of a quadrangle inscribed in a conic is the line connecting the other two diagonal points.*" Consider the curve

$$\mathscr{C}_0: x_1 x_2 - x_3^2 = 0,$$

and the following four points lying on it:

$$Q_1: (\lambda_1^2, \mu_1^2, \lambda_1\mu_1), \qquad Q_2: (\lambda_2^2, \mu_2^2, \lambda_2\mu_2),$$
$$Q_3 = A_1^\circ: (1, 0, 0), \qquad Q_4 = A_2^\circ: (0, 1, 0).$$

Then it is a straightforward calculation to see that

$$Q_1 Q_2 \cap Q_3 Q_4 = D_1: (\lambda_1\lambda_2, \mu_1\mu_2, 0),$$
$$Q_1 Q_3 \cap Q_2 Q_4 = D_2: (\lambda_1\lambda_2, \mu_1\mu_2, \lambda_1\mu_2),$$
$$Q_1 Q_4 \cap Q_2 Q_3 = D_3: (\lambda_1\lambda_2, \mu_1\mu_2, \lambda_2\mu_1).$$

Consider now the line

$$d: [\mu_1\mu_2, \lambda_1\lambda_2, 0],$$

which passes through the nucleus $(0, 0, 1)$, i.e. d is a tangent; moreover, it passes through all the diagonal points D_1, D_2, D_3. The coordinates of the point of contact D of the line d with $\mathscr{C}_0$ are given by:

$$\varrho x_1 = \lambda_1\lambda_2, \quad \varrho x_2 = \mu_1\mu_2, \quad \varrho x_3 = \sqrt{\lambda_1\lambda_2\mu_1\mu_2} = v$$

($v \in \mathbf{K}$ because every element of $\mathbf{K}$ is a square).

We know already that the tangent to the curve $\mathscr{C}_0$ at any point is the polar of this point with respect to $\mathscr{C}_0$ and one and only one tangent passes through every point, except the nucleus. Thus the line d is the polar of every one of the points D_1, D_2, D_3.

Hence,

Theorem: *The diagonal points of a quadrangle, inscribed in a curve of the second order on a Galois plane of order two, are collinear on a tangent to the curve; this tangent being the polar of each of the diagonal points.*

The point of contact of the tangent containing the diagonal points of a quadrangle is said to be the *point associated to the quadrangle*.

---

* As is well known, there is an analogous theorem on the classical projective plane.

2.11 Point configurations of order 2 on a Galois plane of odd order

Throughout this section the coordinate field **K** is assumed to be of odd characteristic.

Detailed proofs of the results in this section will not always be given, since in many cases we can copy exactly the arguments used in classical projective geometry.

We can write the general equation of a figure of the second order in the following form:

(2.11.1) $$f(x) = 0: a_{11}x_1^2 + a_{22}x_2^2 + a_{33}x_3^2 +$$
$$+ 2a_{12}x_1x_2 + 2a_{23}x_2x_3 + 2a_{31}x_3x_1 = 0.$$

If we agree to write $a_{jk} = a_{kj}$, then the symmetrirmical detenant

(2.11.2) $$A = \begin{vmatrix} a_{11} & a_{12} & a_{13} \\ a_{21} & a_{22} & a_{23} \\ a_{31} & a_{32} & a_{33} \end{vmatrix}$$

is said to be the *determinant of the figure of the second order*.

The quadratic form $f(x)$ is the product of two linear forms if and only if $A=0$; thus *the condition of the irreducibility is* $A \neq 0$. If $A \neq 0$, then the set of points defined by the equation $f(x)=0$ is said to be *curve of the second order* or a *quadratic curve*. Here the word "curve" indicates that the figure contains no lines. In what follows, a quadratic curve will be denoted by $\mathscr{C}$.

Let two points, determined by the non-equivalent coordinate sequences (x_1, x_2, x_3) and (x_1, y_2, y_3), be denoted by X and Y, respectively. The point Z generated by a linear combination of the two points lies on the curve $\mathscr{C}$, if

(2.11.3) $$f(z) = f(\lambda x + \mu y) = f(x) \cdot \lambda^2 + 2\varphi(x, y\lambda) \cdot \lambda\mu + f(y) \cdot \mu^2 = 0,$$

where

(2.11.4) $$\varphi(x, y) = (a_{11}x_1 + a_{12}x_2 + a_{13}x_3)y_1 +$$
$$+ (a_{21}x_1 + a_{22}x_2 + a_{23}x_3)y_2 + (a_{31}x_1 + a_{32}x_2 + a_{33}x_3)y_3.$$

Because $a_{jk} = a_{kj}$ it is obvious that

(2.11.5) $$\varphi(x, y) = \varphi(y, x) \quad \text{and} \quad \varphi(x, x) = f(x).$$

We say that the point Y is *conjugate* to the point X with respect to the curve $\mathscr{C}$, if

(2.11.6) $$\varphi(x, y) = 0.$$

From (2.11.5) we see that conjugacy is a symmetrical relationship between the two points; moreover, the points of the curve $\mathscr{C}$ are *self-conjugate* points.

In what follows we shall apply some well-known theorems of the theory of Galois fields. An element of the field will be called *square,* if it is the square of at least one of the elements of the field. Among the $q-1$ non-zero elements of a field of *odd* order there are $(q-1)/2$ squares and the same number of *non-squares.* Every squares is the square of exactly two elements and these two elements are the additive inverses of each other. There exists always an element $S \neq 1$ of the field, called a *primitive element,* such that

$$S^{q-1} = 1 \quad \text{and} \quad S^n \neq 1$$

for any n dividing $q-1$. Thus every non-zero element of the field can be written as a power of S; also the squares are the even powers of S and 0.

As an example consider the element 3 on the table of operation given in Fig. 44: $3=3$, $3^2=7$, $3^3=8$, $3^4=2$, $3^5=6$, $3^6=5$, $3^7=4$, $3^8=1$, Thus the squares of $GF(9)$ are 0, $3^2=7$, $3^4=2$, $3^6=5$, $3^8=1$, which can also be seen in the main diagonal of the multiplication table.

Finally, consider the solutions of the quadratic equation $ax^2+bx+c=0$ in a finite field, namely

$$(-b+\sqrt{b^2-4ac})/2a \quad \text{and} \quad (-b-\sqrt{b^2-4ac})/2a.$$

Let $d=b^2-4ac$ if d is non-zero square then there exist two (distinct) solutions; if $d=0$, then there exists one solution, if d is a non-square, then the equation has no solution. If $b=0$ and $-c/a$ is a square (but $c \neq 0$), then there exist two solutions and the sum of the two solutions is equal to zero.

In the possession of these algebraic facts we can now continue our discussion concerning the curve $\mathscr{C}$.

The common points of the curve $\mathscr{C}$ and a line can be determined by means of (2.11.3) and (2.11.4). The number of common points will be 2, 1, or 0 and the line is said to be *secant, tangent,* or *skew (exterior line),* respectively.

If X and Y are points, not on the curve $\mathscr{C}$, which are conjugate with respect to the curve and the line XY cuts the curve in the points P and Q, then $(XYPQ) = = -1$. This follows by (2.11.3).

The set of points conjugate to a fixed point X with respect to the curve $\mathscr{C}$ forms a line which is said to be the *polar of the point.* This follows by the relation (2.11.4), since only the elements (y_1, y_2, y_3) are variable in it. — If X is on the curve $\mathscr{C}$, then its polar is the tangent to the curve $\mathscr{C}$ at the point X.

If the diagonal points of a quadrangle $Q_1 Q_2 Q_3 Q_4$ inscribed in the curve $\mathscr{C}$ are D_1, D_2, D_3, where

$$Q_1 Q_2 \cap Q_3 Q_4 = D_1, \quad Q_1 Q_3 \cap Q_2 Q_4 = D_2, \quad Q_1 Q_4 \cap Q_2 Q_3 = D_3$$

then the points form a triangle in which each of the vertices has the opposite side as its polar. A triangle like this is a selfpolar figure with respect to the curve $\mathscr{C}$ and it is called a *polar triangle*.

The one-to-one mapping which assigns to every point its polar can be defined, as is easily seen from (2.11.4) by the following linear transformation:

(2.11.7) $$\pi(x) = [u] \begin{cases} \varrho u_1 = a_{11}x_1 + a_{12}x_2 + a_{13}x_3, \\ \varrho u_2 = a_{21}x_1 + a_{22}x_2 + a_{23}x_3, \\ \varrho u_3 = a_{31}x_1 + a_{32}x_2 + a_{33}x_3, \end{cases}$$

where u_1, u_2, u_3 are the line coordinates of the polar of the point (x_1, x_2, x_3). The determinant of this transformation is the determinant of the equation of the curve $\mathscr{C}$ because $A \neq 0$ the transformation can be inverted, i.e. every line is the polar of one and only one point. The *pole* of a line is the point which has the line in question as its polar. The mapping $\pi(x)=[u]$ is called the polarity established by the curve $\mathscr{C}$.

If the polar of a point cuts the curve in two points, then the lines connecting these two points with the point in question are tangents to the curve.

Consider now the following non-singular linear transformation:

(2.11.8) $$\gamma: \begin{cases} \varrho x_1 = c_{11}y_1 + c_{12}y_2 + c_{13}y_3, \\ \varrho x_2 = c_{21}y_1 + c_{22}y_2 + c_{23}y_3, \\ \varrho x_3 = c_{31}y_1 + c_{32}y_2 + c_{33}y_3. \end{cases}$$

As before, the reference point A_i° in the coordinate system (y_1, y_2, y_3) is the image of the point A_i given by the original coordinates $(c_i) = (c_{1i}, c_{2i}, c_{3i})$ $(i=1, 2, 3)$.

Then we find that $f(x)$ is transformed by γ into the quadratic form $g(y)$ where

$g(y) \equiv$
$\equiv f(c_1)y_1^2 + f(c_2)y_2^2 + f(c_3)y_3^2 + 2\varphi(c_1, c_2)y_1y_2 + 2\varphi(c_2, c_3)y_2y_3 + 2\varphi(c_3, c_1)y_3y_1.$

The form of $g(y)$ will depend on the relationship of the triangle $A_1A_2A_3$ with the curve $\mathscr{C}$.

The following three cases are of particular interest:

1° The triangle $A_1A_2A_3$ is a triangle inscribed in the curve $\mathscr{C}$.
2° $A_1A_2A_3$ is a polar triangle with regard to the curve $\mathscr{C}$.
3° The sides A_1A_3 and A_2A_3 of the triangle $A_1A_2A_3$ are tangents of the curve $\mathscr{C}$ in the points A_1 and A_2 respectively.

The corresponding equations $g(y)=0$ will be given by

$$b_{12}y_1y_2 + b_{23}y_2y_3 + b_{31}y_3y_1 = 0,$$
$$b_{11}y_1^2 + b_{22}y_2^2 + b_{33}y_3^2 = 0,$$

and
$$b_{12}y_1y_2 + b_{33}y_3^2 = 0,$$
respectively, where $b_{jj}=f(c_j)$ and $b_{jk}=2\,(c_j,\,c_k)$ if $j\neq k$.

The third equation can be simplified further by another linear transformation. Namely, consider the linear transformation
$$\beta: \varrho z_1 = \beta_1 y_1, \quad \varrho z_2 = \beta_2 y_2, \quad \varrho z_3 = y_3$$
where $\beta_1=b_{12}$ and $\beta_2=b_{33}^{-1}$; this transforms the third equation into the *canonical form*

(2.11.9) $$z_1 z_2 - z_3^2 = 0.$$

We arrived at this canonical equation by a linear transformation $\omega = \gamma\beta$ from the original equation $f(x)=0$. According to the geometrical interpretation of the linear transformation ω this can be expressed in two different ways.

1° The equation of the curve $\mathscr{C}$ takes on the canonical form (2.11.9) for a suitably chosen coordinate system.

2° The linear mapping ω which transforms the quadrangle $A_1 A_2 A_3 E$ of the original coordinate system into the quadrangle $A_1^\circ A_2^\circ A_3^\circ E^\circ$ carries the curve $\mathscr{C}$ into the curve $\mathscr{C}_0$. — We called the curve $\mathscr{C}_0$, in the original coordinate system, the point figure with the equation $x_1, x_2 - x_3^2 = 0$. — On the basis of this, if we recall the parties of $\mathscr{C}_0 = \Omega_0$ we can say that *every curve $\mathscr{C}$ is an oval.* — Later we shall show that the converse statement is also true, namely: On a Galois plane of odd order every oval is a curve of the second order.

Combining the results of this section and section **2.10** we see that any curve $\mathscr{C}$ in a Galois plane, of *arbitrary* characteristic, has a canonical equation $x_1 x_2 - x_3^2 = 0$. Therefore, in the next section we shall not distinguish between the cases of odd and even characteristic.

2.12 Correspondences between two pencils of lines

Let A and B be two distinct points of a Galois plane. Suppose a correspondence is given between the lines of the pencil through A and the lines of the pencil through B. For each pair of corresponding lines we can consider their point of intersection. The figure generated by these points of intersection is called the *projectively generated locus* established by the mapping. Of course, if the mapping has the special property that the line AB corresponds to itself, i.e. it is a united line of the correspondence, then the line AB will be a component of the locus.

In general only the points A and B of the line AB will be contained in the locus, because a line through A corresponding to AB meets it in the point A, whilst a line through B and corresponding to AB meets it in the point B.

If the correspondence in question is a projectivity and the line AB is self corresponding, then the locus consists of the line AB and another line; as can readily be seen from the theorem of Pappus. In this particular case the projectivity between the two pencils is said to be a *perspectivity*.

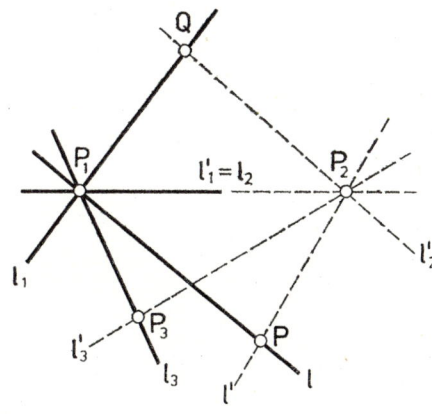

Figure 54

In what follows, we shall deal with the locus formed by the points of the intersection of the corresponding lines of two projectively related pencils, where the projectivity is not a perspectivity.

Let the proper quadrangle $P_1P_2P_3Q$ be given (Fig. 54). The pencil of lines with centre P_1 is carried into the pencil of lines with centre P_2 by a projectivity defined by means of the quadrangle as follows: (We assume the Galois plane is of order >2).

Let

$$P_1Q = l_1 \to P_2P_1 = l_1', \quad P_1P_2 = l_2 \to P_2Q = l_2', \quad \text{and} \quad P_1P_3 = l_3 \to P_2P_3 = l_3'.$$

To each line l of the remaining $q-2$ lines P_1 we assign the line l' through P_2 for which the cross ratio condition

$$(l_1 l_2 l_3 l) = (l_1' l_2' l_3' l') = \varrho, \text{ say.}$$

By intersecting l and l' we obtain a point P of the projectively generated locus. Thus the locus consists of $q+1$ points $P_1, \ldots, P_{q+1}$. We can obtain the equation of the projectively generated locus $\{P_1, P_2, \ldots, P_{q+1}\}$ so determined by applying the linear mapping ω determined by the correspondence

$$P_1 \to A_1^\circ, \quad P_2 \to A_2^\circ, \quad P_2 \to A_3^\circ, \quad P_3 \to E^\circ.$$

Since this correspondence is cross ratio preserving, we have for the image point $\omega(P) = P^\circ$ the relation

$$A_1^\circ(A_1^\circ A_2^\circ E^\circ P^\circ) = A_2^\circ(A_1^\circ A_2^\circ E^\circ P^\circ) = \vartheta$$

and from this the set of the points P° is determined by the equation $x_1 x_2 - x_3^2 = 0$. If this equation is transformed by the linear transformation $\omega^{-1}(x) = y$, then we obtain an irreducible second-degree equation $f(y) = 0$ which is the equation of the original locus. Thus *the equation of the locus of the points of intersection of the corresponding lines of two projectively (non-perspectively) related pencils is an irreducible second degree equation.*

Furthermore, we know that a curve $\mathscr{C}$ defined by an irreducible equation of the second degree can always be mapped onto an oval Ω_0. Hence *a curve defined by an irreducible equation of the second degree can always be generated as the locus of the points of intersection of corresponding lines of two projectively (non-perspectively) related pencils.*

We can now easily prove the following

Theorem: *On a Galois plane of order >3, there is one and only one curve of the second order which passes through five given points, no three of which are collinear.*

Suppose that the five points P_1, P_2, P_3, P_4, P_5 are common to the two distinct curves of the second order $\mathscr{C}_1$ and $\mathscr{C}_2$. Since each of these curves consists of $q+1$ points, they can only be distinct if the curve $\mathscr{C}_1$ has at least one point X which is not a point of the curve $\mathscr{C}_2$ as well.

Since a curve of the second order possesses all properties of the oval Ω_0 that are invariant under a linear mapping, it has the following property: the cross ratio of the four lines connecting any four points of the curve with any other points P of the curve is independent of the choice of P (provided the four points are always taken in the same order). Thus with respect to the curve $\mathscr{C}_1$ we have

$$P_1(P_2 P_3 P_4 X) = P_5(P_2 P_3 P_4 X).$$

Now if we consider the curve $\mathscr{C}_2$ which does not pass through the point X, this curve is cut by the line $P_1 X$ in a point P_1 and in a point X^* distinct from X, so that $P_5 X$ and $P_5 X^*$ are distinct and

$$P_1(P_2 P_3 P_4 X^*) = P_5(P_2 P_3 P_4 X^*).$$

The conditions $X^* \in P_1 X$, $X^* \neq X$ imply

$$P_1(P_2 P_3 P_4 X^*) = P_1(P_2 P_3 P_4 X) = P_5(P_2 P_3 P_4 X) = P_5(P_2 P_3 P_4 X^*).$$

From this follows $P_5 X = P_5 X^*$, which contradicts our assumption. This contradiction proves the statement of the theorem. —

We could prove the following theorems by similar reasoning:

A line passing through a vertex of a proper quadrangle determines (as a tangent) together with the other vertices of the quadrangle one and only one curve of the second order.

The vertices of a proper triangle and two lines which pass through different vertices and are not sides of the triangle determine (as tangents) one and only one curve of the second order.

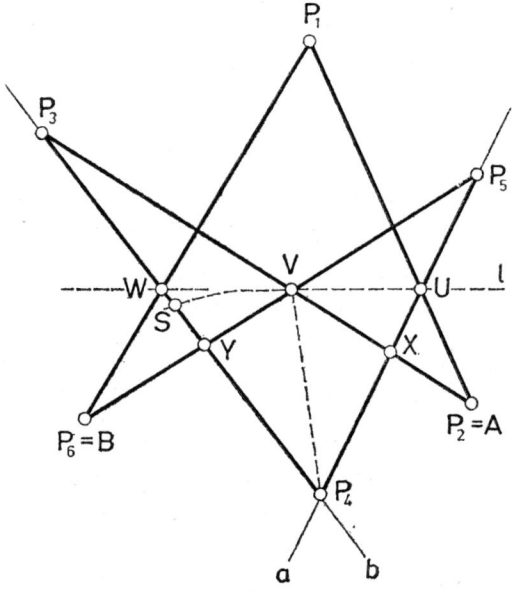

Figure 55

As an application of the previous statements let us discuss the *linear generation* of the irreducible curve of the second degree from five points no three of which are collinear.

Consider Fig. 55: the points $P_1, P_2(=A), P_3, P_4, P_5, P_6(=B)$ are such that no three are collinear and we assume further that these six points lie on a curve of the second order $\mathscr{C}$. It follows from these assumptions that $q > 4$, and further that the five points

$$P_1P_2 \cap P_4P_5 = U, \quad P_2P_3 \cap P_5P_6 = V, \quad P_3P_4 \cap P_6P_1 = W;$$
$$P_2P_3 \cap P_4P_5 = X, \quad P_3P_4 \cap P_5P_6 = Y$$

are all distinct from each other and the given points P. Consider the lines projecting the points P_1, P_3, P_4, P_5 of the curve $\mathscr{C}$, from the points A and B, respectively. It follows from the properties of the curve $\mathscr{C}$ and from the theorem of Pappus that

$$A(P_1P_3P_4P_5) = B(P_1P_3P_4P_5), \quad (P_1P_3P_4P_5) = (UXP_4P_5),$$
$$B(P_1P_3P_4P_5) = (WP_3P_4Y).$$

By comparing all these we have

$$(UXP_4P_5) = (WP_3P_4Y).$$

The projections of the points P_5, X, P_4 of the the line $B_4B_5 = a$ from point V onto the line $B_3B_4 = b$ are in turn the points Y, P_3, P_4 and if we assume further that the projection of the point U is the point S, then by the theorem of Pappus we have

$$(UXP_4P_5) = (SP_3P_4Y).$$

From this and from the previous relation follows

$$(WP_3P_4Y) = (SP_3P_4Y),$$

which proves the coincidence of the points S and W. Thus the points U, V and W lie on a line l. Therefore we can say that if a

hexagon $P_1P_2P_3P_4P_5P_6$ *is inscribed in a curve of the second order, the three pairs of opposite sides (taken in the given order) meet in collinear points.*

If a pentagon $P_1P_2P_3P_4P_5$ determining the curve $\mathscr{C}$ given by the figure shown in Fig. 55, then any of the further $q-4$ points of the curve $\mathscr{C}$ can be taken as the point P_6 in the above theorem. Moreover, we have a construction of the remaining $q-4$ points of the curve $\mathscr{C}$.

Namely, the given five points determine the points of the line b in which it is cut by the sides of the triangle $P_1P_2P_5$ and the points P_3 and P_4 themselves, i.e. five points in all. Any of the remaining $q-4$ points of the line b can play the role of the point W in the above. A point W chosen in this manner leads to a single point P_6 by the following procedure:

$$UW = l, \quad P_2P_3 \cap l = V, \quad P_5V \cap P_1W = P_6.$$

The point P_6 generated in this manner is a point of the curve $\mathscr{C}$, since if it were not on the curve we would be led to a consequence which contradicts the theorem above. By varying the point W we obtain $(q-4)$ district points P_6.

2.13 A theorem of Segre

In this section our discussion refers to a plane of odd order. It is easy to prove that every curve of the second order possesses the following property. *A triangle inscribed in a curve of the second order and the triangle which consists of the tangents at its vertices are in perspective.* This property is said to be the π-*property*. It was perceived by Segre, that an oval with the π-property is a curve of the second order, that is the investigation of the ovals of the Galois planes of odd order can be transformed into an algebraic question.

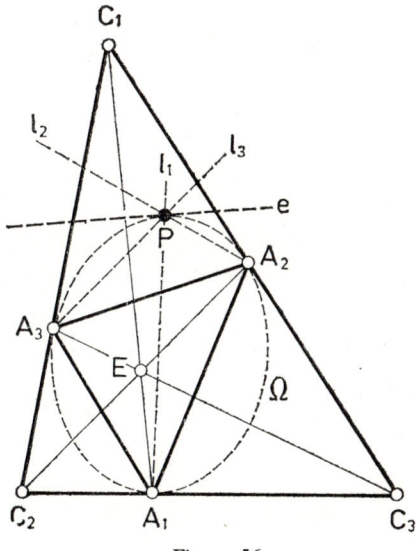

Figure 56

The lines touching an arbitrary oval at the vertices of an inscribed triangle $A_1 A_2 A_3$ are the sides of a triangle $C_1 C_2 C_3$ (Fig. 56). If the oval Ω has the π-property, then the lines $A_k C_k$ ($k=1, 2, 3$) meet in a common point E. Now let $A_1 A_2 A_3$ be the fundamental triangle and E the unit point, then the coordinates of the point C_k are found to be

(2.13.1) C_1: $(-1, 1, 1)$; C_2: $(1, -1, 1)$; C_3: $(1, 1, -1)$.

Namely with a suitably chosen element ξ_k we have $C_k = \xi_k A_k + E$, i.e. the k-th coordinate of the point C_k is $\lambda_k = \xi_k + 1$, the other two being both 1. Since the points C_1, C_2, A_3 are collinear, we have

$$\begin{vmatrix} \lambda_1 & 1 & 1 \\ 1 & \lambda_2 & 1 \\ 0 & 0 & 1 \end{vmatrix} = 0,$$

i.e. $\lambda_1\lambda_2=1$. Thus, by symmetry, we obtain the equations

$$\lambda_1\lambda_2 = 1, \quad \lambda_2\lambda_3 = 1, \quad \lambda_3\lambda_1 = 1.$$

From this it follows that $\lambda_1^2=\lambda_2^2=\lambda_3^2$ and since $C_k \neq E$, we have $\lambda = -1$, thus we proved our statement (2.13.1).

We choose now a point P, say, of the oval Ω distinct from the points A_k, and take the tangent e through it.
Suppose

$$P: (a_1, a_2, a_3) \quad \text{and} \quad e: a_1x_1 + a_2x_2 + a_3x_3 = 0.$$

Since the points A_k and C_k ($k=1, 2, 3$) do not lie on the line e, it follows that e_1, e_2 and e_3 are non-zero.

(2.13.2) $\quad -e_1+e_2+e_3 = c_1 \neq 0, \quad e_1-e_2+e_3 = c_2 \neq 0,$

$$e_1+e_2-e_3 = c_3 \neq 0.$$

Let jkl be an arbitrary permutation of the elements 123 and let us determine the coordinates of the point of intersection $E_j = e \cap C_k C_l$.
It is easy to see that

$$C_2C_3: [0, 1, 1], \quad C_3C_1: [1, 0, 1] \quad \text{and} \quad C_1C_2: [1, 1, 0].$$

Thus we obtain the following coordinates for the points E_1, E_2 and E_3:

$$E_1: (e_2-e_3, -e_1, e_1),$$

(2.13.3) $\qquad E_2: (e_2, e_3-e_1, -e_2),$

$$E_3: (-e_3, e_3, e_1-e_2).$$

The tangent to the oval Ω at the point P is the line e and therefore we have

(2.13.4) $\qquad e_1a_1 + e_2a_2 + e_3a_3 = 0.$

Also, it is a straightforward calculation to see that

$$C_1P: [a_3-a_2, \; a_3+a_1, \; -a_1-a_2],$$

(2.13.5) $\qquad E_2A_3: [e_3-e_1, \quad -e_2, \qquad 0 \quad],$

$$E_3A_2: [e_2-e_1, \quad 0, \qquad -e_3 \;\;].$$

Since the oval is assumed to the π-property, the inscribed triangle PA_3A_2 and the triangle $C_1D_2E_3$ formed by the tangents of the oval at the vertices of this inscribed triangle are in perspective, and therefore the lines C_1P, E_2A_3

and E_3A_2 are concurrent. Therefore the determinant formed from the line coordinates of the three lines must vanish, i.e.

$$\begin{vmatrix} a_3-a_2 & a_3+a_1 & -a_1-a_2 \\ e_3-e_1 & -e_2 & 0 \\ e_2-e_1 & 0 & -e_3 \end{vmatrix} = \begin{vmatrix} 0 & a_3+a_1 & -a_1-a_2 \\ c_1 & -e_2 & 0 \\ c_1 & 0 & -e_3 \end{vmatrix} = 0.$$

(In the first determinant we subtract the sum of the second and third columns from the first in order to obtain the second determinant.) We therefore obtain the equation $(a_1+a_2)e_2=(a_3+a_1)e_3$, as $c_1 \neq 0$. Replacing the tangent triangle $E_2E_3C_1$, in turn, by the triangles $E_3E_1C_2$ and $E_1E_3C_3$ in the argument above we obtain the following equations:

(2.13.6) $$\begin{cases} (a_1+a_2)e_1 = (a_2+a_3)e_3, \\ (a_2+e_3)e_2 = (a_3+a_1)e_1, \\ (a_3+a_1)e_3 = (a_1+a_2)e_2. \end{cases}$$

The fact that all of the elements e_k are non-zero implies that the factors (a_j+a_k) are non-zero, as well; since if one of them was 0, then the other two would be also, so we would have $a_1=a_2=a_3=0$, which is a contradiction.

Thus, from (2.13.6) there exists a non-zero $\varrho \in \mathbf{K}$ such that

(2.13.6′) $$\begin{cases} \varrho e_1 = 0 + a_2 + a_3, \\ \varrho e_2 = a_1 + 0 + a_3, \\ \varrho e_3 = a_1 + a_2 + 0. \end{cases}$$

This non-singular linear transformation defines a polarity which carries the point (a_1, a_2, a_3) into the line e_1, e_2, e_3. This polarity is established by the curve of the second order having the equation

$$\mathscr{C}: x_1x_2 + x_2x_3 + x_3x_1 = 0,$$

namely

$$\begin{cases} \varrho u_1 = 0 + x_2 + x_3, \\ \varrho u_2 = x_1 + 0 + x_3, \\ \varrho u_3 = x_1 + x_2 + 0. \end{cases}$$

Thus $q+1$ points of the curve $\mathscr{C}$ are identical with those of the oval Ω, moreover their tangents with respect to $\mathscr{C}$ and Ω are identical, thus $\mathscr{C}=\Omega$. The result so obtained can be expressed in the following.

Theorem i: *On Galois planes of an odd order every oval with the π-property is a curve of the second order.*

Let **K** be an arbitrary Galois field, let the product of all the elements of the set $\mathbf{K}^* = \mathbf{K} - 0$ be denoted by $\Pi(a)$ and the additive inverse of the unit element by -1. Then we have the following well-known relation:

(2.13.7) $$\Pi(a) = -1$$

expresses a well-known relation. (If the number of the elements of the field **K** is a prime, then (2.13.7) expresses the theorem of Wilson.) The proof of the following theorem rests upon the relation (2.13.7).

Theorem ii: *On a Galois plane of an odd order every oval has the π-property.*

Proof: Let the fundamental triangle $A_1 A_2 A_3$ be an inscribed triangle of an oval Ω. The equation of a line passing through the vertex A_j of the fundamental triangle but distinct from the sides $A_j A_k$ and $A_j A_l$ can be expressed in the form

$$x_k = c x_l \quad (c \in \mathbf{K}^*)$$

(where *jkl* is a cyclic permutation of the elements 123).

Let now the equations of the tangents in the points A_1, A_2 and A_3 be

$$x_2 = \alpha_1 x_3, \quad x_3 = \alpha_2 x_1 \quad \text{and} \quad x_1 = \alpha_3 x_2,$$

respectively (clearly $\alpha_1 \alpha_2 \alpha_3 \neq 0$). If the point $P: (a_1, a_2, a_3)$ of the oval is distinct from the fundamental points, then the equations of the lines $l_1 = A_1 P$, $l_2 = A_2 P$, $l_3 = A_3 P$ are in turn

$$x_2 = \lambda_1 x_3, \quad x_3 = \lambda_2 x_1, \quad x_1 = \lambda_3 x_2$$

where $\lambda_i \neq 0$, $\lambda_i \neq \alpha_i$ ($i = 1, 2, 3$).
As P lies on each of the lines l_1, l_2, l_3 we have

$$a_2 = \lambda_1 a_3, \quad a_3 = \lambda_2 a_1, \quad a_1 = \lambda_3 a_2;$$

from which

$$\lambda_1 = a_2 a_3^{-1}, \quad \lambda_2 = a_3 a_1^{-1}, \quad \lambda_3 = a_1 a_2^{-1},$$

and so

(2.13.8) $$\lambda_1 \lambda_2 \lambda_3 = 1.$$

For each of the $q-2$ points of Ω, distinct from A_1, A_2 and A_3, we will obtain 3 elements λ_1, λ_2 and λ_3 such that $\lambda_1 \lambda_2 \lambda_3 = 1$. Let S denote the product of the $q-2$ products $\lambda_1 \lambda_2 \lambda_3$. Clearly λ_i will take each value, except α_i, only and once only ($i = 1, 2, 3$). Thus in the product $\alpha_1 \alpha_2 \alpha_3 S$ each element of $\mathbf{K}^*$ occurs precisely three times, i.e.

$$\alpha_1 \alpha_2 \alpha_3 S = (\Pi(a))^3.$$

But, from (2.13.8) we have $S = 1$ and from (2.13.7) $\Pi(a) = -1$, consequently

(2.13.9) $$\alpha_1 \alpha_2 \alpha_3 = -1.$$

We can now determine the vertices of the triangle formed by the tangents of Ω at the fundamental points. Let the tangents of the oval at the point A_j be t_j. The line coordinates are given by

$$t_1: [0, -1, \alpha_1], \quad t_2: [\alpha_2, 0, -1], \quad t_3: [-1, \alpha_3, 0].$$

From this the coordinates of the points $C_1 = t_2 \cap t_3$, $C_2 = t_3 \cap t_1$ and $C_3 = t_1 \cap t_2$ are

$$C_1: (\alpha_3, 1, \alpha_2\alpha_3), \quad C_2: (\alpha_2\alpha_1, \alpha_1, 1) \quad \text{and} \quad C_3: (1, \alpha_1\alpha_2, \alpha_2),$$

respectively.

Thus the equations of the lines A_1C_1, A_2C_2 and A_3C_3 are

$$x_3 = \alpha_2\alpha_3 x_2, \quad x_1 = \alpha_3\alpha_1 x_3 \quad \text{and} \quad x_2 = \alpha_1\alpha_2 x_1,$$

respectively, and it can be readily seen, by considering (2.13.9) that these three lines pass through the point

(2.13.10) $\qquad Q: (\alpha_3\alpha_1, -\alpha_1, 1).$

Hence the triangles $A_1A_2A_3$ and $C_1C_2C_3$ are perspective from the point Q and Theorem ii is proved.

Now by combining Theorems i and ii we have

Theorem iii: *Every oval of a Galois plane of an odd order is a curve of the second order.*

Finally, to demonstrate the contrast between Galois planes of odd and even order, we give a trivial example which shows that Theorem iii, above, is not valid on a Galois plane of even order. Namely on a plane like this $q+1$ points of a curve of the second order together with the nucleus form a hyperoval. Let us delete a point of the hyperoval belonging to the curve of the second order. The remaining $q+1$ points form an oval, but this is not a curve of the second order, if $q>4$, since otherwise it would have a number $q>4$ of points in common with the original curve; this contradicts the theorem that a curve of the second order is uniquely determined by 5 of its points.

2.14 Supplementary notes concerning the construction of Galois planes

In this and the following few sections we give some supplementary information, mainly of an algebraic nature. This could have been introduced at the beginning of the chapter but was delayed until now so that the interesting and surprising *geometrical* properties of the linear figures, ovals, and quadratic figures could be dealt with as soon as possible.

We recall from section 1.6 that a plane $S_{2,q}$ was explicitly constructed from a pair of operation tables $[A, M]$ of the field $GF(q)$. However, is the pair of tables $[A, M]$ determined uniquely by the number q and if it is not, what kind of relation exists between the planes determined by different pairs of tables? We shall deal now with this problem.

As usual, let the elements of the finite field $GF(q)$, where $q=p^r$, be denoted by the integers $0, 1, 2, \ldots, q-1$ in such a manner that the zero element of the field is denoted by 0 and the unit element by 1; furthermore, let the elements of the prime subfield $GF(p)$ be denoted by

(2.14.1) $$0, 1, \ldots, p-1$$

where $1+1=2$, $1+1+1=3$, etc.

Then the operation tables of $GF(p)$ are uniquely determined, and so in any addition (resp. multiplication) table of $GF(q)$ the array of p squares $\times p$ squares in the top left hand corner will contain the addition (resp. multiplication) table of the residue class field mod p. We shall mention now some concepts and theorems of the theory of finite fields.

The number of elements in $\mathbf{K}=GF(q)$ is $|\mathbf{K}|=p^r$, the number of elements in $\mathbf{P}=GF(p)$ is $|\mathbf{P}|=p$. The *degree* of the field $\mathbf{K}$ over $\mathbf{P}$, denoted by $|\mathbf{K}|:|\mathbf{P}|$ is the integer r.

Clearly, $\mathbf{K}$ can be considered as a vector space over the subfield $\mathbf{P}$ and then $|\mathbf{K}|:|\mathbf{P}|$ is just the dimension of this vector space. Every element of the field $\mathbf{K}$ satisfies the equation $x^q-x=0$, i.e. the set of elements of the field is formed by the zeros of the polynomial x^q-x. The field $\mathbf{K}$ contains an element s such that $S^q=S$ but $S^n \neq S$ for any $n<q$. S is said to be a primitive root of the equation. Clearly the sequence

$$0, 1, s, s^2, \ldots, s^{q-2}$$

must contain every element of $\mathbf{K}$.

As an example we mention the element $s=3$ of the field $GF(q)$ (Fig. 44). We can readily check, by means of the table M, that the aforementioned sequence is now

$$0, 1, 3, 3^2 = 7, \quad 3^3 = 8, \quad 3^4 = 2, \quad 3^5 = 6, \quad 3^6 = 5, \quad 3^7 = 4$$

and this comprises all elements of the field.

Let two fields $\mathbf{K}$ and $\mathbf{L}$ be given, both of order q, furthermore let $x \to x'$ be a one-to-one mapping of $\mathbf{K}$ onto $\mathbf{L}$, if this mapping has the property that $(x+y)'=x'+y'$ and $(x \cdot y)'=x' \cdot y'$ (where $x, y \in \mathbf{K}$, $x', y' \in \mathbf{L}$), i.e. it is *operation-preserving*, it is called an *isomorphism;* moreover, if $\mathbf{K}=\mathbf{L}$ it is called an *automorphism*. Obviously, the identity mapping $x \to x$, denoted by E, of a field onto itself is an automorphism. Can other automorphisms exist? If $q=p$ is a prime, then there exists no further automorphism, if, however, $q=p^r$ and $r>1$, then there exist r automorphisms (including ε).

In fact, the mapping $x \to x^p$ is an automorphism, denoted by α, and the automorphisms of the field are precisely

$$\varepsilon, \alpha, \alpha^2, \ldots, \alpha^{r-1}.$$

α is called the Frobenius automorphism.

As an example we compile the automorphisms of the field $GF(8)$ (Fig. 51). Here we have

$$\varepsilon = \begin{pmatrix} 0 & 1 & 2 & 3 & 4 & 5 & 6 & 7 \\ 0 & 1 & 2 & 3 & 4 & 5 & 6 & 7 \end{pmatrix} \quad \alpha = \begin{pmatrix} 0 & 1 & 2 & 3 & 4 & 5 & 6 & 7 \\ 0 & 1 & 4 & 5 & 6 & 7 & 2 & 3 \end{pmatrix}$$

$$\alpha^2 = \begin{pmatrix} 0 & 1 & 2 & 3 & 4 & 5 & 6 & 7 \\ 0 & 1 & 6 & 7 & 2 & 3 & 4 & 5 \end{pmatrix}.$$

Every automorphism of a field **K** induces the identity automorphism on its prime field **P**. As an example we mention the automorphism of the field $GF(q)$ corresponding to Fig. 44. In this case, because $r=2$, there exist only two automorphisms, namely ε and

$$\alpha = \begin{pmatrix} 0, & 1, & 2, & 3, & 4, & 5, & 6, & 7, & 8 \\ 0, & 1, & 3, & 8, & 6, & 7, & 4, & 5, & 3 \end{pmatrix}$$

and these leave invariant the elements 0, 1 and 2, i.e. elements of **P**.

Finally we shall give a method for operation tables of a field **K** of $q=p^r$ elements where $r>1$.

We have already assigned the integers $0, 1, \ldots, p-1$ to the elements of the prime subfield **P** of **K**. Let us choose any other element of **K** and denote it by the integer p. Now each element of $\mathbf{K} \setminus \mathbf{P}$ is a zero of a unique irreducible monic polynomial of degree r over **P**. Suppose $f(x)$ is the polynomial of this type having p as a zero.

We can establish a 1—1 correspondence between the elements of **K** and the polynomials of degree at most $(r-1)$ over **P**. Namely, to the polynomial

$$a_{r-1}x^{r-1}+\ldots+a_1x+a_0 \quad (a_0, \ldots, a_{r-1} \in \mathbf{P})$$

assign the element

$$a_{r-1}p^{r-1}+\ldots+a_1p+a_0 \quad \text{of} \quad \mathbf{K}.$$

Now if the elements s and t correspond to the polynomials $g(x)$ and $h(x)$ respectively then $s+t$ is the element corresponding to $g(x)+h(x)$ reduced mod $(p, f(x))$. In this way we can build up a pair of operation tables $[A, M]$ for the field **K**. This procedure has the property that for every choice of the element p the addition table A is the same, whilst the multiplication table M will clearly depend on the polynomial $f(x)$ that has p as a zero.

As an example, consider the field $\mathbf{K} = GF(q)$. Here the only irreducible monic polynomials of degree 2 over $\mathbf{P} = GF(3)$ are

$$f_1: x^2 + 1, \quad f_2: x^2 + x + 2 \quad \text{and} \quad f_3: x^2 + 2x + 2.$$

The six elements in $\mathbf{K} \setminus \mathbf{P}$ occur as the roots of these polynomials. Thus we have three choices for the polynomial $f(x)$ in the construction above.

Thus we obtain three pairs of tables $[A, M^{(1)}], [A, M^{(2)}]$ and $[A, M^{(3)}]$ (Fig. 57). Let the corresponding field be denoted by $\mathbf{K}^{(1)}, \mathbf{K}^{(2)}$ and $\mathbf{K}^{(3)}$, respectively.

It can easily be checked whether each of the permutations

$$\varrho: \mathbf{K}^{(1)} \to \mathbf{K}^{(2)}: \begin{pmatrix} 0 & 1 & 2 & 3 & 4 & 5 & 6 & 7 & 8 \\ 0 & 1 & 2 & 7 & 8 & 6 & 5 & 3 & 4 \end{pmatrix}$$

$$\sigma: \mathbf{K}^{(1)} \to \mathbf{K}^{(3)}: \begin{pmatrix} 0 & 1 & 2 & 3 & 4 & 5 & 6 & 7 & 8 \\ 0 & 1 & 2 & 4 & 5 & 3 & 8 & 6 & 7 \end{pmatrix}$$

gives an isomorphism.

In fact this is the case in general. *Any* two fields of q elements are isomorphic.* We shall not prove this.

In the example above let the planes constructed from $[A, M^{(1)}], [A, M^{(2)}]$ and $[A, M^{(3)}]$ be $S_{2,g}^{(1)}(1), S_{2,g}^{(2)}(2)$ and $S_{2,g}^{(3)}(3)$, respectively.

How are the planes $S_{2,g}^{(1)}$ and $S_{2,g}^{(2)}$ related to each other? We cannot establish a 1—1 correspondence between them by letting the point with coordinate sequence (x_1, x_2, x_3) correspond to the point with the same coordinate triple. For example, the points $(2, 6, 5)$ and $(5, 1, 8)$ both represent the point with normal coordinates $(8, 7, 1)$ on the plane $S_{2,g}^{(1)}$, but on $S_{2,g}^{(2)}$ they represent *different* points, namely $(3, 4, 1)$ and $(6, 7, 1)$, respectively.

However, the non-zero triplet of elements (x_1, x_2, x_3) of the field $\mathbf{K}^{(1)}$ determines a point P of the plane $S_{2,g}^{(1)}$. The substitution ϱ carries x_1, x_2 and x_3 into x_1', x_2' and x_3', respectively of the field $\mathbf{K}^{(2)}$, and the triplet (x_1', x_2', x_3') determines a point P' of the plane $S_{2,g}^{(2)}$. In this way we get a one-to-one mapping

$$\varrho^*: S_{2,g}^{(1)} \to S_{2,g}^{(2)} \quad \text{induced by} \quad \varrho: \mathbf{K}^{(1)} \to \mathbf{K}^{(2)};$$

ϱ^* preserves the linear dependence or independence of two points and assigns a linear combination of the images to a linear combination of two points. Hence the mapping ϱ^* assigns a collinear triplet of points of the plane $S_{2,g}^{(2)}$ to a collinear triplet of points of the plane $S_{2,g}^{(1)}$ and only to a collinear one, i.e. the mapping ϱ^* *is line-preserving.*

---

* However we could still use any pair of tables $[A, M]$ of the field of order q to construct a plane $S_{2,q}$, according to the method of section **1.6**

x^2+1

X	0	1	2	3	4	5	6	7	8
0	0	0	0	0	0	0	0	0	0
1	0	1	2	3	4	5	6	7	8
2	0	2	1	6	8	7	3	5	4
3	0	3	6	2	5	8	1	4	7
4	0	4	8	5	6	1	7	2	3
5	0	5	7	8	1	3	4	6	2
6	0	6	3	1	7	4	2	8	5
7	0	7	5	4	2	6	8	3	1
8	0	8	4	7	3	2	5	1	6

$M^{(1)}$

x^2+x+2

X	0	1	2	3	4	5	6	7	8
0	0	0	0	0	0	0	0	0	0
1	0	1	2	3	4	5	6	7	8
2	0	2	1	6	8	7	3	5	4
3	0	3	6	7	1	4	5	8	2
4	0	4	8	1	5	6	2	3	7
5	0	5	7	4	6	2	8	1	3
6	0	6	3	5	2	8	7	4	1
7	0	7	5	8	3	1	4	2	6
8	0	8	4	2	7	3	1	6	5

$M^{(2)}$

x^2+2x+2

X	0	1	2	3	4	5	6	7	8
0	0	0	0	0	0	0	0	0	0
1	0	1	2	3	4	5	6	7	8
2	0	2	1	6	8	7	3	5	4
3	0	3	6	4	7	1	8	2	5
4	0	4	8	7	2	3	5	6	1
5	0	5	7	1	3	8	2	4	6
6	0	6	3	8	5	2	4	1	7
7	0	7	5	2	6	4	1	8	3
8	0	8	4	5	1	6	7	3	2

$M^{(3)}$

Figure 57

It is readily seen that the construction of a mapping ϱ^* from an isomorphism ϱ can be carried out for planes of any order.

A line-preserving one-to-one mapping of a finite plane onto a finite plane is said to be an *isomorphism* or a *collineation* and in the case of the mapping of the plane onto itself we call the isomorphism an *automorphism*.

Thus we can say that *the pair of tables realising the field of q elements determines up to an isomorphism the Galois plane of order q*.

2.15 Collineations and homographies on Galois planes

As we indicated in the last section, the Galois plane of order q are all isomorphic to each other. We know also that a linear mapping transforming a Galois plane onto itself is cross ratio preserving, it is a collineation. We know that the one-to-one mapping of the classical projective plane onto itself is cross ratio preserving if and only if it is line preserving. On the classical projective plane the linear mapping (arithmetic definition) and the collineation (geometric definition) are equivalent concepts. On a Galois plane this is not so.

On a Galois plane a collineation is not necessarily cross ratio preserving; if it is cross ratio preserving, the collineation is then called an *homography*. Consider again the Galois plane of order 9 represented by the table A of Fig. 44 and table $M^{(2)}$ of Fig. 57; let this plane be denoted by S. We determined already in Section **2.14** the unique automorphism α of this field distinct from the identity, we rewrite it here:

$\alpha\colon x \to x^3$; i.e. in the form of a permutation

$$\alpha\colon \begin{pmatrix} 0 & 1 & 2 & 3 & 4 & 5 & 6 & 7 & 8 \\ 0 & 1 & 2 & 8 & 6 & 7 & 4 & 5 & 3 \end{pmatrix}.$$

We know that the automorphism α induced an automorphism α^* of the plane, i.e. a collineation of the plane. To put it in algebraic form:

$$\alpha^*\colon \begin{cases} \varrho x_1' = x_1^3 + 0 + 0, \\ \varrho x_2' = 0 + x_2^3 + 0, \\ \varrho x_3' = 0 + 0 + x_3^3, \end{cases} \quad (0 \neq \varrho \in \mathbf{K}).$$

But, α^* does not preserve cross ratios:
Consider

$$A\colon (1, 1, 0), \quad B\colon (4, 8, 1), \quad C\colon (1, 5, 1), \quad D\colon (2, 3, 1)$$

and their image under α^*, the quadruple

$$A'\colon (1, 1, 0), \quad B'\colon (6, 3, 1), \quad C'\colon (1, 7, 1), \quad D'\colon (2, 8, 1).$$

It is easy to check by the tables of operation of the field that

$$6 \cdot A + B = C, \quad 7 \cdot A + B = D, \quad 4 \cdot A' + B' = C', \quad 5 \cdot A' + B' = D'.$$

Let us add to this that the automorphism α carries the elements 1, 6, 7 into the elements 1, 4, 5. If we consider now the cross ratios, we obtain the following

$$\vartheta = (ABCD) = \frac{1}{6} : \frac{1}{7} = 6, \quad \vartheta' = (A'B'C'D') = \frac{1}{4} : \frac{1}{5} = 4.$$

Obviously, we can compose any linear mapping φ with α^* to obtain a collineation of the plane S: this will not, however, be a homography.

$$\alpha^*\varphi: \begin{cases} \tau x_1' = a_{11}x_1^3 + a_{12}x_2^3 + a_{13}x_3^3, \\ \tau x_2' = a_{21}x_1^3 + a_{22}x_2^3 + a_{23}x_3^3, \\ \tau x_3' = a_{31}x_1^3 + a_{32}x_2^3 + a_{33}x_3^3. \end{cases}$$

In general, we can generate a family of collineations of a plane by composing linear mappings and mappings induced from field automorphisms. However, we cannot assume that every collineation of plane can be obtained in this way. We shall return to this problem in section **2.17**. Let the number of the elements of the field $\mathbf{K}$ be $q=p^r$. Let φ be any non-singular linear mapping with coefficients $a_{jk} \in \mathbf{K}$, then the mappings

(2.15.1) $$\psi_t: \begin{cases} \tau x_1' = a_{11}x_1^{p^t} + a_{12}x_2^{p^t} + a_{13}x_3^{p^t}, \\ \tau x_2' = a_{21}x_1^{p^t} + a_{22}x_2^{p^t} + a_{23}x_3^{p^t}, \\ \tau x_3' = a_{31}x_1^{p^t} + a_{32}x_2^{p^t} + a_{33}x_3^{p^t} \end{cases}$$

where $0 \leq t \leq r-1$, are collineations. We shall prove that of these only the linear mapping $\psi_0 = \varphi$ is an homography. Let the image of a point P under the collineation ψ_t be denoted by P'. Then if A, B, C and D are four points we have

(2.15.2) $$(ABCD)^{p^t} = (A'B'C'D').$$

Thus, $(ABCD) = (A'B'C'D')$ if and only if $t=0$.

2.16 The characteristic of a finite projective plane

We shall now interpret the characteristic of $GF(q)$ by a closure-property of a certain figure in the plane $S = S_{2,q}$. This figure also plays an important role in the structure of finite planes.

Consider the points

$$O = Q_0: (0,0,1), \quad X: (1,0,0), \quad Y: (0,1,0), \quad P_0: (1,1,1)$$

of the planes S (Fig. 58), i.e. $OXYP_0$ is the fundamental figure of the plane S. Put $Z = OP_0 \cap XY$, hence $Z = (1,1,0)$. The lines x, z and e of Fig. 58 represent the point sets with equations $x_2 = 0$, $x_3 = 0$ and $x_2 = x_3$, respectively; the line y of the figure represents the point set with equation $x_1 = 0$.

We shall now construct a polygon of $2p$ sides, where p is the characteristic of the coordinate field K, as follows:

We project the origin Q_0 from Z onto e, the image P_0 so obtained is projected from Y onto x, we project the point Q_1 so obtained from Z onto e, the point P_1 so obtained is projected from Y upon x, and so on, projecting alternately from Z and Y onto e and x, respectively. The coordinates of each point of the

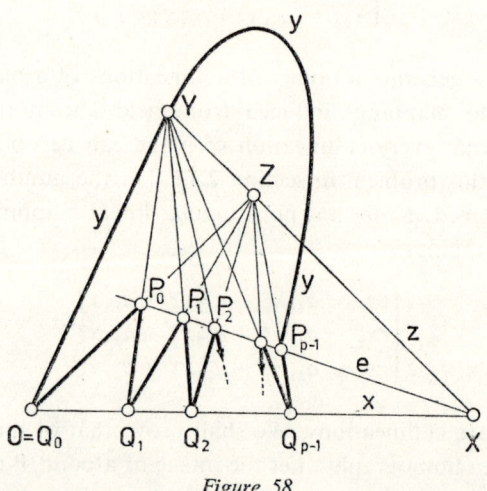

Figure 58

sequence can easily be determined from the coordinates of previous point. Thus we eventually obtain the points

$$P_{p-2}:\ (p-1, 1, 1) \quad \text{and} \quad Q_{p-1}:\ (p-1, 0, 1).$$

But then consider $P_{p-1}: XP_0 \cap ZQ_{p-1}$, i.e. $P_{p-1}: (p, 1, 1) = (0, 1, 1)$, as $\mathbf{K}$ is of characteristic p. Therefore, $Q_{p-1} = XQ_0 \cap YP_{p-1} = Q_0$. Thus we can say that *the process of projection in question is closed and consists of $2p$ projections.* The special polygon of $2p$ sides so derived will be denoted by $\mathbf{P_0}$. — In the figure we emphasized it by bold lines. —

We could perform a similar construction starting with any proper quadrangle $O^* X^* Y^* P_0^*$ and we would still arrive at a polygon with a similar property. This can easily be seen since there exists a collineation ψ (cf. **2.6**), which transforms the quadrangle $O^* X^* Y^* P_0^*$ into the reference quadrangle $OXYP_0$ and the polygon we would obtain is the image of P_0 under ψ^{-1}.

For the sake of brevity let the generating quadrangle be denoted by $\mathbf{N}$, let the construction in question starting from the quadrangle be denoted by $\pi(\mathbf{N})$ where no confusion can occur, we denote this by π and let the closed polygon generated by the process be denoted by $\mathbf{P}$, i.e. $\pi(\mathbf{N}) = \mathbf{P}$. It is clear from the generation of the polygon $\mathbf{P}$ that the quadrangle $\mathbf{N}$ can uniquely be reconstructed

from a given polygon **P**. A given sequence of $2p$ points $(P_0, Q_0, P_1, Q_1, \ldots, P_{p-1}, Q_{p-1})$ determines a polygon **P**, if it satisfies the following conditions:
  (a) The sequences of points $P_0, P_1, \ldots, P_{p-1}$ and $Q_0, Q_1, \ldots, Q_{p-1}$ are on the same lines e and x, respectively.
  (b) The lines $P_0Q_0, P_1Q_1, \ldots, P_{p-1}Q_{p-1}$ meet in a point Z and the lines $P_0Q_1, P_1Q_2, \ldots, P_{p-1}Q_0$ meet in a point Y.
  (c) The point $e \cap x = x$ together with the points Y and Z form a collinear triplet of points.

The polygon **P** so defined is obviously generated by the quadrangle Q_0XYP_0. Given any projective plane T (not necessarily a Galois plane) we could look for polygons **P** satisfying the conditions above. Suppose for a given quadrangle **N** there exists a number p such that a polygon of $2p$ sides exists satisfying the conditions (a), (b) and (c), and p is the least such number, then we say that *on the plane T the condition "$p=0$" is satisfied with respect to the quadrangle* **N** and if this condition is satisfied for each proper quadrangle of the plane, with the same number p, then we say that the "$p=0$" theorem hold on the plane T. If the "$p=0$" theorem holds on a plane, then we say that the plane has *characteristic* p.

If the condition "$2=0$" is satisfied for a quadrangle **N**, then **N** is a Fano quadrangle. If the "$2=0$" theorem holds on a plane T, then T is called a Fano plane. The structure of the Fano planes was completely determined by Gleason. Several mathematicians (notably Lombardo Radice and Zappa) have discussed planes of characteristic 3. Lombardo Radice conjectured that the only planes of characteristic p are the Galois planes.

Let us return to the case of the Galois planes. (We refer again to Fig. 58.) If we delete the points P_i and Q_j ($i, j = 0, 1, \ldots, p-1$) from the lines e and x of quadrangle $Q_0XYP_0 = N_0$ in the plane S, then on each of the lines e and x there remain $q-p+1$ points. Let us choose any one of the remaining points on the line x. Let the chosen point be denoted by Q_0^s, where $s > p-1$. Let P_0^s be the projection of Q_0^s from point Z onto e. Continuing the construction as before we obtain a sequence of points $(P_0^s, Q_0^s, P_1^s, Q_1^s, \ldots, P_{p-1}^s, Q_{p-1}^s)$ where $P_i^s \in x$ and $Q_i^s \in e$ ($i = q, \ldots, p-1$), denote the polygon so formed by $\mathbf{P}_{sp}$. In this way we can decompose the points of the lines e and x, distinct from the point x, into a set of p^{r-1} polygons, with p vertices on each of the lines e and x. Let these polygons be denoted by

$$\mathbf{P}_0, \mathbf{P}_p, \mathbf{P}_{2p}, \ldots, \mathbf{P}_{(p^{r-1}-1)p}.$$

The aforegoing discussion makes it clear that *every vertex of the polygon* $\mathbf{P}_0$ *is a fixed point under the collineation* α^* (*induced by the Frobenius automorphism* α) *which maps every one of the points* Q_0, X, Y, P_0 *forming the coordinate system* *onto itself.* For the time being we can say nothing in general about the images of the polygons $\mathbf{P}_{sp}$ ($s\neq 0$) under the collineation α; but we shall mention a particular case. Namely α^* of $S_{2,q}$ induced by the automorphism $\alpha: x \to x^3$ of $GF(q)$, interchanges polygons $\mathbf{P}_3$ and $\mathbf{P}_6$, while every point of $\mathbf{P}_0$ is carried into itself.

As a further example of a collineation α^* we mention the plane $S_{2,4}$. If we use the tables of operation given in Section **1.6** we find that α^* is given by the following permutation:

$$\alpha^*: \begin{pmatrix} 1 & 2 & 3 & 4 & 5 & 6 & 7 & 8 & 9 & 10 & 11 & 12 & 13 & 14 & 15 & 16 & 17 & 18 & 19 & 20 & 21 \\ 1 & 2 & 3 & 4 & 5 & 6 & 7 & 8 & 9 & 10 & 11 & 13 & 12 & 18 & 19 & 21 & 20 & 14 & 15 & 17 & 16 \end{pmatrix}$$

where the indices are those of the point (and of the lines) in Figs 7 and 10. Thus, for example,

$$l_{15} = \{P_4, P_7, P_{13}, P_{16}, P_{18}\} \to l_{19} = \{P_5, P_7, P_{12}, P_{21}, P_{14}\}.$$

2.17 The set of collineations mapping a Galois plane onto itself

First we shall determine which collineations of a Galois plane of characteristic p have $(1, 0, 0)$, $(0, 1, 0)$, $(0, 0, 1)$, $(1, 1, 1)$ as fixed points. Obviously, the identical mapping ε is such a collineation. Moreover on the classical projective plane these four points can only be a fixed point of the identity. Furthermore we know from our earlier examples that each of the mappings

(2.17.1) $$\omega^t: \varrho x_1' = x_1^{p^t}, \quad \varrho x_2' = x_2^{p^t}, \quad \varrho x_3' = x_3^{p^t}$$

$$(t = 0, 1, ..., r-1)$$

satisfies this requirement. In fact ω^t, the t-th power of the automorphism $\omega: x \to x^p$ of the field $GF(p^r) = \mathbf{K}$ is again an automorphism which induces a collineation of S_{2,p^r}.

We claim that there are *no other line-preserving mappings satisfying the requirements above apart from* the mappings

(2.17.2) $$\omega^0 = \varepsilon, \ \omega^1, \omega^2, ..., \omega^{r-1}.$$

Before we prove this statement we emphasize the fact that the plane $S_{2,q}$ over the prime field $\mathbf{P}$ of the field $\mathbf{K}$ is a (closed) subplane of the plane $S_{2,q}$ over the field $\mathbf{K}$ and every automorphism $x \to x^{p^t}$ of the field $\mathbf{K}$ induces a collineation ω^t mapping the plane $S_{2,q}$ onto itself for which every point of the subplane $S_{2,p}$ is a fixed point. This follows from the fact that an automorphism of the field $\mathbf{K}$ maps any element of the prime field $\mathbf{P}$ onto itself.

We shall refer to Figs 59 and 60 during the proof of our statement (2.17.2) and we shall use the notation introduced at the beginning of the last section.

Consider the point $A: (a, 0, 1)$ and $B: (b, 0, 1)$ on the line x, these determine, in an arithmetic sense, the point sum

(2.17.3) $\qquad A+B = C: (a+b, 0, 1).$

Now we can construct the point C from the given points A and B in the following manner: let $ZA = [1]$ and $[1] \cap e = U$; let $BE = [2]$ and $[2] \cap Z = S$ so if $US = [3]$, we have $[3] \cap x = C$. Let this linear construction be called the construction α.

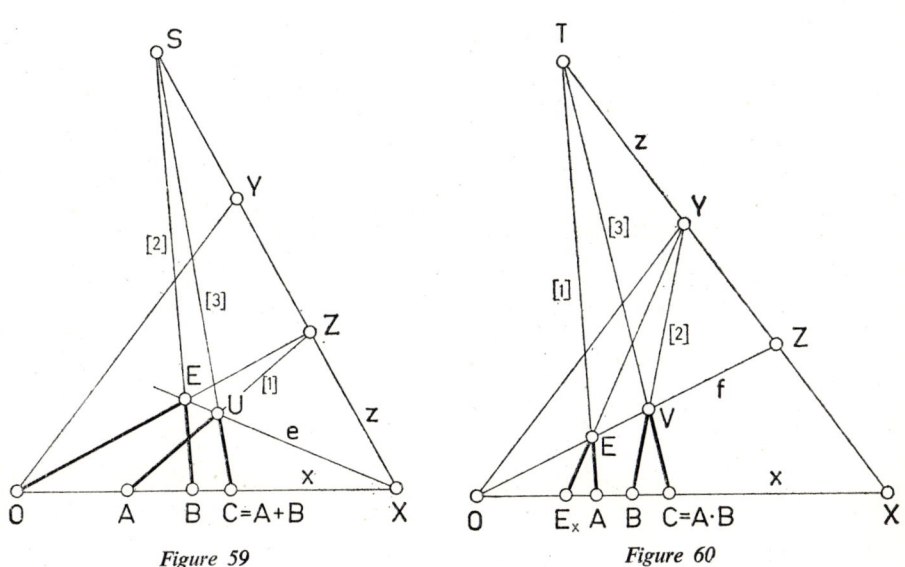

Figure 59 Figure 60

This can be verified by determining the coordinates of the points occurring in the construction by means of linear combinations of points with known coordinates.

The line e contains all the points having the form $(\xi, 1, 1)$ and thus

$\qquad U: (a+1, 1, 1),\quad$ since $\quad Z+A: (1, 1, 0)+(a, 0, 1) = (a+1, 1, 1).$

The line z contains all the points having the form $(\xi, 1, 0)$ and thus

$\qquad S: (1-b, 1, 0) \quad$ since $\quad E-B: (1, 1, 1)-(b, 0, 1) = (1-b, 1, 0).$

All the points of the line x, except the point X, can be written in the form $(\xi, 0, 1)$, hence

$\qquad C: (a+b, 0, 1), \quad$ since $\quad U-S = (a+1, 1, 1)-(1-b, 1, 0) = (a+b, 0, 1).$

11 Introduction

Figure 60 shows a construction which starting with the points A and B, above, determines the point

(2.17.4) $\qquad A \cdot B = C : (a \cdot b, 0, 1).$

This construction can be described as follows:

Let $AE = [1]$, and $[1] \cap z = T$; let $BY = [2]$, $OE = f$, and $[2] \cap f = V$; then if $VT = [3]$, we have $[3] \cap x = C$. Let this construction be denoted by μ.

Again, we verify this by determining the coordinates of the point occurring during the construction.

Now, $T: (1-a, 1, 0)$ since $E - A: (1, 1, 1) - (a, 0, 1) = (1-a, 1, 0)$. The line f contains all the points of the form $(\xi, \xi, 1)$, therefore

$\qquad V: (b, b, 1),$ since $B + b \cdot Y: (b, 0, 1) + (0, b, 0) = (b, b, 1).$

Also,

$\qquad C: (a \cdot b, 0, 1),$ since $V - b \cdot T: (b, b, 1) - (b - b \cdot a, b, 0) =$
$\qquad\qquad\qquad\qquad\qquad = (b \cdot a, 0, 1) = (a \cdot b, 0, 1).$

If we consider the coordinates of the point C obtained by a construction μ, we see immediately that we arrived at the (arithmetically defined) point $A \cdot B = C$.

We obtain now the following statements from the description of the operations α and μ. Let a collineation ω^* be given satisfying the conditions

$$O' = O, \quad X' = X, \quad Y' = Y, \quad E' = E,$$

where we denote $\omega^*(O)$ by O', etc.

Then we have $(A+B)' = A' + B'$ and $(A \cdot B)' = A' \cdot B'$. Consequently, *the effect of the collineation ω^* with the fixed points O, X and E_x: $(1, 0, 1)$ on the line x*

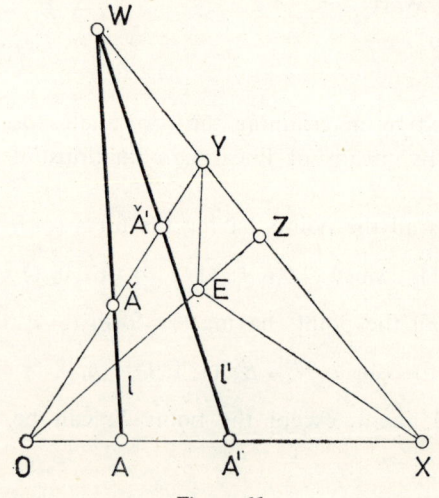

Figure 61

is identical with the effect of a collineation induced by an automorphism ω^t of the field of coordinates **K**. It is easy to see that the point $W: (-1, 1, 0)$ is obtained from the fundamental figure by a linear construction, namely we obtain it as the common point of the line XY and the line connecting the points $XE \cap OX$ and $YE \cap OX$. (Fig. 61). From this it is clear that we must have $\omega^*(W) = W$. If we consider this fixed point, we see that the image of the line $l = AW$ under ω^* is clearly the line $A'W = l'$. Consequently the point $A = OY \cap l$ has for its image the point $OY \cap l' = A'$. If the coordinates of the points A and A' are $(a, 0, 1)$ and $(a', 0, 1)$ then we can obtain the coordinates of the points A and A' in the following manner:

$$a \cdot W + A: \ (-a, a, 0) + (a, 0, 1) = (0, a, 1): \ \check{A};$$

$$a' \cdot W + A': \ (-a', a', 0) + (a', 0, 1) = (0, a', 1): \ \check{A}'.$$

Thus we can say that *the effect of the collineation ω^* upon the line* $OY = y$ *is identical with the effect of the collineation induced by the automorphism ω^t mentioned in the statement above.*

Now the point $P: (a, b, 1)$ is the common point of the line projecting the point $A: (a, 0, 1)$ from the point Y and the line projecting the point $B: (0, b, 1)$ from the point X. Let $A': (a', 0, 1)$ and $B': (0, b', 1)$, then the two statements above imply that the point $P': (a', b', 1)$ is the image of the point P under the collineation ω^* and also under the collineation induced by the automorphism in question. Therefore the collineation ω^* is induced by an automorphism of the field **K**. Thus we have proved the statement (2.17.2).

Consider now the linear mapping λ carrying the vertices of the proper quadrangle $N = OXYE$ into the vertices of a given proper quadrangle $N' = O'X'Y'E'$ and collineation φ satisfying the same conditions but not necessarily identical with λ. Obviously, the mapping

(2.17.5) $$\psi = \varphi \lambda^{-1}$$

is a collineation having each of the points O, X, Y, E as fixed points i.e. the mapping ψ is a collineation induced by an automorphism of the field **K**. From this it follows that any collineations mapping the plane $S_{2,q}$ onto itself can be expressed in the following form:

(2.17.6) $$\begin{cases} \varrho x_1' = a_{11} x_1^{p^t} + a_{12} x_2^{p^t} + a_{13} x_3^{p^t}, \\ \varrho x_2' = a_{21} x_1^{p^t} + a_{22} x_2^{p^t} + a_{23} x_3^{p^t}, \quad (t = 0, 1, \ldots, r-1) \\ \varrho x_3' = a_{31} x_1^{p^t} + a_{32} x_2^{p^t} + a_{33} x_3^{p^t}, \end{cases}$$

where the determinant formed from the coefficients a_{jk} does not vanish.

Finally, we have to determine the number of collineations mapping the plane $S_{2,q}$ onto itself. In order to obtain this number we shall not use the formula (2.17.6), instead we shall use a simple combinatorial argument. There is a unique linear transformation carrying a given (ordered) proper quadrangle $OXYE$ into any other given (ordered) proper quadrangle $O'X'Y'E'$. Thus, in order to find the number of linear transformations it suffices to determine how many proper ordered quadrangle can be chosen from the q^2+q+1 points of the plane.

The point O' can be chosen in (q^2+q+1) different ways. X' can then be chosen in only $q^2+q=(q+1)q$ different ways. For Y' we can choose any point of the plane which does lie on the line $O'X'$, i.e. there are only q^2 choices for Y' once $O'X'$ is fixed. Finally for E' we can choose any point which does not lie on any of the sides of the triangle $O'X'Y'$ and this requires the exclusion of $3(q+1)-3=3q$ points. Consequently we can choose E' in $q^2+q+1-3q=(q-1)^2$ distinct ways. Thus the number of the proper ordered quadrangles of our plane and so the number of its linear mappings is

$$(q^2+q+1)(q+1)q^3(q-1)^2.$$

As there are r automorphisms of the field $\mathbf{K}$ we see that the number of collineations of $S_{2,q}$ is

(2.17.7) $$r(q^2+q+1)(q+1)q^3(q^2-1).$$

It is easy to see that the set of collineations of $S_{2,q}$ forms a group. The number of collineations increases very rapidly with the order of the plane. For instance, for the Galois plane or order 9, we find that the order of the collineation group (given by (2.17.7)) is 106, 142, 400.

2.18 Desarguesian finite planes

In this section we shall deal with the planes for which the axiom system *ID* incorporating the axiom system I and the theorem of Desargues, is valid. For the time being a plane of order q satisfying **ID** will be denoted by $D_{2,q}$. Our aim is to show the equivalence of the notions of the plane $S_{2,q}$ introduced arithmetically, and the plane $D_{2,q}$ introduced geometrically.

1° According to the axioms I_1, I_2, I_3 there exist four points, namely the points

$$A_1(=0), \quad A_2(=X), \quad A_3(=Y) \quad \text{and} \quad E,$$

such that no three of them lies on a line. — In Fig. 62 this is visualized by a Euclidean figure (extended with ideal elements), where X and Y are ideal points. —

Let $OE=e$ and $Z=e \cap xy$. By means of this quadrangle $OXYE$ we shall introduce the line XY two *point operations*, called addition and multiplication, on the set $e^* = e \setminus z$.

We understand by the *addition* α of the points A and B (in this order) the following construction of a point called $A+B$:

(2.18.1) Let $BX \cap OY = U$ and $UZ \cap AY = V$, then $VX \cap OZ = A+B$.

We understand by the multiplication μ of the points A and B (in this order) the following construction of a point called $A \cdot B$:

(2.18.2) Let $BX \cap EY = U$ and $UO \cap AY = V$, then $VX \cap EO = A \cdot B$.

Let us compare the operations α and μ. Starting with the given points, the elements (in the notation of the figure) [1], U, [2], [3], V, [4], S occur in both constructions. The construction μ involves both of the points O and E whilst the construction α involves only the point O. The roles of the points Z and O in the construction α are taken over in the construction μ by the points O and E, respectively.

Our program is now the following: We shall derive from the axiom system **ID** that the operations α and μ impose a skew field structure on the point set e^*. Firstly we shall discuss those axioms of a skew field whose validity on e^* follows from I_1, I_2 and I_3 and then those which require also the theorem of Desargues.

The following facts are easy to see. If in the construction α we have $A=O$, then $V=U$, [4]=[1] and hence $O+B=B$. Similarly, if $B=O$, then [1]=OX, $U=O$,

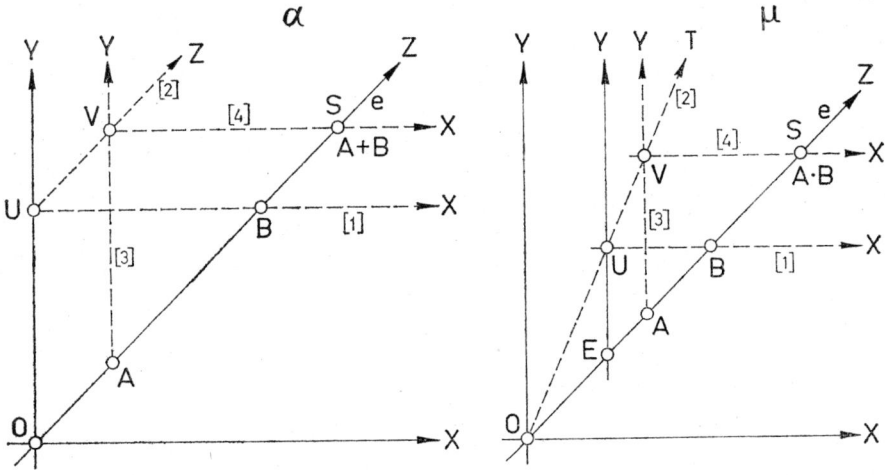

Figure 62

hence $[2]=e$ and $V=A$, which implies $A+O=A$. If $A=O=B$, then $U=O=V$ and $[1]=OX=[4]$, $[3]=OY$, consequently $O+O=O$. Summing up:

(A_1) If $P \in e^*$, then $O+P=P+O=P$.

We can verify in a similar manner the following two basic properties of the operation μ:

(M_0) If $P \in e^*$, then $O \cdot P = P \cdot O = O$.

(M_1) If $P \in e^*$, then $E \cdot P = P \cdot E = P$.

If the point S and if A respectively B is given, then by changing the order of the construction α we can get back to B respectively to A. In the case of the construction μ the same can be done only if we start with S and $A \neq 0$ or with S and $B \neq 0$. Thus we have the following properties:

(A_2) If $C, S \in e^*$, there exists uniquely a point P and a point Q such that $P+C=S$ and $C+Q=S$.

(M_2) If $C \neq 0$, and $C \in C^*$, there exists uniquely a point P and a point Q such that $P \cdot C=S$ and $C \cdot Q=S$.

In order to verify the following properties we shall use the theorem of Desargues or a special case of it.

(A_3) given any three points A, B and C of e^*, then the relation $(A+B)+C=$ $=A+(B+C)$ is valid.

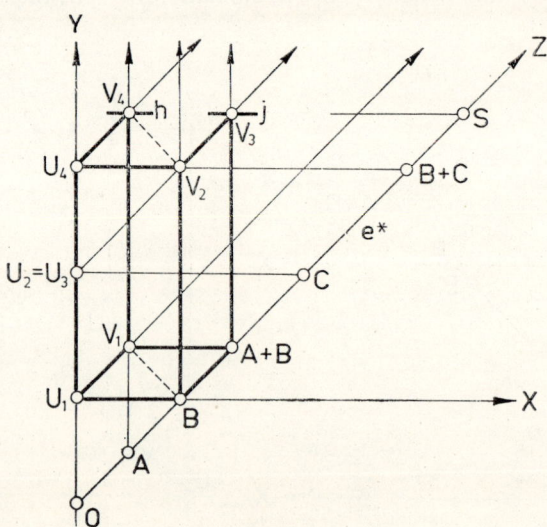

Figure 63

We shall use Fig. 63 in the proof of this statement.

The construction α occurs in the figure in four ways: as the construction of the points $A+B$, $B+C$, $(A+B)+C$ and $A+(B+C)$. The latter two constructions are not drawn completely in the figure.

In order to prove (A_3) consider the triangles BU_1V_1 and $V_2U_4V_4$ which are in perspective from the point Y. According to the theorem of Desargues, the

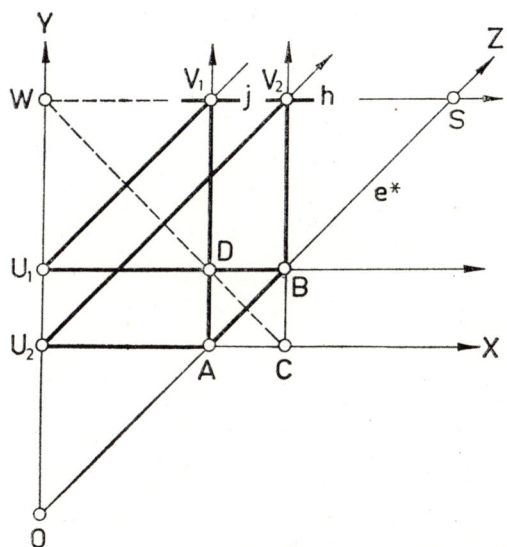

Figure 64

pairs of corresponding sides cut each other in the collinear points X, Z and W, where $W = BV_1 \cap V_2V_4$. Thus the points X, Y, Z, W are all collinear. (In the figure the point W is an ideal point.) Consider now the triangles $V_1B(A+B)$ and $V_4V_2V_3$ which are also in perspective from the point Y. Two of the corresponding lines meet in the point W, two in the point Z i.e. on the line $XYZW$ which is cut by the line $V_1(A+B)$ in the point X. According to the theorem of Desargues the corresponding line V_4V_3 must also pass through the point X. Thus the triplet of the points V_3V_4X is collinear, and (A_3) follows.

The proof of the next property rests again upon a special case of the theorem of Desargues. The proof appeals to the constructions shown in Fig. 64.

(A_4) For any pair of points A, B of e^* we have $A+B = B+A$.

In order to prove this consider the points $U_2X \cap BY = C$, $U_1X \cap AY = D$ and $V_1X \cap DY = W$. The triangles ASC and V_1U_1W are in perspective from the axis XY, thus according to the theorem of Desargues the lines U_1A_2, V_1V_2, DC

connecting the corresponding points meet in the same point, namely in the point $U_1 U_2 \cap DC = W$. And this means that the points $XV_1 W$ and V_2 are collinear, i.e. $V_1 X$ coincides with $V_2 X$ and consequently $A + B = V_1 X \cap e^* = V_2 X \cap e^* = B + A$.

Thus, as the elements of the point set e^* satisfy A_1, A_2, A_3 and A_4 under the operation of addition α, e^* has the structure of a commutative group.

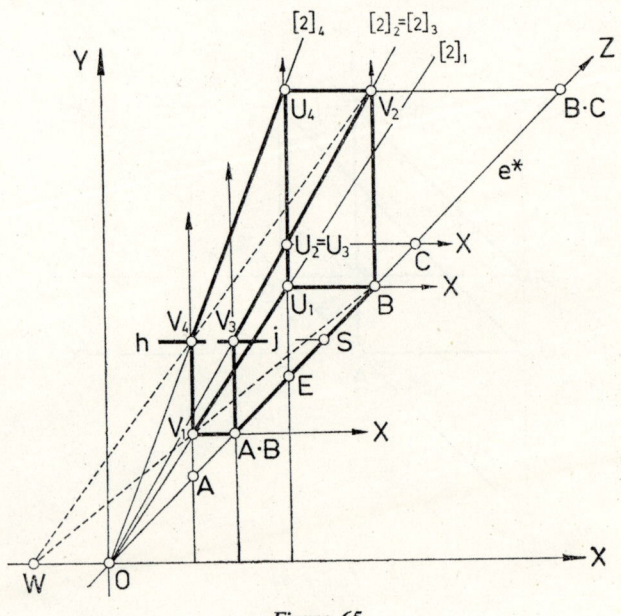

Figure 65

(M_3) Given any three points A, B and C of e^*, then $(A \cdot B) \cdot C = A \cdot (B \cdot C)$.

The proof of this statement, using Fig. 65, is similar to the proof of (A_3). In this figure, the construction μ is made four times, namely for the points

$$A \cdot B, \quad B \cdot C, \quad (A \cdot B) \cdot C, \quad A \cdot (B \cdot C).$$

Consider the triangles $BU_1 V_1$ and $V_2 U_4 V_4$ which are in perspective from the point Y. According to the theorem of Desargues the corresponding sides meet in the points of the same line; thus the point $W = BV_1 \cap V_2 V_4$ lies on the line OX. Consider now the triangle $(A \cdot B) BV_1$ and $V_3 V_2 V_4$ which are also in perspective from the point Y, according to the theorem of Desargues two from the corresponding lines meet in the point O, two meet in W, the line OW is cut by the line $(A \cdot B) V_1$ in the point X and thus the line $V_3 V_4$ corresponding to the line $(A \cdot B) V_1$ passes also through the point X, i.e. $V_3 V_4$ and X are collinear and the result follows.

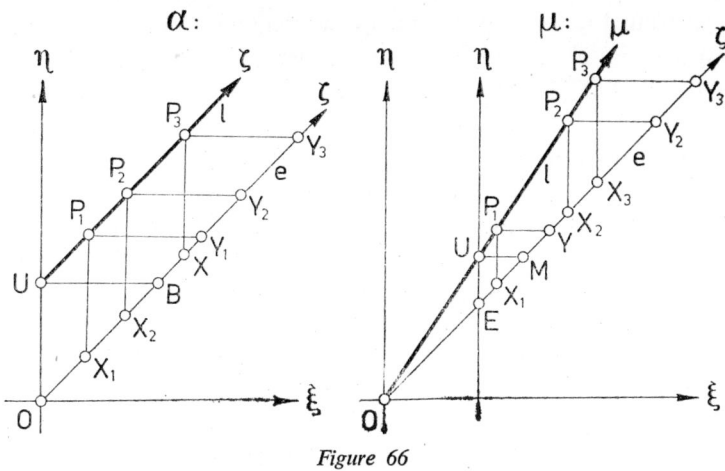

Figure 66

Thus, as the elements of the point set $e^* \setminus O$ satisfy (M_1), (M_2) and (M_3) under the operation μ, $e^* \setminus O$, has the structure of a group.

If we compare the axiom system defining the (not necessarily commutative) field and the present system of properties

$$A_1, A_2, A_3, A_4, \; M_0, M_1, M_2, M_3$$

we notice that if the axioms of right and left distributivity are satisfied then the structure (e^*, α, μ) is a skew field.

For the time being we shall merely formulate the missing properties, their proofs will be given only after the discussion of certain consequences of the properties A and M.

We have for any three points A, B, C of point the set e^*

$$(D_1) \; A \cdot (B+C) = A \cdot B + A \cdot C$$

and

$$(D_2) \; (B+C) \cdot A = B \cdot A + C \cdot A.$$

We note further that in the proof of the properties (A) and (M) we did not consider the special cases coincident points; however, our statements are valid in these cases and the proof of this fact is left here to the reader.

2° Already the properties (A), (M) make possible the introduction of a coordinate system on the plane and, to a certain extent, the developing of an analytic approach to the geometry of the plane.

As in Fig. 66, the fundamental points X, Y of the coordinate system will be denoted by ξ and η and, similarly, the other points of the fundamental line X, Y will be denoted by Greek letters.

By comparing Fig. 62 accompanying the definition of the operation α with Fig. 66 we see that if we fix the point B and let the point X range over the point set e^*, then the line $[2]=l$ involved in the construction of the point $X+B=Y$ remains unaltered. Further more, the point P playing the role of the auxiliary point V will range over all the points of the line l; except the point ζ. Thus we can establish a one-to-one relation between the points of the line $U\zeta=l$, the points P and the ordered pairs of points (X, Y) of the line $0\zeta=e$ satisfying the equation $y=X+B$. Obviously, under this correspondence $U=P_0 \leftrightarrow (D, U)$. We can speak in this sense of the equation of the line l without the point ζ:

(2.18.3) $\qquad\qquad\qquad l^*: Y = X + B.$

Obviously, when the point B ranges over the elements of the point set e^*, then the point sets l^* induced by it, cover all points of the plane except the points of the line $\xi\eta$.

Similarly, we can discuss the part of the figure illustrating the operation μ. If we fix the point M and let the point X range over the elements of the point set e^*, then the line $[2]=l$ involved in the construction of the point $X \cdot M = Y$ remains unaltered. The point P playing the role of the auxiliary point P ranges over the elements of the point set l^* formed by the line l without the point μ. Thus we can establish a one-to-one relation between the points of the point set $OU-\mu=l^*$, the points P and the ordered pairs of points (X, Y) of the point set e^* satisfying the equation $Y=X \cdot M$. Obviously, under the correspondence we have $P=E \leftrightarrow (E, M)$. Thus we can speak of the equation of the point set l^*:

(2.18.4) $\qquad\qquad\qquad l^*: Y = X \cdot M.$

If the point M ranges over the elements of the point set e^*, the point sets l^* will cover all points of the plane except the point of the lines $O\eta$ and $\xi\eta$.

We see from this discussion that the correspondence between point P and the ordered pair of points (X, Y) is analogous to the assignment pairs of coordinates (x, y) in the ordinary plane. In this sense we may speak about the coordinates (X, Y) of the point P with respect to the coordinate system $O\xi\eta E$. Thus we can make the following

Definition: We shall understand by the point coordinates (X, Y) of the point $P \notin \xi\eta$ of the plane with respect to the coordinate system $O\xi\eta E$ the ordered pair of points of the set E^* the ordered pair of points of the point set e^* for which holds

(2.18.5) $\qquad P = X\eta \cap Y\xi, \quad \text{where} \quad X = P\eta \cap e \quad \text{and} \quad Y = P\xi \cap e.$

Obviously, as the points of the line $\xi\eta$ were omitted we have coordinatized only q^2 the points of the projective plane of order q.

So far we have described only $2q-1$ lines of the plane by an equation, namely $Y=X+B$ and $Y=X\cdot M$. The points of any line passing through the point η but distinct from the line $\xi\eta$ are characterized by the fact that their first coordinate is a fixed point $x=A$ of the point set e^* and their second coordinate ranges over all points of e^*. We can say in this sense that

(2.18.6) $\qquad\qquad P\eta: X = A \quad (A\in e^*)$

is the equation of the line in question. Thus we can assign equations to q further lines.

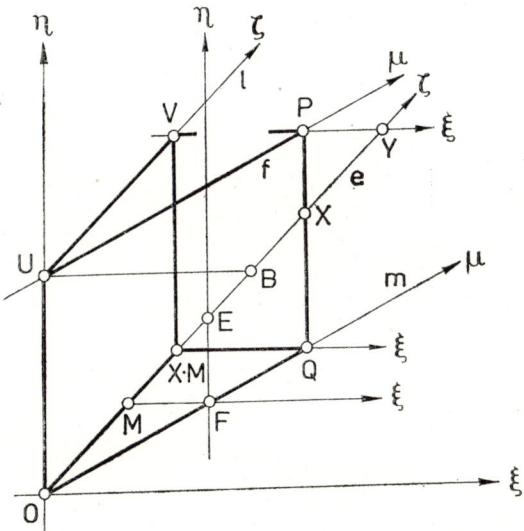

Figure 67

We shall now introduce line coordinates by ordered pairs of points.

Definition: We shall understand by the line coordinates of a line f not through the point η of the plane with respect to the coordinate system $O\xi\eta E$ and ordered pair of points $[M, B]$ of the point set e^*; the points M and B are obtained by the following construction (Fig. 67):

(2.18.7) $\qquad \begin{cases} M = F\xi \cap e, & \text{where} \quad F = E\eta \cap m, \quad m = O\mu, \\ \mu = f\cap \xi\eta; \\ B = U\xi \cap e, & U = f \cap O\eta. \end{cases}$

Obviously, this construction can be inverted, i.e. f can be reconstructed uniquely from a given pair $[M, B]$. By this construction line coordinates can only be assigned to those lines — q^2 in number — which do not pass through the point η. The arithmetic relation between the line coordinate $[M, B]$ of a line and the point coordinates (X, Y) of a point $P(\xi\eta)$ incident with it is expressed by the *equation* of the line f:

(2.18.8) \qquad (L): $Y = X \cdot M + B$.

The validity of the relation (L) can be seen easily as follows (Fig. 66): the line f and the point P are determined by the pairs of points $[M, B]$ and

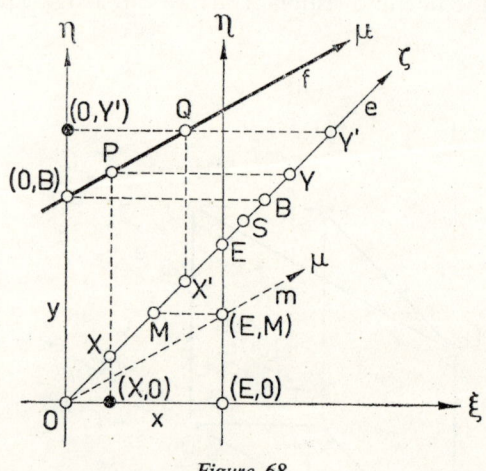

Figure 68

(X, Y) respectively. In the construction of the point $X \cdot M$ from the points X and M the roles of the point U, line [2] and point V are played by the point F, the line m, and the point Q respectively. In the construction the point sum $X \cdot M + B$ the role of the line [2] is taken over by the line e. Therefore the only thing we have to prove is that the point $V\xi \cap e$ coincides with the coordinate-point Y of the point P i.e. that the line VP passes through the point ξ.

In order to do this consider the perspective triangles $(X \cdot M)OQ$ and VUP. Two of the corresponding sides meet in the point ζ and two in the point μ, i.e. on the line $\xi\eta$. Thus, according to the theorem of Desargues, the line $\xi\eta$ is the axis of the perspectivity. Since this axis is cut by the line $(X \cdot M)Q$ in the point ξ, line VP must also pass through the point ξ and this is what we had to prove.

It is easy to see that (2.18.3) and (2.18.4) are special cases of (2.18.8). The first occurs in the case of $M=E$, the second in the case of $B=Q$.

3° By means of the properties of the point operations A and M we can introduce the notion of the *linear point function* (Fig. 68).

If X, M, B are arbitrary elements of the point set e^*, then the line f determined by the fixed pair $[M, B]$ generates a linear point function $Y=f(X, M, B)$, where

$Y \in e^*$ and the explicit form of $f(X, MB)$ is $X \cdot M + B$; conversely, we can say that a line f generates a linear point function $f(X, M, B)$, if M and B are fixed.

Following classical analytic geometry we call the lines $O\xi$ and $O\eta$ and the point E the *x-axis*, the *y-axis* and the *unit point* of the coordinate system, respectively. Given a line f and the projection of one of its points onto x from η or on y from ξ, then the point itself can uniquely be reconstructed. Thus for instance in our figure the point P is determined by (X, O) and the point Q by (O, Y').

Now we return to the proof of the properties (D_1) and (D_2).

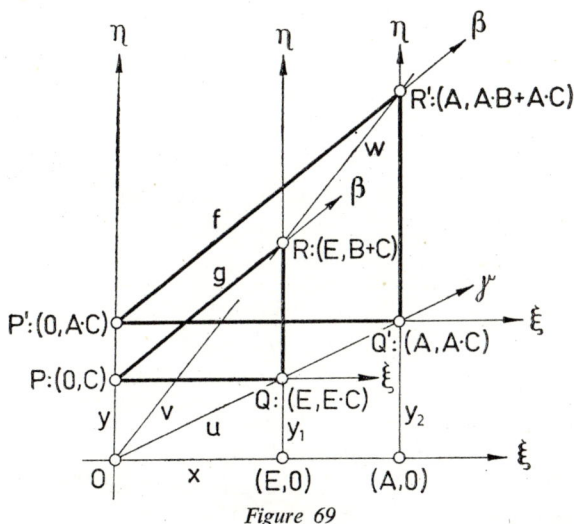

Figure 69

In order prove the property D_1 we start form four lines determined by four equations (Fig. 69). Let these be the lines

$u: Y = X \cdot C,$
$v: Y = X \cdot (B+C),$
$g: Y = X \cdot B + C,$
$f: Y = X \cdot B + A \cdot C$

Obviously, the coordinates of the points $u \cap y_1 = Q$ and $u \cap y_2 = Q'$ are (from the equation of the line u) $(E, E. C) = (E. C)$ and $(A, A. C)$ respectively. Similarly, we obtain from the equations of the lines g and f the coordinates of the points

$P = g \cap y, P' = f \cap g, R = g \cap y_1$ and $R' = f \cap y_2,$

these are written down on the figure. The first coordinates of the lines g and f are both equal to B, this implies that the point $g \cap f$ lies on the line $\xi\eta$. (In the figure this point is denoted by β.) Thus the triangles PQR and $P'Q'R'$

are in perspective from the axis $\xi\eta$ and so, according to the theorem of Desargues, the line $RR' = w$ passes through the point O. Consequently, the equation of the line w has the form $Y = X \cdot M$. The point R lies on this line and thus $E \cdot M = B+C$ i.e. $M = B+C$. Hence the equation of w is $Y = X \cdot (B+C)$ and so $w = v$. Thus the coordinates of the point R' are $(A, A \cdot (B+C))$ but R' lies on the line f so from the equation of f, $R': (A, A \cdot B + A \cdot C)$. Hence $A \cdot (B+C) = A \cdot B + A \cdot C$, which was to be proven.

Similarly, in order to prove the property D_2 we start from four lines given with their equations (Fig. 70). Let these lines be

$$e: Y = X,$$
$$c: Y = X + C,$$
$$l: Y = X \cdot A,$$
$$f: Y = X \cdot A + B \cdot A.$$

If we remember the constructions α and μ, we can easily determine by means of these equations the coordinates of the points

$$C' = l \cap y_2, \quad D' = c \cap y_3, \quad U' = C'\xi \cap y, \quad \text{and} \quad V' = f \cap y_1.$$

— We have also that $D = B+C$, as is illustrated in the figure. — Thus we obtain

$$C': (C, C \cdot A), \quad D': (B+C, (B+C) \cdot A),$$
$$U': (O, C \cdot A), \quad V': (B, B \cdot A + C \cdot A).$$

By comparing the first terms of the equations of e and c with the first terms of the equations of l and f we can readily see that both of the points $e \cap c = \zeta$ and $l \cap f = \lambda$ are on the line $\xi\eta$. Thus the triangles VUU' and DCC' are in perspective from the point ξ and so by the theorem of Desargues η, ζ and $d_1 \cap d_2 = \delta$ (where

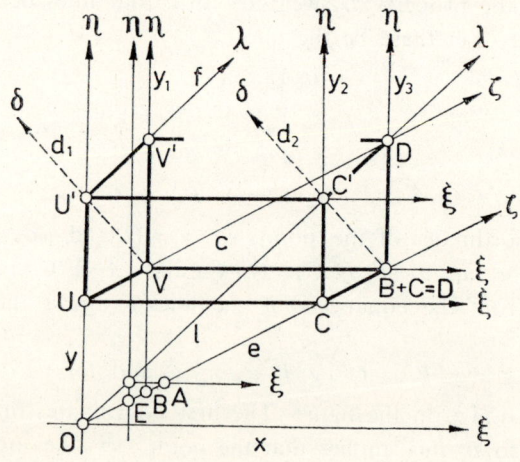

Figure 70

$d_1 = U'V$ and $d_2 = DC'$) are collinear. Therefore the triangles $VU'V'$ and $DC'D'$ are in perspective with respect to the axis $\xi\eta$, since the corresponding lines meet in the points δ, λ, η of this axis. And this implies, according to the theorem of Desargues — since $VD \cap U'C' = \xi$ — that the line $V'D'$ passes through the point ξ. But then the second coordinates of D' and V' are the same, i.e. $(B+C) \cdot A = = B \cdot A + C \cdot A$, which was to be proved.

The foregoing discussion can be summarized in the following

Theorem: *The point operations α, μ with respect to the proper quadrangles $O\xi\eta E$ endow the point set e^* with a skew field structure.*

In fact, there is a well-known theorem due to Wedderburn which states that this field is commutative i.e. *the point multiplication is a commutative operation.* Consequently (2.18.8) can be written in the form

$$Y = M \cdot X + B.$$

Thus we arrived at an expression analogous to the equation $y = mx + b$ of ordinary analytic geometry.

4° Consider the Galois plane $S_{2,q}$ over the coordinate field of order q, **K**. Let us establish a correspondence between the points

$$A_3: (0, 0, 1), \quad E: (1, 1, 1), \quad A_1: (1, 0, 0), \quad A_2: (0, 1, 0)$$

of this plane and the points

$$O, E, \xi, \eta$$

of the plane $D_{2,q}$. The line f with the equation $u_1 x_1 + u_2 x_2 + u_3 x_3 = 0$ in the plane $S_{2,q}$, which does not pass through the point A_2, satisfies the condition $u_2 \neq 0$. If a point $P: (x_1, x_2, x_3)$ is not on the line $A_1 A_2$, it satisfies the condition $x_3 \neq 0$. If we write

$$-\frac{u_1}{u_2} = m \in \mathbf{K}, \quad -\frac{u_3}{u_2} = b \in \mathbf{K}; \quad \frac{x_1}{x_3} = x \in \mathbf{K}, \quad \frac{x_2}{x_3} = y \in \mathbf{K},$$

then every point of the line f, except the point with $x_3 = 0$ satisfies the equation

$$y = mx + b.$$

In particular, there exists a one-to-one correspondence Q between the elements of the field **K** and the points of the line $y = x$ of the plane $S_{2,q}$, namely:

$$Q(x) = (x, x) \quad \text{for} \quad x \in \mathbf{K}.$$

It is easy to see that the point operations α and μ induced by the quadrangle $A_3 E A_1 A_2$ form the point set (x, x) into a field $\mathbf{K}^*$ isomorphic with the field **K**.

Let us follow the constructions shown in Fig. 62.

In the case of α, $A: (a, a)$, $B: (b, b)$, the equation of [1] is $y = b$ and $U: (0, b)$; the equation of [2] is $y = x + b$ and the equation of [3] is $x = a$, hence $V^*: (a, a+b)$, the equation of [4] is $y=a+b$ and thus the line $y=x$ is cut by it in the point $S: (a+b, a+b)$.

In the case of μ, $E: (1, 1)$, $A: (a, a)$, $B: (b, b)$, the equation of [1] is $y=b$ and $U: (1, b)$; also the equation of [2] is $y=xb$, the equation of [3] is $x=a$, $V: (a, ab)$ and the equation of [4] is $y=ab$, thus the line $y=x$ is cut by the line [4] in the point $S: (ab, ab)$.

Hence there corresponds to the addition and the multiplication in the field **K** the point operations α and μ with respect to the corresponding points in the field **K***, thus the mapping Q is really an isomorphism.

Since the point operations established by the quadrangle $OE\xi\eta$ endow the line e of the plane $D_{2,q}$ with the structure of a field of q elements and fields having the same number of elements are isomorphic, then the field in question is isomorphic with the field **K*** determined above.

So it is obvious that $D_{2,q}$ and $S_{2,q}$ are isomorphic figures and thus *the concepts of the Desarguesian plane of order* **q** *and the Galois plane of order* **q** *are the same*.

Finally let us emphasize that the axiom system **ID** does not require the full force of the theorem of Desargues (more correctly the axiom of Desargues), but only its validity in certain special cases; indeed we only used the validity of the theorem in a few special cases in the proofs given in this section. But validity in these special cases implies the validity of the theorem of Desargues in general, moreover, by means of the Wedderburn theorem it follows that the axiom P (the Pappus—Pascal theorem) which is stronger than the theorem of Desargues, is also valid. The importance of these remarks will appear when we investigate planes on which exist pairs of triangles in perspective from a point but in perspective from an axis. Such planes are said to be non-Desarguesian. On a non-Desarguesian plane there can, of course, occur particular pairs of triangles for which the theorem of Desargues is valid. (An example of this occurs in Exercise 14).

2.19 Problems and exercises to Chapter 2

17. Let six distinct points $P_1 P_2 P_3 P_4 P_5 P_6$ of a Galois plane be given such that $P_1 P_3 P_5$ lie on a line u and $P_2 P_4 P_6$ lie on a line v, but the point $u \cap v = Q$ does not coincide with any of the six points above. Prove that the points $P_1 P_2 \cap P_4 P_5 = Q_1$, $P_2 P_3 \cap P_5 P_5 = Q_2$ and $P_3 P_4 \cap P_6 P_1 = Q_3$ and distinct from the points P_i ($i=1, \ldots, 6$), from the point Q, and from each other and further that the three points $Q_1 Q_2 Q_3$ lies on a line w.

18. Extend the conditions of exercise 17 by requiring that the three points $Q Q_1 Q_2$ be collinear. Prove then, that the collinearity of $Q_1 Q_2 Q_3$ can be deduced from the axiom system **ID**.

19. Consider on a Fano plane (defined by the axiom system I and the validity of the Fano theorem) six points $P_1, \ldots, P_6$ satisfying the conditions of exercise 17. Prove that the collinearity of $QQ_1Q_2Q_3$ follows from the axiom system **IF**.

The phrase "finite pascal plane" occurs in the literature. We shall understand by this a plane defined as follows: The axiom system **I** and the axiom **P**, i.e. the axiom system **IP**, is valid. Adding the axiom **P** implies the following: every triple of points $Q_1Q_2Q_3$ established by the six points $P_1, \ldots, P_6$ defined in exercise 17 is collinear.

20. Prove that the following theorem **T** is valid on a Pascal plane: If not more than two of the six distinct points $P_1P_2P_3P_4P_5P_6$ are collinear and further the lines P_1P_2, P_3P_6 and P_4P_5 meet in a point W_1 and the lines P_1P_4, P_2P_3 and P_5P_6 meet in a point Q_2, prove that the lines P_1P_6, P_2P_5, P_3P_4 also met in one point.

It follows from exercise 20 that theorem **T** is just a re-formulation of theorem **P**. In some cases it is easier to consider theorem **T**.

21. If we consider the points and lines of the projective plane of order five with respect to an oval of the plane, we can see that the plane has 10 "interior" points and 10 "exterior" lines which do not cut the oval. Prove that these 10 points lie by threes on each of the "exterior" lines and are the ten points of a Desarguesian configuration, i.e. 6 of the 10 points form the vertices of a pair of triangles in perspective; 1 point is the centre of the perspectivity; and the 3 remaining points are the three points of the axis in which corresponding sides meet.

22. Consider the points

$$B_1: (0, b_1, 1), \quad B_2: (1, 0, b_2), \quad B_3: (b_3, 1, 0);$$
$$C_1: (0, c_1, 1), \quad C_2: (1, 0, c_2), \quad C_3: (c_3, 1, 0)$$

of a Galois plane of order greater than 4 if

$$b_j \neq c_j \quad \text{and} \quad b_j c_j \neq 0 \quad (j = 1, 2, 3).$$

Prove that these six points satisfy a second degree equation if and only if

$$a_1 a_2 a_3 b_1 b_2 b_3 = 1.$$

23. Prove that on the plane $S_{2,q}$ the Pappus—Pascal theorem is valid.

24. Notice that the statement of the last exercise implies that the theorem **T** is valid on the plane $S_{2,q}$.

25. Prove that the point operation μ is commutative on the plane $D_{2,q}$.

CHAPTER 3

GEOMETRICAL CONFIGURATIONS AND NETS

In this short chapter we shall deal with concepts which were original but later became abstract and obtained an important role in the formulation of the theory of finite planes.

3.1 The concept of geometrical configurations

Let an array of $m \times n$ squares be given with m rows and n columns. An array which consists of empty squares and of squares containing incidence signs •, is said to be an *incidence table*. — The configurations $\Sigma_1, \Sigma_2, \Sigma_3$ of Fig. 71 are incidence tables. —

If we consider the set of columns in an incidence table then the rows can be identified with certain subsets of this set. Namely, let each row be identified with the subset of columns that contain an incidence sign in that row. The indices denoting the rows and the columns are taken in the order $1, 2, ..., m$ and $1, 2, ..., n$; respectively, unless otherwise stated. Thus the incidence table Σ_1 of Fig. 71 represents in a perspicuous form the system formed by the subsets $\{1, 2, 3\}, \{1, 2, 5\}, \{1, 3, 4\}, \{2, 3, 6\}, \{2, 5, 6\}, \{3, 4, 6\}, \{1, 4, 5\}, \{4, 5, 6\}$ of the set $\{1, 2, 3, 4, 5, 6\}$.

Now we shall compare the three incidence tables of Fig. 71. If we permutate the columns of table Σ_1 as follows:

$$\omega = \begin{pmatrix} 1 & 2 & 3 & 4 & 5 & 6 \\ 1 & 2 & 3 & 5 & 4 & 6 \end{pmatrix}$$

and then permutate the rows according to

$$\sigma = \begin{pmatrix} 1 & 2 & 3 & 4 & 5 & 6 & 7 & 8 \\ 1 & 7 & 5 & 6 & 2 & 3 & 4 & 8 \end{pmatrix}.$$

The composition of these two permutations is an example of an *elementary transformation* and this transformation τ carries Σ_1 into Σ_2: $\tau(\Sigma_1) = \Sigma_2$. It is easy to see that Σ_3 cannot be obtained in this manner from Σ_1 or from Σ_2. Clearly, the subset {1, 2, 4} of the eight subsets of Σ_3 has at least one element in common with each of the remaining seven subsets. And this property remains invariant under any elementary transformation. However, each subset of Σ_2 is disjoint from at least one other subset.

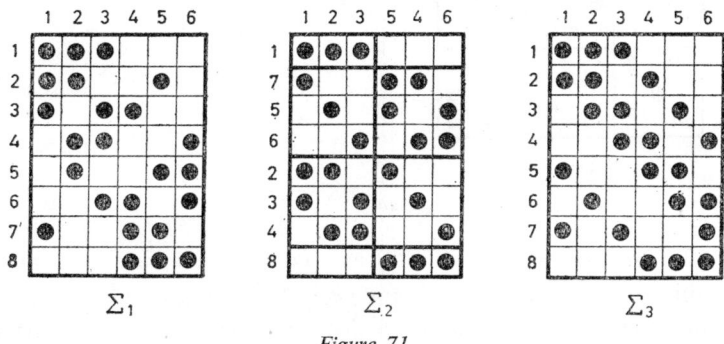

Figure 71

Two incidence tables derivable from each other by an elementary transformation are said to be *isomorphic*; otherwise they are said to be *heteromorphic*. Isomorphic incidence tables are said to define the same *incidence structure*, whereas different structures belong to heteromorphic incidence tables.

If an incidence table Σ consists of n rows and if each of its rows contains a number v and each of its columns a number of the incidence sign •, the incidence table is then said to be a *configuration table* or more briefly a *configuration*. We understand by the *type of the configuration* the matrix

$$\begin{pmatrix} n & m \\ \mu & v \end{pmatrix}.$$

Clearly, $mv = n\mu = J$ is the number of the incidence sign • occurring in the incidence table. In our figure all the three configurations have the type

$$\begin{pmatrix} 6 & 8 \\ 4 & 3 \end{pmatrix}.$$

Hence the type does not tell us much about the structure of the configuration.

The configuration is *symmetric*, if $m = n$, which implies $v = \mu$. — A configuration table like this is represented in Fig. 72. On the other hand, the configuration tables shown in Fig. 71 are not symmetric. —

The Euclidean figure shown in Fig. 72 and the configuration table are identifiable, if we assign to the points the columns of the table having the same indices and to the lines the rows of the table. A configuration table like this cannot

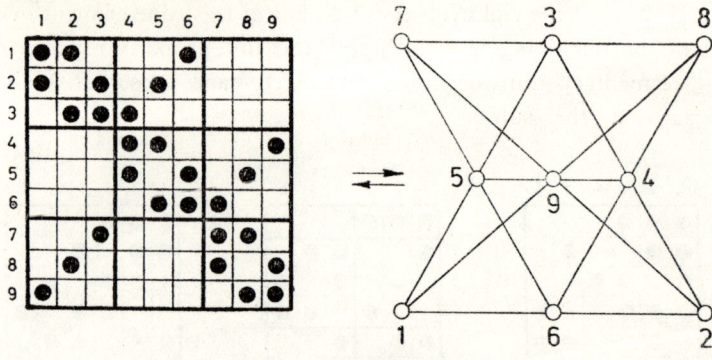

Figure 72

contain a 2×2-sub-table without at least one empty square. A configuration satisfying this condition is said to be a *geometric configuration*. Of course, the existence of a geometric configuration does not mean that it can be realized on an Euclidean or even on a classical projective plane. — In order to see this we have only to consider the finite projective planes. Every finite projective plane is a geometric configuration satisfying even further hypotheses (equivalent to the axioms I_1 and I_2). We know, however, that even the plane of order two cannot be realized on the classical projective plane. —

Let the configuration table Σ be an incidence table of m and n columns and let us choose arbitrarily $m^*(<m)$ rows and $n^*(<n)$ columns. The m^*n^* squares of intersection of these rows and columns together with the incidence sign arrangement in them need not form a configuration Σ^*. — An example of a sub-table which is a configuration table is Σ^* formed by the intersection of the rows 4, 5, 6 and the columns 4, 5, 6 of Fig. 72. However, let us consider the sub-table given by the intersection of the rows 1, 3, 9 and the columns 1, 2, 6. It is easy to see that this is not a configuration. — If a sub-table * of a configuration table is itself a configuration table it is then said to be a *sub-configuration* of Σ.

One of the fundamental problems of the investigation of the finite projective plane is the establishing of the existence of certain sub-configurations and, in the case of their existence, the determination of their total number.

3.2 Two pentagons inscribed into each other

For the sake of curiosity, we shall introduce now some configurations on the classical projective plane. These may perhaps seem to be quite trivial, however, they often lead to an interesting and simple interpretation of an abstract geometrical idea.

We understand by the *sides* of a polygon $\mathbf{A}=A_1 A_2 \ldots A_n$ and of a polygno $\mathbf{B}=B_1 B_2 \ldots B_n$ the lines $A_1 A_2, A_2 A_3, \ldots, A_n A_1$ and $B_1 B_2, B_2 B_3, \ldots, B_n B_1$, respectively. The polygon is said to be *ordinary*, if it does not have three collinear vertices, otherwise it is called *special*. A line joining two vertices of an ordinary polygon, not a side, is said to be a *diagonal line,* more briefly a diagonal.

If there exists a one-to-one correspondence between the sides of the polygon $\mathbf{A}$ and the vertices of the polygon $\mathbf{B}$ such that every side passes through the vertex corresponding to it, then $\mathbf{A}$ is said to be a *circumscribed* polygon with respect to $\mathbf{B}$ and $\mathbf{B}$ is said to be an *inscribed* polygon with respect to $\mathbf{A}$. This relationship will be denoted by $\mathbf{A} \succ \mathbf{B}$.

Figures 72 and 73 both express the incidence table of a geometric configuration of the type

$$\begin{pmatrix} 9 & 9 \\ 3 & 3 \end{pmatrix}$$

as well as a geometrical realization of it; in comparing the two, however, we can see immediately that they are heteromorphic. Consider in each figure the triangles

$$\mathbf{A} = \{1, 2, 3\}, \quad \mathbf{B} = \{4, 5, 6\}, \quad \mathbf{C} = \{7, 8, 9\}.$$

For both figures we have

$$\mathbf{A} \succ \mathbf{B} \succ \mathbf{C} \succ \mathbf{A}.$$

The incidence tables cannot be carried into each other by an elementary transformation: this fact characterizes the difference of the two configurations from a combinatorial point of view. If we represent them on the classical projective plane, we can realize a geometrical difference as well. In fact, if the triangles $\mathbf{A}$ and $\mathbf{B}$ satisfying the condition $\mathbf{A} \succ \mathbf{B}$ are given, then one vertex of the triangle

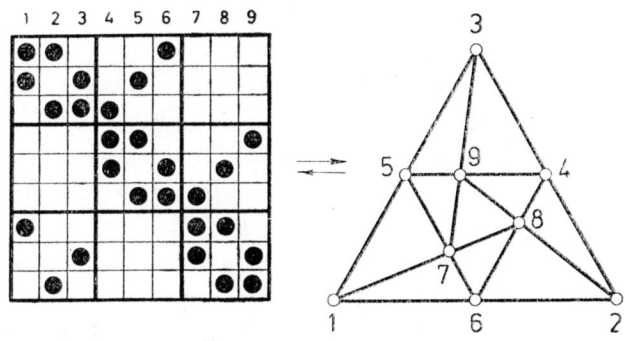

Figure 73

C satisfying the condition $B \succ C \succ A$, in the case of Fig. 72 vertex 7 on the line $\{5, 6\}$ can be chosen arbitrarily and then the points 8 and 9 are uniquely determined by applying Pascal's theorem to the hexagon 1 9 2 4 6 5. However, the position occupied by the point 7 on line $\{5, 6\}$ in Fig. 73 is fixed. — In fact, if the point 7 moves on the line towards point 5, then the point $45 \cap 37 = 9$ also moves towards point 5, but the point $46 \cap 17 = 8$ moves towards point 4. Consequently, by displacing point 7 the line 8, 9 will vary and will not contain the point 2. —

Let **A** and **B** be two distinct triangles the vertices of which are six distinct points of the classical projective plane. It is easy to prove that $A \prec B$ and $A \succ B$ cannot both hold. Surprisingly, however, two pentagons **A** and **B**, can satisfy both $A \succ B$ and $A \prec B$, i.e. *there exist two pentagons such that each is inscribed in the other.*

We shall prove this statement by appealing to a figure in three dimensional space. Let five points 1, ..., 5 space be given such that no more than three of them lie in the same plane (Fig. 74). Consider the skew pentagons 1 2 3 4 5 and $1'2'3'4'5' = 1\ 3\ 5\ 2\ 4$. Let the sides of the former be $a_1, ..., a_5$, and those of the latter be $b_1, ..., b_5$; these pentagons will be denoted by **A** and **B**, respectively. The sides of each of these pentagons are diagonals of the other pentagon. The ten sides of the two pentagons form ten triangles. Five of these triangles have two sides in common with **A** and one in common with **B** and the remaining five triangles have two sides in common with **B** and one in common with **A**. These triangles span ten different planes. The solid configuration so defined consisting of 5 points, 10 lines and 10 planes is called a Θ-configuration. Our aim is now to study a plane section of the Θ-configuration by a plane Σ which does not contain any of the vertices of the configuration.

Let the lines a_j and b_j meet this plane in the points A_j and B_j, respectively $(j=1, ..., 5)$. If we consider the way in which the planes of the configuration meet the cutting plane, we see immediately that two A-points and one B-point are collinear and furthermore two B-points and one A-point are collinear, too. A figure like this is shown in Fig. 75.

This figure proves that the positional relation $A \succ B \succ A$ is indeed realizable for two pentagons. Furthermore, it can be proved easily by means of Fig. 75 that a configuration like this and the Desargues configuration are equivalent figures. — Thus, for instance, the axis of the pair of triangles $A_2 A_3 B_5$, $B_2 A_4 A_5$ in perspective from the point B_1 is the line $B_4 A_1 B_3$. —

The vertices of the *double pentagon configuration* are denoted, in Fig. 76, merely by integers and thus Fig. 77 is an incidence table of the configuration. The latter expresses in an easily understandable manner the relation between the two pentagons.

Two pentagons inscribed

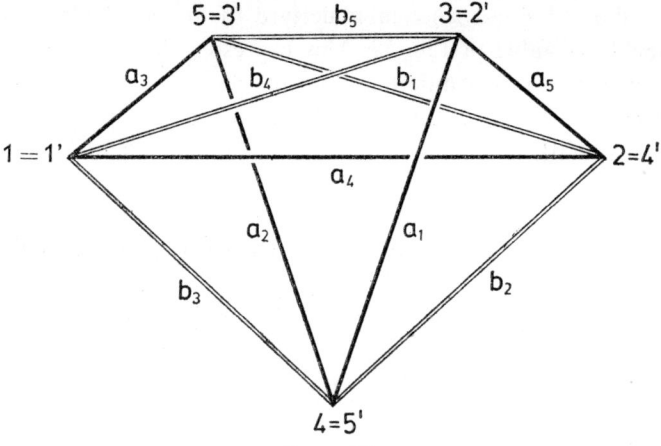

Figure 74

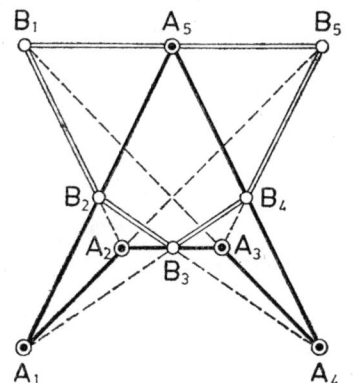

Figure 75

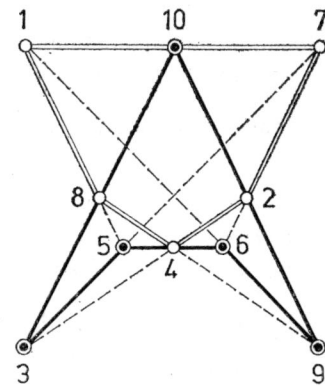

Figure 76

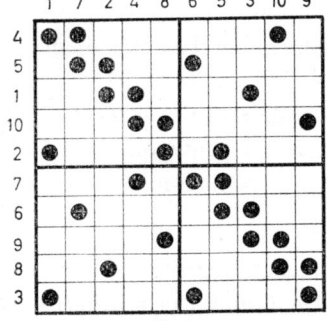

Figure 77

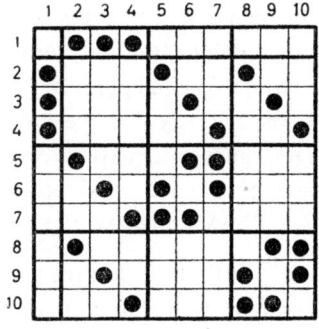

Figure 78

Figure 78 shows the incidence table derived by an elementary transformation from the incidence table of Fig. 77. This expresses clearly the structure of the configuration from a rather different point of view; namely it is a figure formed by pair of triangles {5, 6, 7} and {8, 9, 10} in perspective from the point 1 and from the axis {2, 3, 4}.

3.3 The pentagon theorem and the Desarguesian configuration

1° In connection with the double pentagon configuration we obtain a closure theorem equivalent to the theorem of Desargues. Let us denote the side opposite to the vertex A_k of the pentagon **A** by a_k and similarly for **B**. Further let the *projection* (A, a) mean that if $A \notin a$ and $X \neq A$, then the image of the point X under (A, a) is $X' = AX \cap A$. By a *chain of projection* we mean the composition of a sequence of projections in which at each stage the point projected is the image of the preceding projection. A chain of projections is said to be *closed*, if it leads back to the original point; the number of projections in a closed chain is said to be its *length*. We can now formulate the

Pentagon theorem: *The chain of projections*

$$(A_5, a_4), \; (A_3, a_2), \; (A_1, a_5), \; (A_4, a_3), \; (A_2, a_1)$$

starting from an arbitrary point B_1 *of the side* a_1 *of the pentagon* **A** *leads back to the point* B_1, *if the theorem of Desargues hold on the plane.* (Fig. 75.)

Proof: By the first projection we obtain point B_5 from point B_1, by the second B_4 from B_5, by the third B_3 from B_4, by the fourth B_2 from B_3. This derivation implies that the triangle $B_2 A_2 B_3$ is in perspective with the triangle $A_5 B_5 B_4$ from the centre A_1. But then the axis of perspectivity is the line $A_3 A_4 = a_1$ and this is cut by the side $B_5 A_5$ of the triangle $A_5 B_5 B_4$ in a point B_1 and if Desargues' theorem holds the side $A_2 B_2$ of the triangle $B_2 A_2 B_3$ must cut the axis in the same point. But this implies that the projection (A_2, a_1) carries the point B_2 onto the original point B_1.

We may say that the pentagon theorem is a re-formulation of the theorem of Desargues into a theorem of closure. Moreover, given a pentagon **A** we can assign to each point the line a_1 one and only one pentagon **B** satisfying $A \succ B \succ A$ and have the points of a_1 as a vertex.

2° With respect to the sign-configuration of the incidence table of the Desargues configuration shown in Figs 77 and 78, we speak of a Γ and a Θ arrangement, respectively; or more briefly of a table Γ and a table Θ. 30 incidence signs occur in each of these tables. A configuration which is not geometrical, i.e. one whose incidence table contains a 2×2 sub-table filled with incidence signs cannot be realized on any kind of projective plane because this would require the existence of two distinct lines with two distinct points in common.

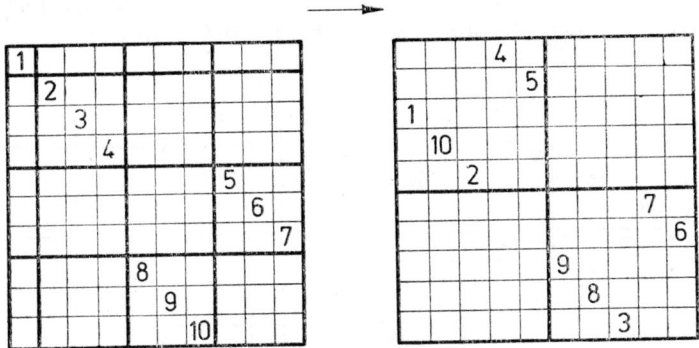

Figure 79

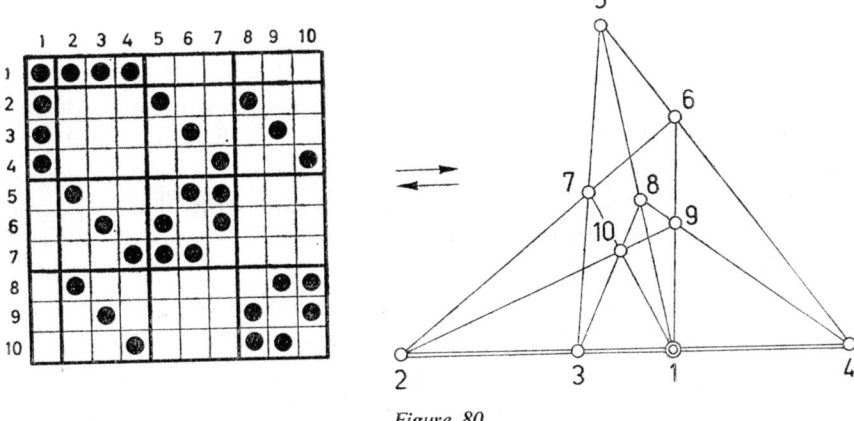

Figure 80

In what follows we shall refer to a Desargues configuration (having a table containing 30 incidence signs) as an *ordinary Desargues configuration*.

Now if we consider the table Γ in Fig. 78, we see that we can insert a thirty first sign • into its upper left hand corner square and the *extended* incidence table still represents a geometrical configuration. However, if we insert, for instance, into the square in the lower right hand corner a 31st sign of incidence, the extended configuration is not geometrical. In this sense we can speak of *free* and *fixed* squares of the original table.

In Fig. 79 we indicate the free squares of the incidence table in Figs 77 and 78. For the time being we investigate only the possibility of extension by a single incidence sign. Notice that every row and every column of the table has just one free square.

By extending the original Γ-table by putting an incidence sign in the square emphasized by the sign "1" in Fig. 80 we obtain a *special Desargues configuration*.

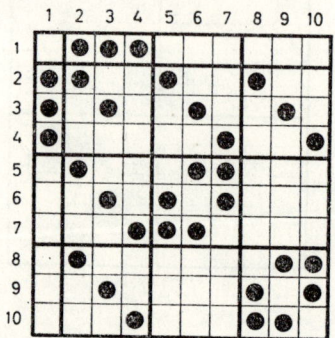
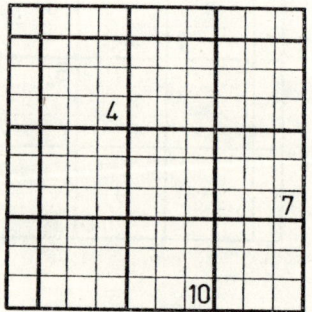

Figure 81

For the sake of brevity we shall call this a D_1-configuration. By comparing the models of the ordinary D-configuration and the special D_1-configuration we arrive at the following conclusion. In the former case there are three points of the configuration on every line of the configuration and three lines of the configuration pass through every point of it. In the latter case, the configuration has one point with four lines through it and one line with four points on it (these are emphasized on the model by double lines).

Of the 10 free squares of Fig. 79 we shall use the one numbered "1" for the purpose of extension. Actually, we would obtain the same kind of configuration by using any other free square. In fact, each of the squares numbered 2, 3, ..., 10 is carried in the upper left hand corner by a suitable elementary transformation leaving invariant the pattern of original 30 incidence signs shown in Fig. 78. — Let us mention as an example the case of the square "9". If we put the 31st incidence sign in this square and if we rearrange the rows of the table by the permutation

$$\sigma = (7, 9, 4, 8, 3, 10, 2, 5, 1, 6)$$

and the columns of the table by the permutation

$$\omega = (10, 6, 4, 5, 8, 1, 9, 3, 7, 2),$$

then the sign pattern so obtained is identical with the sign pattern shown in

Fig. 80. — That is, extension by one incidence sign gives a unique D_1-configuration.

Of course, there are fewer free squares in the incidence table of the D_1-configuration of Fig. 80, since by putting an incidence sign in square "1" we have deprived the squares 2, 3 and 4 of their freeness (Fig. 79). Thus further extension can only be accomplished by putting an incidence sign in one of the squares

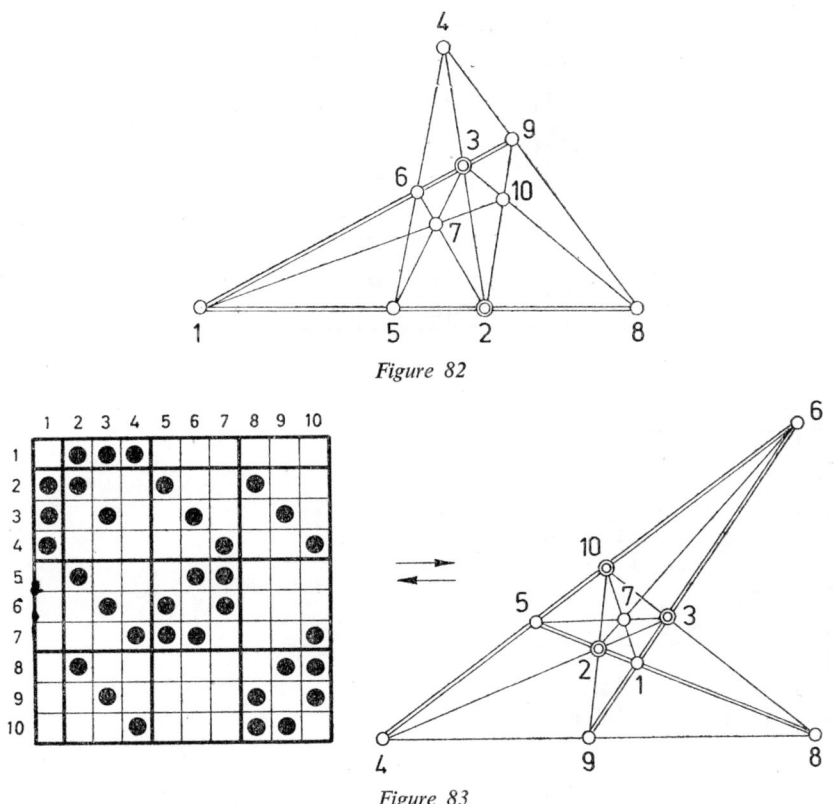

Figure 82

Figure 83

5, 6, 7, 8, 9, 10. Again, it is irrelevant which of the six free squares we choose, since the six cases can be changed by an elementary transformation in the form shown in Fig. 81. We have now a further loss of free squares, only the squares 4, 7 and 10 remain free for a further extension. This doubly extended configuration will be called a D_{-2}*configuration*.

Figure 82 gives a realization of this configuration. This has two points with four lines through each and two lines with four points on each.

However, if we wish to make a further extension the result will depend on which of the squares 4, 7 and 10 we choose. More exactly, the roles of the squares

7 and 10 can be interchanged, but extension by using the square 4 is not equivalent to extension by using squares 7 or 10.

Let us investigate firstly the extension using the square 7. If an incidence sign is placed in square 7 then the freeness of 4 is also lost, and so only square 10 remains for a further extension (Fig. 83). This non-symmetric incidence table can be realized on the classical projective plane.

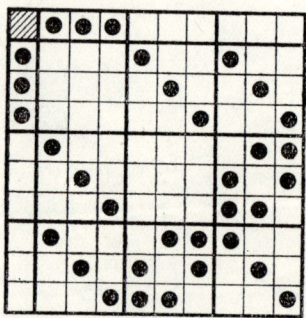

Figure 84

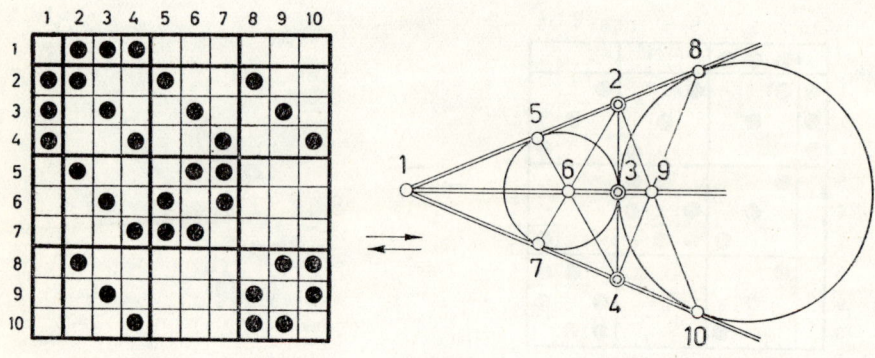

Figure 85

To see the incidence relations more clearly we can transforms the incidence table into a symmetric form.

This can be done by performing permutations of rows and columns as follows.

$$\sigma = (10, 4, 5, 6, 1, 9, 8, 7, 2, 3), \quad \omega = (7, 4, 8, 9, 1, 6, 5, 10, 2, 3).$$

The figure so derived is called a D_3-configuration, it has three special points with four lines through each and three special lines with four points on each. The symmetric incidence table can be seen in Fig. 84.

If we extended the D_2-configuration by putting an incidence sign in square 4 then the freeness of both squares 7 and 10 is lost: we obtain thus a *saturated* configuration. (Saturated in the sense that it cannot be further extended into a geometrical configuration: Fig. 85.) The sub-table formed by the intersection of the first seven rows and columns is nothing else but a Fano configuration (a finite plane of order two), and from this it follows that the configuration cannot be realized by points and lines on the classical projective plane.

This saturated configuration is called a D_3°-*configuration*. This, too, has three lines of four points on each and three points with four lines through each. These, however, form a different figure to points and lines of the D_3°-configuration with the same property. We can also say that the D_3°-configuration is obtained as an extension of the Fano-configuration (F-configuration).

We return now to the incidence table of the D_3-configuration represented in Fig. 84. The upper left hand square of this table is still free. If we put an incidence sign in it, the table becomes saturated (Fig. 86). This D_4-*configuration* possesses four lines with four points on each and four points with four lines through each.

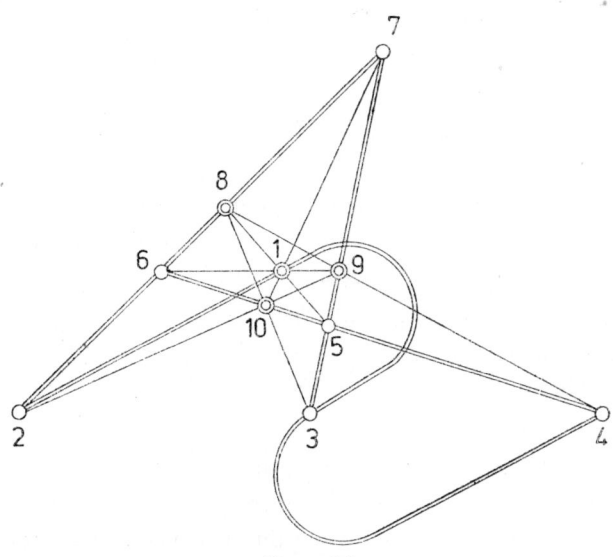

Figure 86

The D_4-configuration cannot be realized on the classical projective plane.

Assume that it is realizable of the plane π and on the ambient 3-dimension space. Let S be any point not lying in π. Project D_4 from S onto a plane parallel to the plane spanned by S and the line $l = \{2, 3, 4\}$. Let the image of the point k of the configuration be denoted by k'. Since, according to our assumption $1 \in l$, each of the images $1', 2', 3', 4'$ will be an ideal point. Thus the sides of the image-triangle $8' 9' 10'$ and the sides of the circumscribed image-triangle $5' 6' 7'$

are pairwise parallel (Fig. 87). Consequently, the common point $1'$ of the lines $5'8'$, $6'9'$, $7'10'$ is the centroid of each of the two triangles, i.e. it is not an ideal point. This contradiction proves our statement.

If we denote the original Desargues configuration by D_0 then the configurations occurring in Figs 78, 80, 82, 83, 86 and 87 are D_0, D_1, D_2, D_3^o and D_4^o, respectively. The lower index signifies the number of incidences in excess of

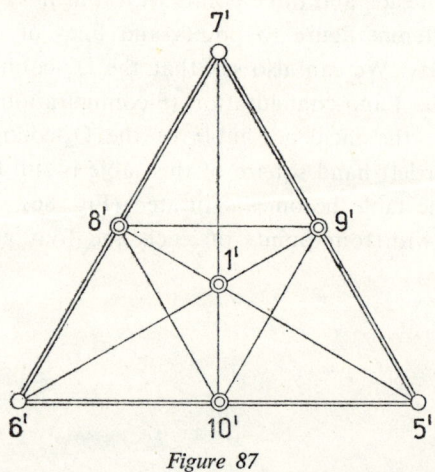

Figure 87

the original thirty and the upper index denotes which configurations are saturated. By a "D-configuration" we shall understand one of the configurations $D_0, D_1, D_2, D_3, D_3^o$ and D_4^o.

3° A line containing four points in a D-configuration is said to be a *singular line* the point through which four lines pass is said to be *singular point* of the configuration. The singular elements themselves form a sub-configuration as follows:

D_1 A point and a line passing through it.

D_2 Two points and one line passing through each.

D_3 Three points and the lines passing through them, which form a perspective pair of triangles with a non-singular point and a non-singular line as centre and axis, respectively.

D_3^o Three points on a line and three concured lines, one through each of the points.

D_4^o The singular elements of the D_3-configuration and their centre and axis, which are also incident.

We have seen already that D_3^o and D_4^o are not realizable on the classical projective plane but does there exist a plane on which they are realizable? We shall show that on a Galois plane of order four there exists a D_3^o-configuration and on a Galois plane of order 3 there exists a D_4^o-configuration.

Consider Fig. 7 (p. 27). Let us choose the rows with indices

$$1, 2, 3, 6, 7, 8, 10, 11, 14, 16$$

and the columns with indices

$$1, 2, 3, 4, 6, 7, 8, 10, 11, 12.$$

Let us permutate these rows and columns as follows:

$$6, 1, 2, 3, 7, 10, 11, 8, 14, 16$$

and

$$1, 2, 6, 10, 3, 7, 11, 4, 8, 12$$

respectively. After this rearrangement, the intersection of the rows and columns forms a 10×10 table having a sign pattern identical with that of Fig. 85.

And if we choose from Fig. 8. *iii* (p. 30) the rows

$$1, 2, 3, 4, 6, 7, 8, 9, 11, 12$$

and the columns

$$1, 2, 3, 4, 5, 7, 9, 10, 12, 13$$

and if we permutate them as follows:

$$1, 2, 3, 4, 7, 9, 12, 6, 8, 11$$

and

$$1, 2, 3, 4, 5, 10, 13, 7, 9, 12$$

respectively, the sign pattern of the 10×10-sub-table formed by the intersection of these rows and columns has precisely the incidence properties of Fig. 85.

3.4 The concept of geometrical nets

In section **1.5** we saw how, in the Euclidean model of a finite plane, the points could be parametrized by two families of lines namely, those parallel to the *x*-axis and those parallel to the *y*-axis. We also introduced a "diagonal" family of lines of the finite plane, which were represented by curves on the Euclidean model. This configuration suggests the following definition:

By a *net* we shall mean a set of n^2 elements, called points, satisfying the following conditions:

(i) There exist three families of subsets ξ, η and ζ each consisting of n subsets called lines, and each line consisting of n points.

(ii) Any two lines of different families have one and only one of the n^2 points in common; any two distinct lines of the same family have no point of the net in common. (So each family contains all of the points of the net.)

(iii) Given any point of the net there exists uniquely a line of the family such that the given point is the common point of these lines.

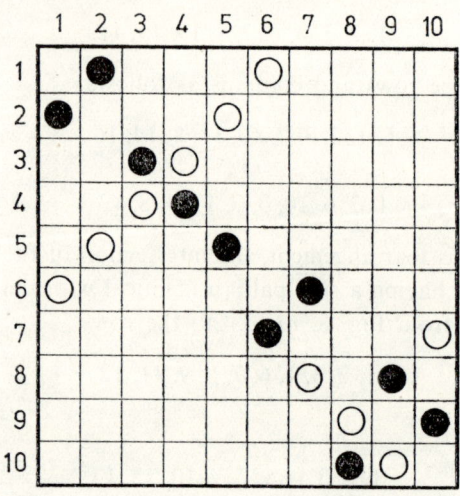

Figure 88

We can represent the structure of a net on an $n \times n$ table as follows: Consider the squares of the table as the points of the net, let its rows be the subsets in the family ξ and its columns be the subsets in the family η, then the subsets in the family ζ can be interpreted as a set of n naturally distinct diagonals of the table. (By a diagonal we mean n squares of the table such that there is precisely one square is each of the rows and each of the columns.) Thus for instance in Fig. 88 the dark and the light signs emphasize two disjoint diagonals.

If we write down the column indices of the squares covered by a diagonal in the order of the rows, we obtain a permutation, describing uniquely the diagonal. — Thus the dark and the light diagonals of Fig. 88 can be expressed by the permutation

$$2, 1, 3, 4, 5, 7, 6, 9, 10, 8$$

and

$$6, 5, 4, 3, 2, 1, 10, 7, 8, 9$$

respectively. If the elements in each family of the net are indexed by $1, 2, \ldots, n$, then by writing in each square of the table the index of diagonal containing it, it is easy to see that we obtain a Latin square.

A numerical table forming a Latin square can be interpreted in various ways as an operation table. By writing in the left hand margin and in the top margin a permutation of the indices $1, 2, \ldots, n$ (not necessarily the same in each case) we obtain an operation table in the following sense: If the index s stands in the square of intersection of the row of index j and the column of index k, then we

∘	1	2	3	4
1	1	2	4	3
2	4	3	1	2
3	2	4	3	1
4	3	1	2	4

U

∘	1	2	3	4	5
1	1	2	3	4	5
2	2	3	1	5	4
3	5	4	2	1	3
4	3	5	4	2	1
5	4	1	5	3	2

V

Figure 89

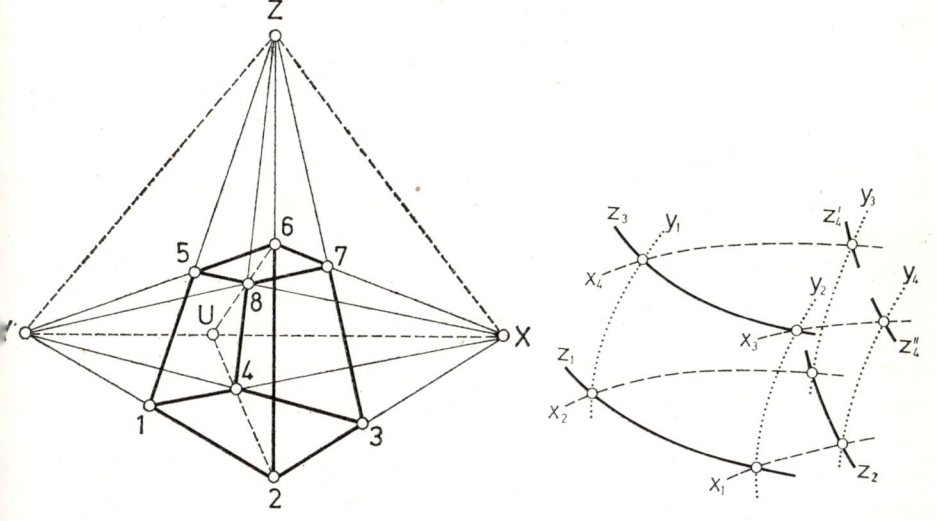

Figure 90

introduce the operation ∘ and write $j \circ k = s$. Thus, consider the tables U and V of Fig. 89. The operation of the table U does not possess either a left or a right neutral element. Table V has a neutral element but only a left neutral element, namely the element 1. If we were to write the permutation 1 2 5 3 4 in the left hand column of this table we would obtain an operation table which has a right neutral element, too, namely the element 1 again.

Consider the configuration shown in Fig. 90 which is similar to a central projection of a parallelepiped. — There is a special case of this figure, namely when the points X, Y, Z are collinear, but we do not give this figure. — If we omit the point U, the three lines meeting in the point U and the lines XZ and YZ, but draw the line 37 passing through point Z, we obtain a configuration consisting of 11 point and 12 lines. This configuration is called a Reidemeister configuration and is denoted by R. If X, Y and Z collinear, we speak of a micro-Reidemeister configuration denoted by R_0.*

If four lines of each of the families ξ, η, ζ of a net can be chosen with the same incidence structure as the three pencils of four lines in the R_0-configuration we speak of an R_0-configuration of the net.

We may now ask, whether *an* R-*configuration or an* R_0-*configuration exists on some projective plane.* This question is closely connected with the validity of the theorem of Desargues and the particular case of it when the centre of the pair of triangles lies on the axis, this special theorem will be called the micro-Desarguesian theorem and will be denoted by MD.

Consider three distinct points X, Y and Z of a plane. Choose any other point 1 that is not collinear with any two of the first three points. Let us choose one further point on each of the lines $1X, 1Y$ and $1Z$. Let these be denoted by 4, 2 and 5, respectively. Let us consider connecting lines and points of intersection as follows:

$$2X, 4Y, \quad 2X \cap 4Y = 3; \quad 5Y, 2Z, \quad 5Y \cap 2Z = 6;$$

$$5X, 4Z, \quad 5X \cap 4Z = 8; \quad 6X, 8Y, \quad 6X \cap 8Y = 7.$$

Now, with exception of the line 37 we have constructed either an R or an R_0 configuration, *provided line 37 passes through the point Z.*

If both the D- and the MD-theorems hold on the plane, then as the triangles 124 and 568 are in perspective from the point Z it follows that the points 4, X and Y are collinear. But now, because the triangles 243 and 587 are in perspective from the axis XY it follows that the point Z lies on the line 37, too. Hence the R-, respectively the R_0-configuration exists on other planes apart from the classical projective plane.

(Reidemeister condition (see Fig. 90): Consider any subsets of nets: $\{x_1, x_2, x_3, x_4\}, \{y_1, y_2, y_3, y_4\}$ and $\{z_1, z_2, z_3\}$; if the points $x_3 \cap y_3$ and $x_3 \cap y_4$ are such that if $z'_4 = x_4 \cap y_3$ and $z''_4 = x_3 \cap y_4$, then $z'_4 = z''_4 (=z_4)$. If the Reidemeister condition is satisfied in a net, then it is called an R-net.)

---

* If in the orthogonal projection of a parallelepiped the sides have 12 distinct image lines, then the image of the parallelepiped forms a R_0-configuration.

We return now to the tables U and V of Fig. 89 and we shall show the essential difference between them, mentioned already above. If we consider the Latin squares established by the operation tables as nets, we may put the question, how can the MR-configuration be interpreted on the table.

The operation table defines over the set $\{1, 2, \ldots, n\} = N$ an operation $x_j \circ y_k = z_l$, where $x_j \in N, y_k \in N$ and $z_l \in N$, i.e. in the square of the intersection of the row of index x_j and the column of index y_k lies the element z_l. The definition of the R net is the case of the operation table can be expressed as follows:

If
$$x_1 \circ y_2 = x_2 \circ y_1, \quad x_1 \circ y_4 = x_2 \circ y_3 \quad \text{and} \quad x_3 \circ y_2 = x_4 \circ y_1,$$
then
$$x_3 \circ y_4 = x_4 \circ y_3.$$

This follows easily from the fact that the line of parameter $c (\in N)$ of the families ξ, η, ζ of the net is formed by the squares belonging to a fixed index $c (\in N)$ and satisfying the conditions $x = c, y = c$ and $x \circ y = c$, respectively.

If we consider the Latin square as a matrix, the property MR can also be expressed in a simple manner as follows:

If the matrix has a submatrix
$$\begin{pmatrix} a & b \\ c & d \end{pmatrix}$$
then in every one of its submatrices
$$\begin{pmatrix} a & b \\ c & * \end{pmatrix}, \quad \begin{pmatrix} c & * \\ a & b \end{pmatrix}, \quad \begin{pmatrix} b & a \\ * & c \end{pmatrix}, \quad \begin{pmatrix} * & c \\ b & a \end{pmatrix}$$
the element in the place of the $*$ is d. We shall denote this fact by speaking of a *matrix with the rectangle-property*.

We can now see by checking the operation tables U and V of Fig. 89 that the former possesses a matrix with the rectangle-property but the latter does not possess any. We can, for instance, form the first and the second rows of V a matrix $\begin{pmatrix} 1 & 4 \\ 2 & 5 \end{pmatrix}$ and from the second and third rows another matrix $\begin{pmatrix} 1 & 4 \\ 2 & 3 \end{pmatrix}$ and $3 \neq 5$.

3.5 Groups and R nets

A finite group can be realized by the so called Cayley table. In general, it is far from easy to decide whether a given operation table does or does not realize a group. Especially, the establishing of the associativity of the operation defined by the table requires a close and lengthy investigation, it is also difficult to determine when two structures given by tables are isomorphic. The existence and uniqueness of inverses and neutral element are more easily verified. If the number of the elements of the underlying set is q, then the verification of the property $(ab)c = a(bc)$ requires $4q^3$ separate searches in the table, since for each of the q^3 triples a, b, c we have to search in turn for the elements $ab, (ab)c, bc$ and $a(bc)$.

We shall now give an application of the idea of the operation table considered as a net, to the problem of the associability of the operation. Let us agree to call the operation multiplication and its neutral element the unit element. Further let the indices of the rows and of the columns be in turn the integers $1, 2, \ldots, q$, where 1 is the unit element. Thus we can speak of the operation table of a *loop*, i.e. of a quasi-group with unit element. Let us see what the meaning of the property (R) is with respect to the operation, i.e. the rectangle property of the table.

1° An operation table satisfying the (R) condition possesses the following property:

If

(3.5.1) $\qquad x_2 = y_2 \quad \text{and} \quad x_3 y_2 = x_4 \quad \text{and} \quad x_2 y_3 = y_4$

then

$$x_3 y_4 = x_4 y_3$$

which follows by putting $x_1 = y_1 = 1$ in the equation defining the property (R) given in the last section.

From (3.5.1) we have

$$x_3(x_2 y_3) = x_3 y_4 = x_4 y_3 = (x_3 y_2) y_3 = (x_3 x_2) y_3.$$

Since the relation $x_3(x_2 y_3) = (x_3 x_2) y_3$ so obtained holds for any triple (x_2, x_3, y_3) of elements of the underlying set $\mathbf{Q} = \{1, 2, \ldots, q\}$ the operation is associative. Thus, if the operation table of the quasi-group $\mathbf{Q}$ with unit element, considered as a net has the property (R), then the quasi-group is a group.

2° Let us now start from the assumption that the operation table defines a group over the underlying set $\mathbf{Q}$. In the case of a group we have

$$(xy)^{-1} = y^{-1} x^{-1}.$$

Suppose that $x_3 y_2 = x_4 y_1$ and $x_1 y_2 = x_2 y_1$ and $x_1 y_4 = x_2 y_3$. From $x_1 y_2 = x_2 y_1$ we obtain, by the uniqueness of inverses.

Therefore from the equations above we have

$$(x_3 y_2)(y_2^{-1} x_1^{-1})(x_1 y_4) = (x_4 y_1)(y_1^{-1} x_2^{-1})(x_2 y_2)$$

and hence

$$x_3 y_4 = x_4 y_3.$$

We deduce that the operation table of the group considered a net has the property (MR).

The results obtained in 1° and 2° are summarized in the following

Theorem: *The operation table of a quasi-group with unit element has as a net the rectangle property, if and only if the quasi-group is a group.*

It is necessary to specify that the quasi-group has a unit element for U consider the example U given in Fig. 89. It is easy to verify that the table has the rectangle property but it does not have a unit element; moreover, it is not associative as is seen from

$$(2 \cdot 3)4 = 1 \cdot 4 = 2, \quad 2(3 \cdot 4) = 2 \cdot 1 = 4.$$

3.6 Problems and exercises to Chapter 3

26. Formulate, what kind of special conditions determine the structure of the pentagon configuration corresponding to the configuration D_0, D_1, D_2, D_3, D_3° D_4°.

27. Determine, by means of the $\Gamma(4)$-table in Fig. 7 how many Desargues configurations there are on the finite projective plane of order four.

28. If we consider the definition of the $\Gamma(q)$-table given in **1.4** and the properties D_1, D_2, D_3, D_4 implied by it, then we can readily see that we obtain a Latin square in the following manner.

In the parcel $C^{1,k}$ replace every incidence sign by the number k ($k=0, 1, \ldots \ldots, q-1$) and then superimpose these parcels one on another. Thus we obtain a Latin square of order q, let it be denoted by **L**.

29. If we write out the indices $1, \ldots, q-1$ in the left hand margin and in the upper margin of the Latin square **L** obtained in the previous example we get the operation table of a quasi-group. But this is not necessarily the operation table of a quasi-group with neutral element (i.e. of a *loop*); let it be denoted by **Q**.

Prove by suitably permuting the columns and the rows that the incidence table can be rearranged into a form which still satisfies the definition of the $\Gamma(q)$-table and such that **Q** belonging to it defines a loop.

30. Consider the operation table of **Q** occurring in the previous exercise. Prove that this defines a group if and only if the theorem of Desargues holds on the plane defined by $\Gamma(g)$ in all cases when the axis is the line l_1 and the centre is one of the points $\{P_1, P_2, \ldots, P_{q+1}\}$.

N.B. Remembering the non-Desarguesian plane corresponding to the $\Gamma^*(9)$-table determined by Fig. 45 (p. 94): and by the table V of Fig. 45, we can readily check by means of the row of parcel $C^{10}, C^{11}, \ldots, C^{18}$ that this is a plane with the property dealt with in Exercise 30. Furthermore, we have shown in the notes added to the Exercises of Chapter 1, that on this plane there exists a perspective pair of triangles for which the D-theorem does not hold. By considering the line l_1 as the ideal line, this plane is a so-called *translation plane*.

CHAPTER 4

SOME COMBINATORIAL APPLICATIONS OF FINITE GEOMETRIES

In this chapter we shall apply some of the properties of finite projective planes to the solution of certain combinatorial problems. We shall deal with some extremum-problems of combinatorial analysis, furthermore with the realization of extremal geometric figures (graphs) giving the solution of these problems. The first section deals only with some background material and can be omitted without affecting the reader's understanding of the later parts of the book. In the section dealing with the notation of graphs we outline only the most necessary facts concerning a graph as a geometric structure.

4.1 A theorem of closure of the hyperbolic space

Let a and b be two lines in the space of Bolyai and Lobachewsky. If a and b are mutually perpendicular we shall denote this fact by writing $a \times b$. If $a \times b$, then we say that the two lines form a *cross*. Let the lines g_1, g_2, g_3, g_4 be pairwise skew and let $n_{j,k}$ denote the normal transversal of the pair of lines g_j, g_k. There are altogether six normal transversals and we have the following

Theorem: *If* $n_{12} \times n_{34}$ *and* $n_{13} \times n_{42}$, *then* $n_{14} \times n_{23}$. *This holds also in the Euclidean space.*

The hypothesis of the theorem speaks of the crosses formed by $4+6=10$ lines, since a normal transversal forms crosses with two g lines and the theorem states that the existence of 14 crosses implies the existence of the fifteenth cross.

The diagram shown in Fig. 91 gives in a clear form the combinatorial structure of the 15 crosses formed by the 10 lines of the figure corresponding to the theorem. The point G_j represents the line g_j of our figure, the point N_{jk} the line n_{jk} of the figure; the *edge*, i.e. the line segment, joining the two points represents the cross-forming relation of the corresponding two lines; the *points* of the diagram are represented by small circles. This figure is the so-called Petersen graph. — It is easier to draw the graph then to describe it. We shall discuss the structure of this graph later, but let us now return to the proof of the theorem. In the proof we shall refer to well-known theorems of classical projective geometry and to theorems concerning Poincare's spherical model of hyperbolic geometry.

We start with the Poincaré model. The surface of the sphere is projected from one of its points onto a plane. The points of this plane represent complex numbers. The model of a line of hyperbolic space is an arc lying inside the sphere and meeting the sphere perpendicularly; the two end points of the arc are not considered as points of the line. Now, if the complex number a_1 and a_2 correspond to the

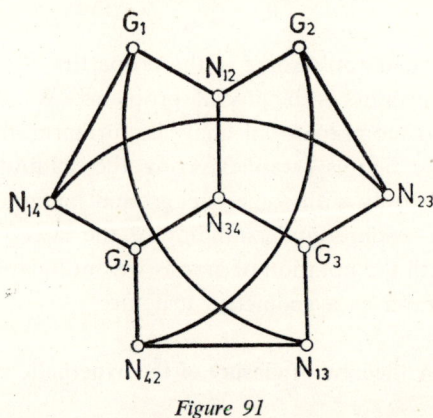

Figure 91

projections of the endpoints of the arc representing a line a, then the unordered pair (a_1, a_2) of complex numbers determines the line a uniquely. The lines a: (a_1, a_2) and b: (b_1, b_2) are said to lie on the same plane of the hyperbolic space if and only if the points of the sphere corresponding to the complex numbers $a_1, a_2; b_1, b_2$ lie on the same circle of the sphere and the two lines are said to cut each other perpendicularly if the following cross ratio condition holds:

$$(a_1 a_2 b_1 b_2) = -1.$$

This can also be written in the form

$$(a_1+a_2)(b_1+b_2) = 2(a_1 a_2 + b_1 b_2).$$

We could also consider a model in the complex Cartesian coordinate plane, i.e. the set of pairs of complex numbers.

Let

$$A: (a_1 a_2, a_1+a_2) \quad \text{and} \quad B: (b_1 b_2, b_1+b_2)$$

be two points of the plane, where a_1, a_2, b_1 and b_2 satisfy the equation above. Consider now on the plane the parabola defined by the equation

$$y^2 = 4x.$$

We understand by a conjugate pair of points with respect to this parabola a pair of points P': (x', y'); p'': (x'', y'') satisfying the equation

$$y'y'' = 2(x'+x'').$$

Obviously, the points A and B, above, are conjugate points with respect to the parabola. Thus in this model a pair A, B of points conjugate with respect to the parabola correspond to the cross formed by the pair of the lines a and b in the first model, and conversely.

The point G_1 corresponding to the given line g_1 forms a conjugate pair of points with each of the points N_{12}, N_{23} and N_{14}, corresponding to the lines n_{12}, n_{13} and n_{14} respectively. Hence the latter three points lie on a line, on the polar of point G_1. Similarly, the triples N_{12}, N_{13}, N_{42}; N_{13}, N_{23}, N_{34}; N_{14}, N_{42}, N_{34} of points each lie on one line, namely on the polars of the points G_2, G_3 and G_4, respectively. Thus the six N_{jk} points are the six vertices of the complete quadrilateral formed by the poles of the lines G_j.

The requirements $n_{12} \times n_{34}$ and $n_{13} \times n_{42}$ of the theorem signify that the diagonal pairs of vertices N_{12}, N_{34} and N_{13}, N_{42} of the complete quadrilateral are conjugate pairs of points. From this, however, it follows by a well-known theorem of Hesse that the third pair of vertices is also a conjugate pair. And this fact implies that the relation $n_{14} \times n_{23}$ is satisfied and the theorem is proved.

The validity of this theorem in Euclidean space can be deduced from the hyperbolic case by a limiting procedure. Hence the theorem is an absolute theorem in the sense of Bolyai.

4.2 Some fundamental facts concerning graphs

In the present section we shall introduce the notion of a graph in the narrower sense, i.e. notion of the so-called geometric graph.

We shall understand by a *graph* a figure consisting of *points* and of *edges* connecting the points and we shall assume that the following conditions are valid:

1° The graph has points.
2° An edge can join two distinct points only.
3° Two points are joined by at most one edge.

A graph like this is shown in Fig. 91.

If a graph has no edge it is said to be *empty*. If any two (distinct) points of a graph are joined by edges, it is said to be *complete*. The latter can be realized, in the case of a finite graph, i.e. if the number of points is finite, by the set of the vertices of a regular n-gon and by the line segments joining the vertices pairwise. The number of points of a finite graph is called the *order* of the graph, the number of the edges will be called the *class* of the graph. — Thus Fig. 91 is a graph of order ten and of class fifteen. If we delete edges from a complete graph of order n, we may arrive at every graph of order n.

Given any two points A and B of the graph. A *path* is said to exist connecting A with B if and only if there exists a set of distinct points $\{P_1, ..., P_s\}$ and edges connecting A with P_1, P_1 with P_2, ..., P_{s-1} with P_s, P_s with B. A graph is said to be *connected* if there exists a path connected any two distinct points of the graph. The number of the edges forming a path is said to be the *length* of the path. A path connecting a point with itself is called a *cycle*. — The shortest cycle of the graph shown in Fig. 91 consists of 5 edges and the longest cycle has 9 edges. If k edges meet in a point of a graph, then the point is said to be of *degree k*. If every point of the graph is of degree k, the graph is said to be a *regular graph of degree k*. — Our example in Fig. 91 is a regular graph of degree three. — The length of its shortest cycle is called the *girth* of a graph. — The girth of the graph in our example is $l=5$. — The girth of a graph is either equal to 0, or is $l \geq 3$ (because of the defining properties 2 and 3). A *tree* is a non-empty connected graph which does not contain a cycle; hence the girth of a tree is 0. If a cycle passes through every point of a graph the cycle is called a Hamilton *circuit* and the graph a Hamilton *graph*.

If we can remove a point of a connected graph so that the remaining graph is not connected, then the point in question is said to be an *articulation point* of the graph. — Our example given in Fig. 91 has no articulation point. — If a tree possesses at least two edges, then it has an articulation point: the points of a tree which are not articulation points are said to be endpoints of the tree. Each point of a tree can be connected to any other by a unique path. In fact, if two paths existed connecting two points then the two paths together would either form a cycle or would contain a cycle as a part. — The length of the path connecting two points is said to be the *distance* between them.

If the points of a graph can be subdivided into two non-empty classes so that an edge joins only points belonging to different classes, then the graph will be called a *bipartite graph* of an *even graph*. — The graph in Fig. 91 is bipartite, the two classes being distinguished by dark and light colouring.

Clearly, a graph may be represented by many different models each of which exhibits the essential properties of the graph, i.e. the discernment of the points and the existence or nonexistence of an edge connecting two points. Consequently, it may be most difficult to decide whether two models represent the same graph or not. Thus, for instance, the two graphs occurring in Fig. 92 are essentially the same as the graph in Fig. 91, i.e. they are Petersen graphs.

The concept of isomorphism is indispensable in graph theory. A transformation φ, mapping the set of the points of a graph Γ upon the set of the points of a graph Γ' in a one-to-one and "connection preserving" manner, is said to be an *isomorphism*, in the case of a mapping φ of a graph onto itself an *automorphism*. — By *connection preserving* we mean that two points are connected

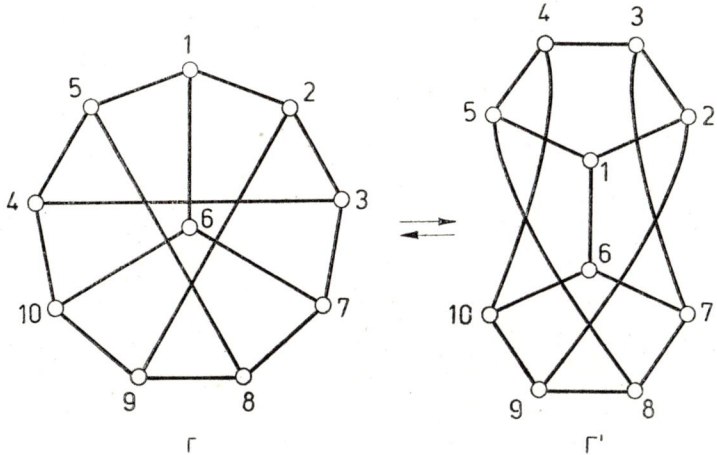

Figure 92

by an edge if and only if their images under are connected by an edge. If a mapping φ exists between two graphs they are said to be *isomorphic*, otherwise they are *heteromorphic*. Clearly, isomorphism is an equivalence relation on the set of graphs. — It is easy to check that the examples given in the figure show isomorphic graphs. — Isomorphic graphs are not considered as different in graph theory.

It is seen readily that the above mentioned numbers and properties (the connectedness of the graph, the regularity, the order, the class, the girth of the graph, the degree of a point of the graph, the length of a path, etc.) are invariants of the graph under isomorphism. Furthermore, it is easy to prove that *the automorphisms of a given graph form a group*. If we index the points of a finite graph of order n by the set of integers $\{1, 2, \ldots, n\}$, then an automorphism of the graph can be expressed as a permutation of $\{1, \ldots, n\}$. — It is an interesting problem to identify the automorphism group of a graph, for instance the Petersen graph. This, however, will not be treated here.

In attempting to identify a particular graph we may so tumble on interesting theorems. For instance, we have already mentioned that the Petersen graph is easier to draw than to describe it. However, let us try to characterize it in terms of the invariants we have introduced.

The Petersen graph (Fig. 91) is a regular graph, its degree is $k=3$, its girth $l=5$. These two properties are common with many graphs; for instance consider the vertices and the edges of a regular dodecahedron as the points and the edges of a graph. This graph is obviously a regular graph of degree 3 and girth 5, just like the graph of our example in the figure but they are clearly heteromorphic

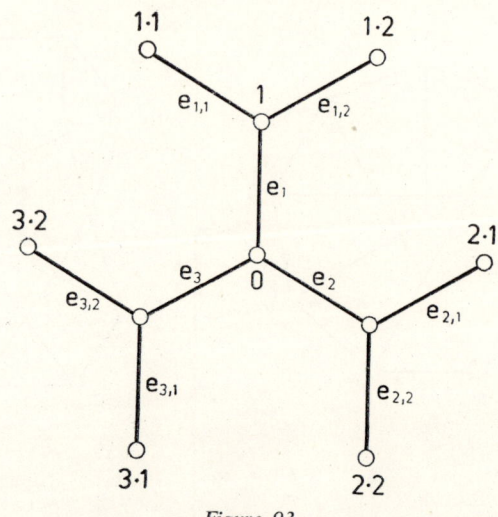

Figure 93

since the order of the one is $n=10$ and that of the other is $n=20$. Let us find the minimal order of a regular graph of degree 3 and girth 5. Let this minimum number be denoted by $f(3, 5)=n$.

Any graph with $k=3$ has a point lying on three edges in Fig. 93 these are the point G and the edges e_1, e_2 and e_3 the other endpoints of these edges being the points 1, 2 and 3, respectively. Every one of these three points must lie on two further edges and the endpoints of these render further six points of the graph, namely the points $1 \cdot 1, 1 \cdot 2; 2 \cdot 1, 2 \cdot 2; 3 \cdot 1$ and $3 \cdot 2$. The 9 edges, above, must exist and the subfigure formed by them must be a tree, since $l=5$. Thus any graph satisfying the conditions $k=3$ and $l=5$ has at least $1+3+6=10$ points, that is $f(3, 5) \geqq 10$.

But the graph in Fig. 91 has order 10, hence $f(3, 5)=10$.

Let us denote the graphs satisfying $f(3, 5)=10$ by $(10; 3, 5)$, let the number of isomorphism classes be denoted by $g(3, 5)=r$. By means of Fig. 93 it is easy to see that $g(3, 5)=1$.

In order to see this consider a point of degree one of the tree occurring in the figure, say the point $1 \cdot 1$. This point is the starting point of two further edges of which one is connected to one of the points $3 \cdot 1$ and $3 \cdot 2$, and the other is connected to one of the points $2 \cdot 1$ and $2 \cdot 2$. It is clear that these four possibilities are essentially the same; for example, if $1 \cdot 1 - 3 \cdot 1$ were an edge then by rotating the broken line through 180° about the axis, which would induce an isomorphism carrying the edge $1 \cdot 1 - 3 \cdot 1$ into the position occupied by an edge $1 \cdot 1 - 3 \cdot 2$. Hence we assume without loss of generality that the edges $1 \cdot 1 - 1 \cdot 1 - -2 \cdot 2$ and $1 \cdot 1 - 3 \cdot 2$ meeting at the point $1 \cdot 1$ belong to the graph in question. Continuing in this way, remembering the assumption $l=5$, we obtain the edges $2 \cdot 2 - 3 \cdot 1, 3 \cdot 2 - 2 \cdot 1, 2 \cdot 1 - 1 \cdot 2$ and $3 \cdot 1 - 1 \cdot 2$. So we arrive at the figure occurring in Figure 91.

Hence, up to isomorphism, *the* Petersen *graph is the only regular graph of degree three of minimal order whose smallest circuit consists of five edges.*

4.3 Generalizations of the Petersen graph

We shall denote by $\Gamma(n; k, l)$ a regular graph of degree k and girth l, having the minimal number $n=f(k, l)$ of points. Of course for an arbitrary pair of integers k and a graph may not exist. However, when a graph exists satisfying the two conditions then the number $f(k, l)=n$ is determined and another problem presents itself, namely the determination of $g(k, l)=r$ (i.e. the number of isomorphism the classes). In general, little is known about the possible values of n and r or about the realization of extremal graphs of this kind. We shall now mention a few particular cases.

1° If the value of (k, l) is either $(3, 7)$ or $(3, 8)$, then an extremal graph exists.
2° For these cases we have $f(3, 7)=24$ and $f(3, 8)=30$.
3° We also gave $g(3, 7)=g(3, 8)=1$.

We shall call a graph $\Gamma(n; k, l)$, by analogy with the graph discussed in the last section, a (generalized) Petersen graph. Before attempting to establish general theorems for these graphs we shall deal with a special case which gives a useful insight into the structure of these graphs.

Consider the question of the existence of a graph satisfying to the assumption $k=3, l=6$ $(n; 3, 5)$.

First of all, if such a graph exists then consider one of its edges (on Fig. 94 the edge 1—1*). Let one of the points joined by this edge be light and the other dark. We start from this edge and progressing to the right and to the left symmetrically we build up according to the assumptions $k=3, l=6$ a tree, forming a part of the graph, until the distance between the "free endpoints" reaches the value $6-1=5$. At each stage of the construction, the endpoint of a new edge is given the opposite colour to the beginning point of the edge. Thus because of $l=6$ it is obvious that the graph in question contains the tree which can be seen on Fig. 94.

This tree has 7 dark and 7 light points, hence $f(3, 6)=n\geq 14$. We can now extend the graph without introducing additional points, e.g. the insertion of the new edge $2\cdot 1 - 3\cdot 2^*$ does not impair the requirement $l=6$, completing as

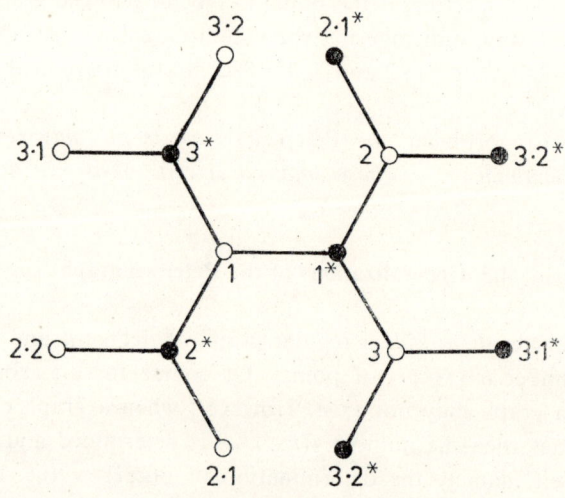

Figure 94

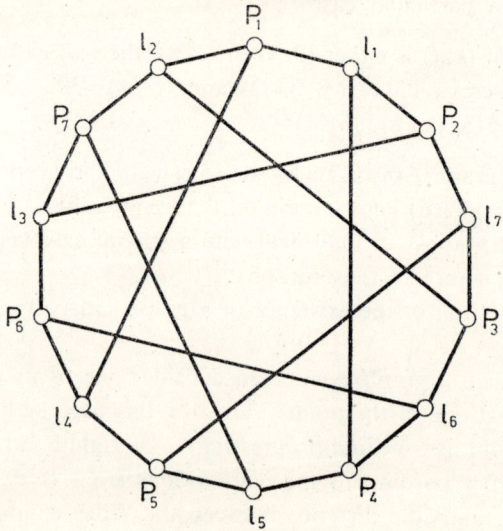

Figure 95

it does a cycle consisting of precisely six edges. Hence we arrive at the graph shown in Fig. 95 which satisfies the definition of $\Gamma(14; 3, 6)$, as can be checked by examining the number of the edges of the closed paths, therefore $f(3, 6) = 14$.

Let the light points be denoted by l_j and the dark points by P_k ($j, k = 1, ..., 7$), thus we can establish a correspondence between Fig. 95 and Fig. 2 (p. 19). If the points P_k of this extremal graph represent the points of a plane and the points l_j of the graph represent the lines of the plane, further if we interpret the connecting edges as the incidence relation, then, as is easily seen, *our graph is a model of the Galois plane of order two*.

Following this example we could conjecture that for $k = q+1$ and $l = 6$, where $q = p^r$ is any power of a prime, there exists an extremal graph $\Gamma(n; q+1, 6)$ and the order of this graph is $n = f(q+1, 6) \geq 2(q^2+q+1)$; furthermore, a graph of this kind is given by a model of the projective plane of order q.

The fact that $f(q+1, 6) \geq 2(a^2+q+1)$ can be derived in a way similar to the case of $k = 2+1 = 3$.

Consider the symmetrical tree constructed as follows: Given an edge 1—1*. Construct q further edges radiating from each of the point 1 and 1*. Thus we obtain $2q$ free endpoints, from each of these let q further edges radiate. The farthest distinct between free endpoints of this tree is 5 and the total number of its points is $2+2q+2q^2$.

Clearly if a graph with $k = q+1$ and $l = 6$ exists then it must contain a tree of the type above, therefore $f(q+1, 6) \geq 2(q^2+q+1)$.

Now we can construct a graph from the incidence table of a projective plane of order q, as follows: Let us take q^2+q+1 "dark" points and index them by the columns of the table and the same number of "light" points and index them by the rows of the table. Let a dark and a light point be connected by an edge if and only if the corresponding row and column intersect in a square occupied by an incidence sign. Obviously, we obtain thus a regular graph of degree $(q+1)$ consisting of $2(q^2+q+1)$ points. In this graph only points of opposite colour are joined, thus it is a bipartite graph. Clearly, a cycle of an odd number of edges cannot exist. Furthermore, a cycle of four edges does not exist either, since otherwise the incidence table would not satisfy axioms l_1 and l_2. Hence the length of the smallest cycle cannot be less than 6. According to axiom l_3, a triangle exists on a finite projective plane on the graph this triangle corresponds to a cycle of six edges. Thus, comparing this with the previous result, we see that $f(q+1, b) = 2(q^2+q+1)$. Thus we have proven the existence of an extremal graph $(2(q^2+q+1), q+1, 6)$.

The graph $(2q^2+2q+2, q+1, 6)$ represented by the table $\Gamma(q)$ is a model of the plane determined by this table. Conversely, we can construct a $\Sigma(q)$-table from any graph $\Gamma(2(q^2+q+1), q+1, 6)$ and this table determines a projective plane of order q. The graph and the table are equivalent models. A particularly regular figure occurs in the case of an *incidence graph* of the Galois plane. (The name "incidence graph" is used to underline the equivalence of the graph with the incidence table.)

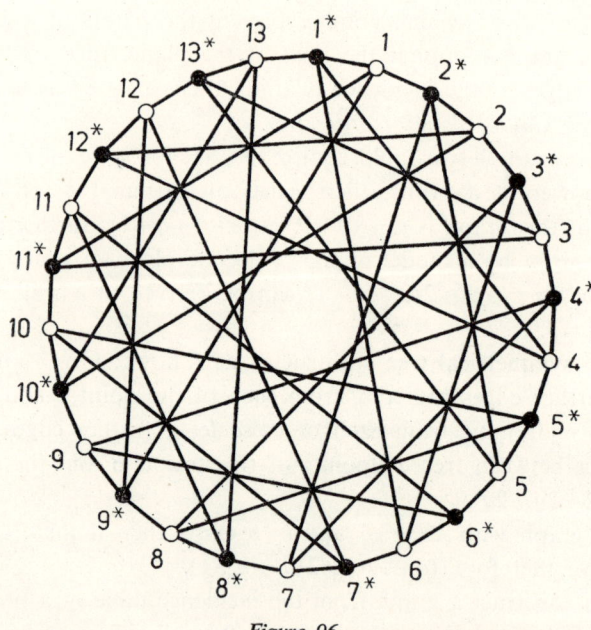

Figure 96

Due to a well-known number theoretic theorem of Singer, the incidence table of any Galois plane can be rearranged into a cyclic sign matrix, as was already mentioned in connection with figures 5 and 6 of Section **1.3**. In this case the incidence graph can be constructed from the Ω-table of the plane. We may make a rearrangement so that the two squares lying in columns P_1 and P_2 of the row l_1 of the table both contain incidence signs. This can be done since otherwise the polygon Λ (cf. Fig. 5, p. 23) would not have a side subtending an angle of ω. — The incidence graph can be constructed from this Ω-table, as follows (Fig. 96):

1° We colour the vertices of a regular $2(q^2+q+1)$-gon alternatively dark and light; these will be the points of the graph.
2° We choose an orientation and index the light points by $1, 2, \ldots, q^2+q+1$ in this order. We denote the dark points in a similar manner, by $1^*, \ldots, (q^2+q+1)^*$, where 1^* is the point falling between the light points of index 1 and 2. We assign to the row l_j of the Ω-table the light points of the graph

with index j, and to the column P_k the dark point of index k^*. An incidence is represented by the edge (chord) connecting the corresponding vertices of the polygon.

This symmetrical form of the extremal graph $(2(q^2+q+1), q+1, 6)$ corresponding to a Galois plane is not only a bipartite graph, it is also a Hamilton graph. Summing up the results of our discussions we can state the following

Theorem: *The incidence graph of a Galois plane is a regular, bipartite Hamilton graph of the minimal order, of degree $(q+1)$, and of the girth 6.*

4.4 A combinatorial extremal problem

We discussed an extremal problem in the preceding section to determine the *minimum number n* of points of a graph with given degree k and girth l. We shall now formulate a similar problem, in which we have to determine the *maximum number of edges* of a graph satisfying the same conditions. This problem is a geometric interpretation of a problem due to Zarenkiewicz, namely the following:

For a given integer n consider the bipartite graphs consisting of n dark and n light points those which do not possess a subgraph consisting of four edges and of two dark and two light points. Determine the greatest number of edges that a graph satisfying this requirement can have; let this number be denoted by $F(n)=e$. Also construct an (extremal) graph corresponding to the numbers n, e; such a graph will be denoted by $\Gamma(e, n)$.

It will be more convenient to use the *incidence matrix* of a graph of the above type, i.e. the following $n \times n$-matrix:

1° The rows of the matrix correspond to the light points of the graph, the columns to the dark points. Let the light points be denoted by P_j, the dark points by P_k^* where $j, k = 1, 2, \ldots, n$.

2° If two points P_j and P_k^* are connected by an edge of the graph let the element a_{jk} of the matrix be 1, otherwise let it be 0.

Our problem, rewritten in matrix language, is to find the maximum number of elements 1 in this matrix, given that a submatrix of the type

$$Z = \begin{pmatrix} 1 & 1 \\ 1 & 1 \end{pmatrix}$$

does not occur; and how can we construct a matrix of this type?

Let the row vectors of the matrix be denoted by $\mathbf{a}_j$, the column vectors by $\mathbf{a}_k^*$. According to the criterion forbidding the existence of the submatrix $\mathbf{Z}$ we have,

(4.4.1)   if $j \neq k$, then $\mathbf{a}_j \cdot \mathbf{a}_k = 0$ or 1 and $\mathbf{a}_j^* \cdot \mathbf{a}_k^* = 0$ or 1,

where $\cdot$ denotes the usual scalar product of vectors. The scalar product of a row or column vector with itself is a non-negative integer equal to the number of the 1's occurring that row or column.

Let these numbers be denoted by

(4.4.2) $\mathbf{a}_k^2 = \varrho_k$ and $\mathbf{a}_k^{*2} = \sigma_k$ $(k = 1, 2, \ldots, n)$.

Let the number of the 1's occurring in the matrix be denoted by η.
Clearly

(4.4.3) $\varrho_1 + \varrho_2 + \ldots + \varrho_n = \sigma_1 + \sigma_2 + \ldots + \sigma_n = \eta$.

We have now transformed the problem into an arithmetic form, namely: Given the number n, find the maximum of the number satisfying the conditions (4.4.1), (4.4.2), (4.4.3) and find an extremal matrix belonging to the number.

A clue to the solution of this arithmetical problem is to be found in the well-known inequality

(4.4.4) $\dfrac{1}{n}(c_1 + c_2 + \ldots + c_n)^2 \leq c_1^2 + c_2^2 + \ldots + c_n^2$,

where the c_k-s are non-negative integers and we have equality if and only if $c_1 = c_2 = \ldots = c_n$. Now

$$(\mathbf{a}_1 + \mathbf{a}_2 + \ldots + \mathbf{a}_n)^2 = (\mathbf{a}_1^2 + \mathbf{a}_2^2 + \ldots + \mathbf{a}_n^2) + 2(\mathbf{a}_1 \mathbf{a}_2 + \ldots + \mathbf{a}_{n-1} \mathbf{a}_n).$$

Obviously the left hand side is equal to

$$\sigma_1^2 + \sigma_2^2 + \ldots + \sigma_n^2 \geq \frac{\eta^2}{n}$$

and the second term, because of (4.4.1) cannot be greater than

$$n(n-1),$$

and is equal to this number if and only if the product of any two different row vectors is 1.

Therefore

$$\frac{\eta^2}{n} \leq \eta + n(n-1),$$

i.e.

$$\eta^2 - n\eta - n^2(n-1) \leq 0,$$

and from this follows for the value of the positive number η

(4.4.5) $$\eta \leq \frac{1}{2}(1 + \sqrt{4n-3})n.$$

Thus from the inequality mentioned above we have

(4.4.6) $$\sigma_1 = \sigma_2 = \ldots = \sigma_n.$$

Similarly we start from the square of the sum of the column vectors. We have the condition

(4.4.7) $$\varrho_1 = \varrho_2 = \ldots = \varrho_n.$$

Thus each of these conditions implies the other.

Of course, the validity of the conditions (4.4.6), (4.4.7) implies $\eta = n(1+\sqrt{4n-3})/2 = e$, if there exists the matrix corresponding to the prescriptions.

Assume that

(4.4.8) $$n = q^2 + q + 1 \quad (q \text{ is an integer}),$$

then from (4.4.5) obviously follows

$$\eta = (q^2 + q + 1)(q + 1).$$

By comparing the conditions (4.4.3), (4.4.6), (4.4.7) we see now that

$$\eta = (q^2 + q + 1)(q + 1) = e$$

can only be fulfilled if $(q+1)$ 1's occur in any of the rows and columns of the incidence matrix. Moreover, if there exists a projective plane of order q, then the incidence matrix derivable from its incidence table is just the one corresponding to the criteria above, when we write a 0 in the empty squares and a 1 the squares containing an incidence sign.

Thus when a plane of order q exists, the problem is completely solved: we have not only found the maximum number $e = (q+1)(q^2+q+1)$ but have also constructed a graph with maximum edge number.

4.5 The graph of the Desargues configuration

Consider now the relation of Fig. 75 to Fig. 92. The former is nothing else but the general Desargues configuration (realized by pentagons inscribed in each other).

Consider the following correspondence:

$$\begin{pmatrix} 1 & 2 & 3 & 4 & 5 & 6 & 7 & 8 & 9 & 10 \\ A_1 & A_2 & A_3 & A_4 & A_5 & A_6 & A_7 & A_8 & A_9 & A_{10} \\ a_1 & a_3 & a_5 & a_2 & a_4 & b_1 & b_2 & b_3 & b_4 & b_5 \end{pmatrix} : (\Gamma \to P \to l).$$

Now the edge joining — for instance — the point 4 with the point 10 of Fig. 92 has a double meaning: on the one hand it signifies the vertex A_4 in the Fig. 75 lies on the line $b_5 = B_2 B_3$, on the other hand that the line $a_2 = A_4 A_5$ passes through the vertex B_4.

In this sense we may say that the *Petersen graph is the incidence graph of the Desargues configuration*. In fact, the endpoints of the edges radiating from a point of the graph indicate precisely lines respectively points are incident with the corresponding points respectively lines of the configuration.

It must be emphasized that our graph is a *twofold* diagram of the configuration: if we consider any endpoint of any edge as an image of a point of the configuration, the other endpoint must then be considered as a line, when we interpret the edge as a model of the incidence.

The collineations mapping D onto itself form a group. Obviously, the investigation of this group can now be performed in terms of the automorphism group of the Petersen graph. Now, a cycle of s edges is mapped by an automorphism onto a cycle of s edges. The study of the cycles formed from the edges of the graph is here restricted to those of five edges, i.e. to the smallest ones. We claim that there are 12 of these, forming six pairs of disjoint pentagons. The vertices of the disjoint pentagons are joined by five pairwise *skew* (disjoint) edges. These five edges of the graph form a *spanning* system of minimal edge number of the graph (every point of the graph belongs at least to one edge of the system).

One of these decompositions is shown in Fig. 92. Suppose that the edges of the pentagon $(1', ..., 5')$ are coloured red, and those of the pentagon $(6', ..., 12')$ are coloured green. Let an edge joining a red point and a green point be coloured white, e.g. the edge $1' - 6'$.

Obviously, any further pentagon to be found in the graph has two white edges. The red ends of the white edges are joined either by a red edge or by a two-edged red path, whence the green ends are joined: either by a green path of two edges or by a green edge, respectively. There are five pentagons of each kind and they form 5 disjoint pairs, each pair consisting of a pentagon of each type. Thus, together with the red and green pentagons the total number of the pairs of pentagons of the graph is six.

We can investigate by similar reasoning the existence of larger cycles, too. It is interesting to note that the Petersen graph and the dodecahedron graph, which both have $k=3$, $l=5$, also have the same number of pentagons, namely 12.

4.6 Problems and exercises to Chapter 4

31. Prove that the theorem dealt with in Section 4.1 is equivalent to the following theorem of closure: Given three mutually skew lines which are not parallel to the same plane such that none of them is perpendicular to the two other, then consider the normal transversal of each line with the normal transversal of the other two then the three lines so obtained have a common normal transversal.

32. The following is also equivalent to the last problem: If each angle of two skew pentagons is a right angle and if no two of the ten sides are parallel, furthermore if we can assign to every side of the one pentagon a side of the other such that non-neighbouring sides should correspond to neighbouring sides, and if four of the pairs of the corresponding sides are lines intersecting each other perpendicularly, then the lines of the fifth pair are also intersect each other perpendicularly.

33. Prove the eixstence of the graph $(n; 3, 7)$; furthermore prove that $n=24$ and $r=1$. Construct a simple figure of this graph.

34. Prove the existence of $\Gamma(n; 3, 8)$ furthermore prove that $n=30$, and $r=1$. Construct a simple figure of this graph.

35. Prove that $\Gamma(86; 7, 6)$ does not exist.

36. Determine the number of cycles of each length occurring in the graph $\Gamma(10; 3, 5)$.

37. Prove that the group of the collineations mapping the general Desargues configuration onto itself is isomorphic to the permutation group on five elements.

We add a few hints as to the solutions of these problems, as some are very difficult and others rather laborious.

33*. (Hint at the Exercise 33). We label the vertices of the regular 24-gon inscribed in a circle by the indices 1, 2, ..., 24 ordered by one of the orientations. We draw the 12-chords connecting the following pairs of points: 1,18; 2,14; 3,10; 4,21; 5,17; 6,13; 7,24; 8,20; 9,16; 11,20; 12,19; 15,22.

These 12 chords and the 24 sides of the polygon furnish all the edges of the extremal graph in question.

34*. We consider the following 15 pairs formed from the vertices of the regular polygon 1, 2, ..., 30 inscribed in a circle:
1,18; 2,23; 3,10; 4,27; 5,14; 6,19; 7,24; 8,29; 9,16; 11,20; 12,25; 13,30; 15,22; 17,26; 21,28.
The 15 chords spanned by these pairs of points and the 30 sides of the polygon exhaust all the edges of the extremal graph in question.

35*. Since there is no projective plane of order six, the graph in question does not exist either.

36*. It is advisable to reason according to the figure Γ' of Fig. 92.

CHAPTER 5

COMBINATORICS AND FINITE GEOMETRIES

The initial development of the theory of finite geometries was by analogy with classical geometry. However, the subject was soon influenced by the combinatorial and algebraic considerations. In this chapter we shall discuss in more detail the connections between the theory of finite geometries and combinatorial analysis mentioned earlier.

5.1 Basic notions of combinatorics

Let $H = \{1, 2, \ldots, v\}$ be a finite set consisting of v elements (denoted by integers). The one-to-one mappings of the set H onto itself are said to be the permutations of H. The set of permutations of set H form a group. This group is called the *complete permutation group* of H and every subgroup of this is said to be a *permutation group*.

Let $c \in H$, a permutation group is said to be *transitive* over the set H if for every element $x \in H$ there exists a permutation such that the image x_c under is the element x.

The notion of transitivity can be made independent of the distinguished element c of the set H. Let x and y be two arbitrary elements of H. According to our assumption there exist $\sigma, \tau \in \Gamma$ such that $c^\sigma = x$, $c^\tau = y$. Let the inverse of σ be denoted by σ^{-1}, furthermore the composition of σ^{-1} and τ by $\sigma^{-1} \cdot \tau = \omega$. Since $\sigma^{-1} \in \Gamma$ and $\omega \in \Gamma$, further $(x^{\sigma^{-1}})^\tau = y$ the group has an element ω such that $x^\omega = y$.

If a permutation group Γ of the set H is not transitive over the set, i.e. it is *intransitive*, then H can be decomposed into subsets such that Γ is transitive over these subsets. Subsets of this kind are said to be *domains of transitivity* of the set H. If $a, b \in H$ then they belong to the same domain if and only if there exists a $\sigma \in \Gamma$ for which $a^\sigma = b$. In this case we may say that a *is joined to b by* Γ. It is easy to see that the relation "is joined to" is reflexive, symmetric and transitive; i.e. the domains of transitivity are *equivalence classes*.

The example in Figs 97 and 97* visualizes a permutation group Γ of the set $H = \{1, 2, \ldots, 13\}$; the elements of the group are $\sigma_1, \sigma_2, \ldots, \sigma_6$. If we consider the element 8 of the set, we can see immediately that the group Γ is intransitive over the set **H**. We can easily establish its domains of transitivity:

$$H_1 = \{8\}, \quad H_2 = \{2, 9, 11\}, \quad H_3 = \{6, 12, 13\},$$
$$H_4 = \{1, 3, 4, 5, 7, 10\}; \quad \text{and} \quad H = H_1 \cup H_2 \cup H_3 \cup H_4.$$

If we delete the columns 2, 6, 8, 11, 12 and 13 from the figure, the remaining table visualizes a group consisting of the 6 permutations of the set H_4. Consider the decomposition $H_4 = H_{41} \cup H_{42} \cup H_{43}$ of this set, where

$$H_{41} = \{1, 4\}, \quad H_{42} = \{3, 5\}, \quad H_{43} = \{7, 10\}.$$

This decomposition into disjoint subsets has the property that any of the permutations transforms each of these subsets into another subset (possibly itself).

	1	2	3	4	5	6	7	8	9	10	11	12	13
σ_1	1	2	3	4	5	6	7	8	9	10	11	12	13
σ_2	7	2	5	10	3	12	1	8	11	4	9	6	13
σ_3	4	9	7	1	10	13	3	8	2	5	11	12	6
σ_4	3	9	10	5	7	12	4	8	11	1	2	13	6
σ_5	10	11	1	7	4	13	5	8	2	3	9	6	12
σ_6	5	11	4	3	1	6	10	8	9	7	2	13	12

Figure 97

	σ_1	σ_2	σ_3	σ_4	σ_5	σ_6
σ_1	σ_1	σ_2	σ_3	σ_4	σ_5	σ_6
σ_2	σ_2	σ_1	σ_4	σ_3	σ_6	σ_5
σ_3	σ_3	σ_5	σ_1	σ_6	σ_2	σ_4
σ_4	σ_4	σ_6	σ_2	σ_5	σ_1	σ_3
σ_5	σ_5	σ_3	σ_6	σ_1	σ_4	σ_2
σ_6	σ_6	σ_4	σ_5	σ_2	σ_3	σ_1

Figure 97*

If in the case of a group Γ transitive over a set **H** the set can be decomposed into at least two disjoint subsets from which at least contains more than one element and any element of the group Γ maps each of the subsets onto another subset, then the group Γ is said to be *imprimitive*. If the set **H** does not have a decomposition like this, then the group Γ is *primitive*. Any transitive group over a set **H** containing a prime number of elements is *primitive*.

We shall understand by the *degree* of a permutation group of a set **H** the number $|\mathbf{H}|=v$. If the permutation group Γ is transitive over the set **H**, then the order of the group, the number $|\Gamma|=g$ is divisible by the degrees of the group. — This can easily be proved and is left to the reader as an exercise. — If $g=v$, the group Γ is then said to be *regular*.

The transitive Abelian group is always regular. — The proof of this is also left to the reader.

We shall now introduce a fairly recent development in combinatorial analysis which is indispensable from the point of view of finite geometries.

Let $\mathbf{B}_1, \mathbf{B}_2, \ldots, \mathbf{B}_b$ be a set non-empty subsets (called *blocks*) of the set $\mathbf{H}=\{1, 2, \ldots, v\}$ satisfying the following:

1° The number of the elements of the set **H** is a given natural number v.
2° The number of the blocks is a given natural number b.
3° Every block consists of the same number of element. Let this number be k.
4° Every element of the set **H** belongs to the same number of blocks, let this number be r.
5° Every element of the set **H** belongs to the same number of blocks, let this number be λ.

A set of blocks $\{\mathbf{B}_1, \ldots, \mathbf{B}_b\}$ satisfying these conditions is called a *block design*. Of course, the existence or uniqueness of a block design with given parameters v, b, k, r and λ is a question which must be investigated. — For instance, let $v=21$, $b=21$, $k=5$, $r=5$ and $\lambda=1$. In this case we can establish not only the existence but also the uniqueness of the block design. Consider namely the projective plane of order 4. Let the set of points of this plane be taken as the set **H** and consider the lines of the plane as the blocks. The given parameter values are the specifications of this and only of this plane. We know that a plane of this kind exists and is unique, namely, the Galois plane of order four (Fig. 7, p. 27).

The five parameters are not independent of each other. In fact, a block design can be represented by an incidence table, as was already mentioned with respect to Fig. 71. We can count the number of the incidence signs contained in the table in two ways; namely be the columns which gives $v \cdot r$ or by the rows which gives $b \cdot k$. Thus

(5.1.1) $$v \cdot r = b \cdot k.$$

Now, the number of pairs of incidence signs in a row of the table is $\binom{k}{2}$; thus the number of pairs of this type, contained in the table is $b \cdot \binom{k}{2}$. There are λ rows, each of which intersect a given pair of columns in squares containing incidence signs, hence enumeration by columns gives $\lambda \cdot \binom{v}{2}$. From these two results we have $bk(k-1) = \lambda v(v-1)$. By comparing this with (5.1.1) we obtain the relation

(5.1.2) $$\lambda(v-1) = r(k-1).$$

The conditions (5.1.1) and (5.1.2) are necessary but not sufficient conditions of the existence of a block design. — In fact, these two conditions are satisfied by $v=43$, $b=43$, $v=7$, $k=7$, $\lambda=1$, though we know that a projective plane of order six does not exist. —

We have seen already, namely by the example of the Galois plane of order four, that a block system exists for which $v=b$ but then because of (5.1.1) we must have $k=r$. In such a case we speak of a *symmetric* (v, k, λ)-*block design*. We may also call this a *symmetric* (v, k, λ)-*configuration*. The necessary condition of the existence of a block design like this is (5.1.2) and, because of the symmetry the equality

(5.1.3) $$\lambda(v-1) = k(k-1).$$

These are satisfied by every projective plane of order q, when $v=q^2+q+1$, $k=q+1$, $\lambda=1$. —

Without going into details we mention that a necessary condition of the existence of a symmetric (v, k, λ)-configuration was known some twenty years ago and it was conjecture that this condition is also sufficient; this conjecture, however, could not be proved until recently. This condition is the following:

Given an even number v, let the equation

(5.1.4) $$z^2 = (k-\lambda)x^2 + (-1)^{\frac{v-1}{2}} \cdot \lambda y^2$$

have a solution in integers, x, y, z. The verification of the conjecture proves to be a most difficult problem.

If we consider the definition of the symmetric (v, k, λ)-configuration, a natural generalization presents itself, i.e. the so-called $t-(v, k, \lambda)$-*block design*. This is defined by the following conditions:

1° The set **H** consists of v elements.
2° Every block is a subset consisting of k elements.
3° Any subsets of t elements is contained in exactly λ blocks.

This generalization is far from being artificial, on the contrary, it occurs quite naturally in the theory of finite planes; as we shall see in the following sections.

5.2 Two fundamental theorems of inversive geometry

We can obtain the so-called Möbius plane (inverse plane) by extending the Euclidean plane with a single ideal point and by considering this new point as incident with every line of the plane. Having done this, we do not distinguish between circles and straight lines, both are now called circles. Thus we arrive at a unified formulation of the theorems of the Euclidean geometry concerning the intersection of lines and circles.

The structures of the Möbius plane and of the sphere are "identical", in the sense that the following *stereographic mapping* of a sphere onto a Möbius plane exists. Let the axis of rotation of the sphere Γ passing through an arbitrary point U of it be the line t and let Σ be an arbitrary Möbius plane which is perpendicular to t; the point $P \neq u$ of the sphere and the point P' of the plane correspond to each other, if and only if the line PP' passes through the point U; and the point U of the sphere corresponds to the ideal point of the plane. This mapping is circle- and angle-preserving — circles touching each other are carried into circles touching each other and the spherical circles passing through point U are carried into the lines of the plane. It is this property of stereographic mappings which suggested that we should consider the lines of the plane as circles, if we extend the Euclidean plane by an ideal point.

In the past century, much was written about the so-called *associated point-octet*, which can be derived as follows: Three quadrics, which do not have a curve in common, can have exactly eight common points; the figure consisting of these eight points is called an associated point-octet, briefly $\mathbf{R}(8)$. In considering the set of the quadrics passing through the points of the figure $\mathbf{R}(8)$, we may find among them a sphere Γ and quadric surfaces degenerated into a pair of planes. The study of this special case led to two fundamental theorems of inversive (or real inversion) geometry. Now we shall given an explicit construction of this special $\mathbf{R}(8)$.

Consider the *closed chain* formed by the circles c_1, c_2, c_3, c_4 of the sphere Γ, by which we understand that

the circles c_1 and c_2 cut each other in the points A_1 and B_1,
the circles c_2 and c_3 cut each other in the points A_2 and B_2,
the circles c_3 and c_4 cut each other in the points A_3 and B_3, and
the circles c_4 and c_1 cut each other in the points A_4 and B_4.

Let the plane of the circle c_k be denoted by γ_k. The four planes are — in general — the four faces of a tetrahedron. Let C_k denote the vertex of the tetrahedron opposite to the face γ_k. In fact, this is an associated point octet, since the eight points can be obtained as the common points of three quadrics, namely: the sphere Γ, the pair of planes γ_1, γ_3, and the pair of planes γ_2, γ_4.

The description of the special $\mathbf{R}(8)$ can be made without the notion of the closed chain of circles in the following: Let a non-coplanar quadruple (C_1, C_2, C_3, C_4) of points have the property that each of the lines C_1C_2, C_2C_3, C_3C_4 and C_4C_1 each cut the sphere Γ in two points. The eight points so obtained form the figure $\mathbf{R}(8)$.

The two kinds of derivation leads to the same point figure on the sphere. But the derivation by means of the closed chain of circles can also be interpreted on the Möbius plane.

The following is a degeneration of the special $\mathbf{R}(8)$: The planes γ_k are distinct and let the quadruple A_1, A_2, A_3 and B_3 lie on a plane γ. This can occur only if the lines A_1B_1, A_3B_3 and A_2B_2 meet in the point C (the common point of the planes γ_2, γ_3 and γ). Since the line A_4B_4 is the line of intersection of the plane γ_1 incident with the line A_1B_1 and of the plane γ_4 incident with the line A_3B_3, it must also pass through the point $C = A_1B_1 \cap A_3B_3$. Thus, the degeneracy consists of the fact that the lines A_kB_k ($k=1, 2, 3, 4$) meet in the point C, which can obviously be taken as the ideal point of the space, thus the tetrahedron C degenerates into four planes meeting at the point C. Because $C \in A_kB_k$ ($k=1, 2, 3, 4$) the quadruple of points (A_2, B_2, A_4, B_4) is, clearly, also coplanar. Naturally, four coplanar points on a sphere on a spherical circle. Thus we proved the following

Theorem: *If we have four point pairs* (A_k, B_k) *determined by a closed chain of circles on a sphere and* A_1B_1, A_3, B_3 *lie on a circle, then the points* A_2, B_2, A_4, B_4 *also lie on a circle.*

This theorem also holds on the Möbius plane, i.e. it is a theorem of inverse geometry, and can easily be proved by means of stereographic mapping. This theorem of inverse geometry will be called briefly as the N-*theorem*.

Let us now return to the special $\mathbf{R}(8)$-configuration described by means of a general tetrahedron C. There is a classical theorem that states that *if the points* A_k ($k=1, 2, 3, 4$) *are coplanar, then the points* B_k *are coplanar.*

By interpreting this statement in terms closed chains of circles we arrive at a theorem of inversive geometry, called the *Theorem of Miguel*, briefly M-theorem.

M-Theorem: *For any four point pairs* (A_k, B_k) *determined by a closed chain, the four points* B_k *are concircular if and only if the four points* A_k *are concircular.*

This theorem follows immediately from a strong result concerning the associated point-octet by applying a stereographic projection. Here, however, we shall give an elementary proof (Fig. 99).

Take an arbitrary sphere and project the figure onto the sphere from the point of the sphere farthest from the plane of the figure. (We have to prove from the existence of the circle c^* containing the points quadruple B_1, B_2, B_3 and B_4.)

Two fundamental theorems

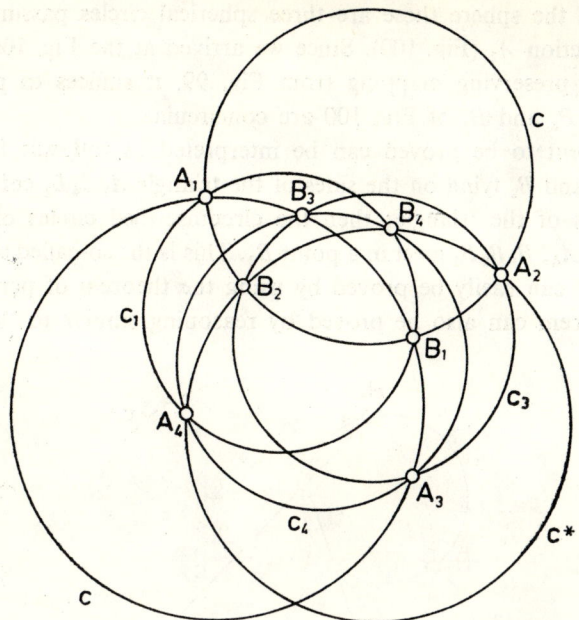

Figure 98

Now the spherical figure is again stereographically mapped from the point A_1 of the sphere onto a plane and thus we again obtain a figure of the same structure as the original figure, but now the "circles" c_1, c_2, c will be straight

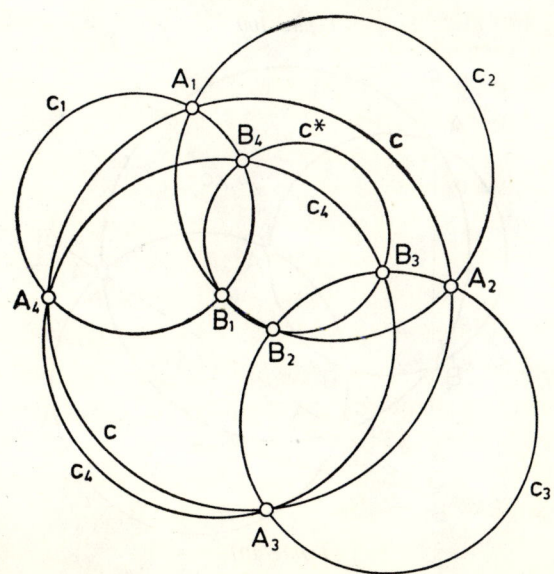

Figure 99

lines, since on the sphere these are three spherical circles passing through the point of projection A_1 (Fig. 100). Since we arrived at the Fig. 100 by two consecutive circle preserving mapping from Fig. 99, it suffices to prove that the points B_1, B_2, B_3 and B_4 of Fig. 100 are concircular.

This statement to be proved can be interpreted as follows: If none of the points A_3, B_4 and B_2 lying on the sides of the triangle $A_2 A_4 B_1$ coincide with one of the vertices of the triangle; then the circumscribed circles of the triangles $B_2 A_2 A_3$, $B_4 A_3 A_4$, $B_1 B_2 B_4$ meet in a point B_3. This is the so-called special theorem of Miguel and can easily be proved by using the theorem of peripheral angles.

The N-theorem can also be proved by reasoning similar to the above.

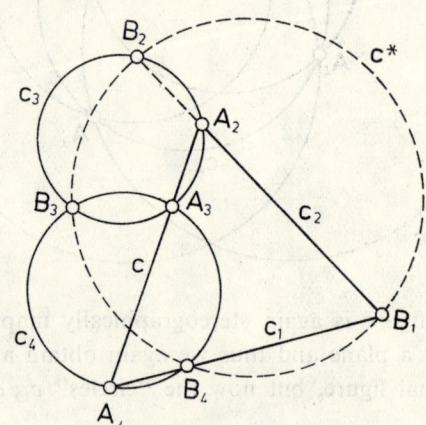

Figure 100

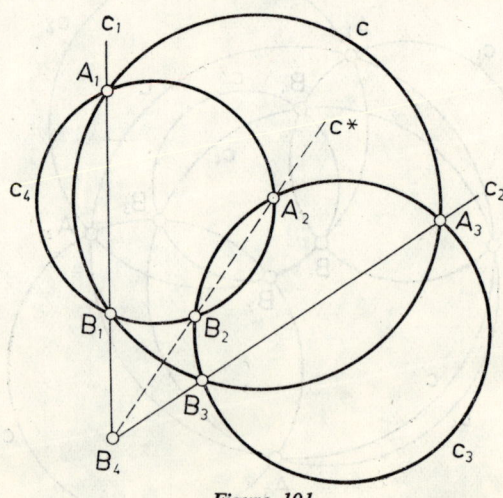

Figure 101

Consider Fig. 98, we have to prove that a circle c^* exists containing the points A_2, B_2, A_4 and B_4. The first step is again a transfer of the structure of this planar figure onto a sphere Γ and then to map it from the point A_4 of the sphere by a second stereographic projection onto a plane. Thus we arrive at the projection shown in Fig. 101, i.e,. in the Euclidean sense, both c_1 and c_2 are straight lines. In considering our plane as a Möbius plane, we have to prove that the points A_2, B_2, B_4 are (in an Euclidean sense) collinear, as A_4 is the ideal point of the plane. But this follows from the theorem concerning the line of powers of three circles: B_4 is the power of the circles c_3, c_4, c.

We emphasize the common idea in the two latter proofs: from the validity of the special M-theorem and of the special N-theorem (Fig. 101) we can deduce the general M- and the general N-theorem by the twofold application of stereographic projection.

We shall now investigate whether finite planes analogous to the inversive plane exist and if theorems analogous to the M-theorem and the N-theorem are valid.

5.3 Finite inversive geometry and the t-(v, k, λ) block design

We have already seen an interesting example of a block design t-(v, k, λ), namely the finite projective plane, where $t=2$, $v=q^2+q+1$, $k=q+1$ and $\lambda=1$. Now we shall introduce the example corresponding to the case $t=3$, $v=q^2+1$, $k=q+1$ and $\lambda=1$ formulated geometrically. As a concrete example, let us choose the case $q=3$. Consider the incidence table shown in Fig. 102, where the columns are called the points $P_0, P_1, \ldots, P_9$, respectively; and the rows are called the circles $c_1, c_2, \ldots, c_{30}$, respectively.

It is easy to check that this incidence structure consisting of 10 points and 30 circles has the following properties:

K$_1$ *Any circle has points.*
K$_2$ ***A*** *unique circle is incident with any given triple of points.*
K$_3$ *If $P_j \in c_r$ and $P_h \in c_r$, then there exists a unique circle $c_s \neq c_r$ for which $P_j \in c_s$ and $P_h \in c_s$.*
K$_4$ *There exist 4 points which are not concircular.*

The axioms K$_1$, K$_2$, K$_3$, K$_4$ together form the *axiom system* **K**.

A plane satisfying the axiom system **K** is called *inversive plane, or Möbius plane*. If the number of the points of the plane is finite — as in the present case — then we speak of a *finite Möbius plane*. Now, the following questions may be asked:

1° How can incidence tables, similar to the above example, be constructed?

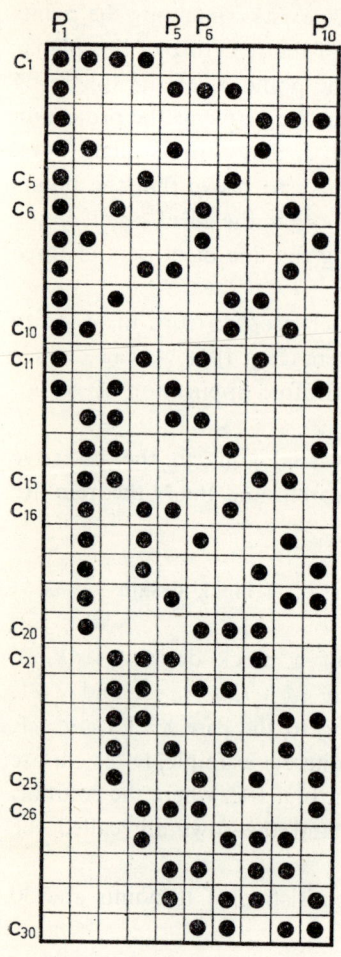

Figure 102

2° What properties do inversive planes have in common?

We shall not give detailed answers to these questions, rather we shall give a program for studying these planes.

1° Let p be an odd prime number. Consider the three dimensional Galois space over the coordinate field $\mathbf{K} = GF(p^r)$ and at the same time consider classical projective space. In the latter the equation

(5.3.1) $$x_1^2 + x_2^2 + x_3^2 = x_4^2$$

corresponds to the unit sphere. If the sphere without the point U: $(0, 0, 1, 1)$ is projected from this point U onto the (fundamental) plane determined by the equation $x_4 = 0$, then we have performed a stereographic mapping of the sphere. If we interpret this equation over the coordinate field $\mathbf{K}$, then we have a quadric in the Galois space, called an ovoid. — We understand by *quadric* the finite geometric figure corresponding to the surface of order two and if the quadric contains no straight lines and consists of more than one point, it is said to be an *ovoid*. The plane sections of an ovoid are ovals. Any point of the ovoid not on the plane $x_4 = 0$ can be used as the point of projection U. We shall call the projections of the plane sections of the ovoid onto the fundamental plane $x_4 = 0$, "*circles*".

We have constructed the example shown in Fig. 102 in this manner. Starting from the ovoid (5.3.1) over the coordinate field $GF(8)$ project the points of the ovoid from the points U: $(0, 0, 2, 1)$ onto the plane formed by the points $(x_1, x_2, x_3, 0)$. The reader can easily check that the points of the ovoid (beside the point U) are the following:

$Q_1(0, 0, 1, 1)$, $Q_2(0, 1, 0, 1)$, $Q_3(0, 2, 0, 1)$, $Q_4(1, 0, 0, 1)$;

$Q_5(1, 1, 1, 0)$, $Q_6(1, 2, 1, 0)$, $Q_7(2, 0, 0, 1)$, $Q_8(2, 1, 1, 0)$,

$Q_9(2, 2, 1, 0)$.

Denoting the projection of the point Q_j by P_i it is obvious that $Q_5=P_5$, $Q_6==P_6$, $Q_8=P_8$, $Q_9=P_9$; and the other points are:

$$P_1(0,0,1,0), \quad P_2(0,1,1,0), \quad P_3(0,2,1,0), \quad P_4(1,0,1,0), \quad P_7(2,0,1,0).$$

If we go over to inhomogeneous coordinates by (x_1, x_2) instead of $(x_1, x_2, 1, 0)$ for the points P_j we have:

$$P_1(0,0), \quad P_2(0,1), \quad P_3(0,2);$$
$$P_4(1,0), \quad P_5(1,1), \quad P_6(1,2),$$
$$P_7(2,0), \quad P_8(2,1), \quad P_9(2,2).$$

Thus we obtain the points of the affine plane of order three. And this extended by a single ideal point P_0 (corresponding to U) is the Möbius plane of order three. (Cf. Fig. 8 *iii*, p.30 and the first 12 rows of Fig. 102).

Next we compute the subsets lying on the same planes of the ovoid $\{U, Q_1, Q_2, \ldots, Q_9\} = \Omega$. Since a line has not more than two common points with an ovoid, it suffices to consider the planes determined by the triplets of points of Ω. Every such plane has four points in common with the ovoid and if we project these from the point U onto the fundamental plane we obtain the sets of four points forming the circles of the Möbius plane. The incidence table of the points and circles of this plane is given in Fig. 102.

2° The notion of an ovoid in the space $A_{3,q}$ can be formulated independently of algebraical considerations. We shall not discuss the case $q=2^r$. — We shall understand by an ovoid a subset of q^2+1 points of the projective space of order q such that no three of its points are collinear. (Cf. Sections 2.7 and 2.13).

The number of points common to a plane and an ovoid cannot be greater than the number of points of an oval of the plane of order q, i.e. it cannot be greater than $q+1$. However, it cannot be less than $q+1$ either. For consider the planes passing through the points Q_j and Q_k ($j \neq k$) of an ovoid; the line $l_{jk}=Q_j Q_k$ and every point of a line of the space which is skew to l_{jk} each determine one plane, hence the number of the planes incident with the line l_{jk} is $q+1$. Let the number of the points of the ovoid be in turn $\pi_1, \pi_2, \ldots, \pi_{q+1}$. With respect to the points Q_j, Q_k clearly

$$\pi_1 + \pi_2 + \ldots + \pi_{q+1} = (q^2+1) + 2q = (q+1)^2.$$

But $\pi_j \leq q+1$, and thus it follows

$$\pi_1 \leq \pi_2 \leq \ldots \leq \pi_{q+1} = q+1.$$

Hence the plane containing two points of the ovoid contains exactly $q+1$ *points of the ovoid.*

The number of planes passing through one point of an ovoid and containing other points of it is $q^2(q+1)/q = q^2+q$. And since the number of the planes incident with a point of the space is q^2+q+1, one through the point in question plane *touches* the ovoid (i.e. it contains a single point of it).

The number of such *tangent* planes to the ovoid is clearly q^2+1. But, the number of the planes containing several points of the ovoid given by the quotient of the number of the point triples of the ovoid by the number of point triples of an oval, namely:

$$\binom{q^2+1}{3} \bigg/ \binom{q+1}{3} = q^3 + q.$$

Hence the number of the planes containing points of the ovoid is $(q^2+1) + (q^3+q) = q^3+q^2+q+1$ which is the total number of planes of the space.

Thus every plane of the space either cuts an ovoid in $q+1$ points or touches it.

This is an unexpected fact, since in the classical projective space only the quadrics which are ruled have the property that every plane has points in common with them.

The following properties can easily be established and are left to the reader:

Every line exterior to an ovoid is incident with two tangent planes and with $q-1$ planes cutting the ovoid in ovals.

A line touching an ovoid in a single point is incident with a single tangent plane and with q planes cutting the ovoid in ovals.

As was already mentioned the line cutting the ovoid in two points in incident with $q+1$ planes cutting the ovoid in ovals.

These three kinds of pencils of ovals show again a close analogy with the elliptic, parabolic, hyperbolic pencils of circles of the sphere of classical geometry.

Any point of the ovoid is contained in $q(q+1)$ ovals of the ovoid. There are $q(q^2+1)$ ovals on the ovoid. The number of the ovals disjoint from a given oval of the ovoid is $q(q-1)(q-2)/2$ and the number of the ovals touching the given oval is q^2-1.

Thus we have given a combinatorial interpretation to the notion of the ovoid and have displayed completely the incidence structure of the points and ovals on it by elementary (combinatorial) arguments. Since any three points of the ovoid determine exactly one oval and since the ovals of the ovoid as point sets can be considered as blocks, an ovoid is the realization of a block design $3-(v, k, \lambda)$, if $v = q^2+1$, $k = q+1$, $\lambda = 1$.

The relationship between inversive geometry, i.e. Möbius plane of order q and the block design $3-(q^2+1, q+1, 1)$ is easily seen by the fact that the Möbius plane with its circles is, because of the stereographic mapping an isomorphic figure with the "sphere" of the equation (5.3.1) of the Galois space of order

q with its "spherical circles". And this sphere with its circles is nothing else but an algebraically derived ovoid with its ovals.

Thus we have infinitely many finite planes satisfying the axiom system **K**, which therefore deserves investigation.

5.4 General theorems concerning the Möbius plane

In comparing the examples of Figs 102 and 101 we see that if we assign the lines c_1, c_2, c_3, c_4, c of the latter to the rows $c_1, c_3, c_{30}, c_{22}, c_{23}$ of the former, then the columns $P_4, P_6, P_0; P_3, P_7, P_9, P_1$ correspond to the points of intersection $A_1, A_2, A_3; B_1, B_2, B_3, B_4$. Furthermore, the row c_2 of Fig. 102 corresponds to the line c^* passing through the three points A_2, B_2 and B_4 that is, a theorem analogous to the N-theorem of classical inversive geometry holds for the figure in Fig. 102.

Now we may ask, whether or not the M-theorem or the N-theorem hold on the inversive plane defined combinatorially. We shall only briefly quote the results known about this problem.

If on the Möbius plane the Theorem of Miguel holds, the plane is said to be a *Miguelian plane* or briefly an M-*plane*.

The finite inversive plane is Miguelian plane, if and only if it is isomorphic with a non-ruled quadric of the Galois space. The points are the points of the quadric, the circles its plane sections which contain more than one points of the quadric.

In a space of odd order the non-ruled quadrics are ovoids, and conversely. In spaces of even order this does not hold. (Cf. the theorem of Segre.)

Thus in the odd order cases we can speak of *ovoidal planes* instead of M-planes.

Up to now only ovoidal Möbius planes are known, and the properties of these planes depend on the properties of the ovoid from which they are derived.

Let us now compile the relations concerning the number of the points and circles characterizing the M-plane of order q:

M_1 *The number of points of the plane is* q^2+1;
M_2 *The number of circles is* $q(q^2+1)$;
M_3 *Each circle is incident with* $q+1$ *points;*
M_4 *Each point is incident with* $q(q+1)$ *circles;*
M_5 *Each hyperbolic pencil consists of* $q+1$ *circles;*
M_6 *Each parabolic pencil consists of* q *circles;*
M_7 *Each elliptic pencil consists of* $q-1$ *circles.*

Note that the number of circles of a hyperbolic pencil is equal to the number of points of a circle.

If we now prepare the incidence table of a plane possessing the properties $M_1 - M_7$ — cf. Fig. 102. —, these properties are visualized in a perspicuous manner. In studying arithmetical relations, the incidence matrix can be of advantage. — The matrix is derived from the incidence table by writing an 0 in the empty squares and replacing the incidence signs by a 1. We show the utility of the incidence matrix in connection with a combinatorial extremum problem. Namely we shall investigate how many 1's are contained in the incidence matrix belonging to the block design consisting of n elements and m blocks, i.e. in the matrix of n columns and m rows, if the columns of the matrix are considered as points, the rows are considered as circles and 1 and 0 signify incidence and non-incidence, respectively. Assuming that the system of the points and circles so defined satisfies the requirements of the axiom system **K** as well. Let the number of 1's be denoted by η.

Axiom K_4 implies that the matrix consists of at least 4 columns. And because of K_2 it consists of at least 4 rows, hence $\eta \geq 12$.

Let the number of the 1's occurring in the rows be denoted in turn by $\lambda_1, \lambda_2, \ldots, \lambda_m$ and those occurring in the columns by $\pi_1, \pi_2, \ldots, \pi_n$. Obviously,

(5.4.1) $$\lambda_1 + \lambda_2 + \ldots + \lambda_m = \pi_1 + \pi_2 + \ldots + \pi_n = \eta.$$

The row vectors, respectively, the column vectors of our matrix are in turn: $\mathbf{x}_1, \ldots, \mathbf{x}_m$, respectively $\mathbf{y}_1, \ldots, \mathbf{y}_n$. We shall now establish the relation of the numbers λ, π to the vectors $\mathbf{x}, \mathbf{y}$.

Clearly, $\mathbf{x}_j^2 = \lambda_j$ and $\mathbf{y}_j^2 = \pi_j$, consequently

(5.4.2) $$\begin{cases} \mathbf{x}_1^2 + \mathbf{x}_2^2 + \ldots + \mathbf{x}_m^2 = \lambda_1 + \lambda_2 + \ldots + \lambda_m = \eta, \\ \mathbf{y}_1^2 + \mathbf{y}_2^2 + \ldots + \mathbf{y}_n^2 = \pi_1 + \pi_2 + \ldots + \pi_n = \eta, \end{cases}$$

and for the sum of the row vectors, respectively of the column vectors we have on the one hand the relations

(5.4.3) $$\begin{cases} (\mathbf{x}_1 + \mathbf{x}_2 + \ldots + \mathbf{x}_m)^2 = \pi_1^2 + \pi_2^2 + \ldots + \pi_n^2 \\ (\mathbf{y}_1 + \mathbf{y}_2 + \ldots + \mathbf{y}_n)^2 = \lambda_1^2 + \lambda_2^2 + \ldots + \lambda_m^2, \end{cases}$$

and on the other hand

(5.4.4) $$\left(\sum_1^m \mathbf{x}_j\right)^2 = \sum_1^m \mathbf{x}_j^2 + \sum_1^m \mathbf{x}_j \mathbf{x}_k \qquad (j \neq k)$$

$$\left(\sum_1^n \mathbf{y}_j\right)^2 = \sum_1^n \mathbf{y}_j^2 + \sum_1^n \mathbf{y}_j \mathbf{y}_k.$$

By comparing (5.4.2)—(5.4.4) we have:

(5.4.5) $$(\lambda_1^2 + \lambda_2^2 + \ldots + \lambda_m^2) - (\lambda_1 + \lambda_2 + \ldots + \lambda_n) = \sum_1^n \mathbf{y}_j \mathbf{y}_k.$$

Let the scalar product $y_j y_k$ be denoted by ω_{jk} ($j \neq k$). Let the largest of these numbers be ω. Since the number of the products of this kind is $n(n-1)$ we have from (5.4.5)

(5.4.6) $\quad (\lambda_1^2 + \lambda_2^2 + \ldots + \lambda_m^2) - (\lambda_1 + \lambda_2 + \ldots + \lambda_m) \leq \omega n(n-1)$,

where the equality holds if and only if every one of the numbers ω_{jk} is equal to ω (Condition (α)).

From the well-known relation between the sum of squares and the arithmetic mean of the positive numbers we have for the positive integers $c_1, c_2, \ldots, c_s$ the inequality

$$c_1^2 + c_2^2 + \ldots + c_s^2 \geq \frac{(c_1 + c_2 + \ldots + c_s)^2}{s},$$

where the equality holds if and only if $c_1 = c_2 = \ldots = c_s$. (Condition (β)). Hence we have from (5.4.6) the following inequality:

$$\frac{\eta^2}{m} - \eta \leq \omega n(n-1).$$

But because of $\eta > 0$ it follows from this that

(5.4.7) $\quad 2\eta \leq m + \sqrt{m^2 + 4\omega mn(n-1)}$,

where the equality holds, if and only if both conditions (α) and (β) are fulfilled.

Let us formulate these two conditions in the language of inversive geometry. (α) requires that circles pass through any two points. (β) requires that every circle has the same number of points; let this number be denoted by v. If we formulate on the incidence table the obtained results, we may say the following:

In the case of given ω, m, n consider all incidence tables of the types $m \times n$ such that the number of the rows intersecting any two columns in squares occupied by the sign • is at most ω and there exists a pair of columns for which there are exactly ω rows of this kind. Let the number of the signs • occurring in a table be denoted by η. Let the tables in question be called $\Theta(\omega; m, n)$-tables. In these tables the number η cannot exceed the upper bound defined by (5.4.7); if we denote this upper bound by $E(\omega, m, n)$, we have

$$\eta \leq E(\omega, m, n).$$

The number η reaches the upper bound E, if and only if any two columns are crossed by exactly ω rows in squares containing an incidence sign; and further that every row contains exactly v incidence signs.

It is easy to see that if the $\Theta(\omega, m, n)$-table satisfies the aforesaid conditions, i.e. it has the extremal properties just formulated, then it is characterized by a certain degree of regularity, with respect to its columns, too: *the same number of incidence signs occur in every one of its columns.*

We shall prove this statement after the introduction of an auxiliary notion. We shall understand by a *link* of a table the pair of incidence signs occurring in the same row of a table and by the terms of the link we shall understand the two columns which contain the incidence signs of the link.

Now let us determine how many links can have an arbitrarily chosen column P_j as a term. The counting can be done in two different ways.

1° π_j of rows cross the column P_j of the table in square containing an incidence sign and each row like this contains $v-1$ further signs. Hence the number of links in question is $\pi_j(v-1)$.

2° The column P_j of the table can be paired with $n-1$ columns distinct from itself. Any pair of columns can be paired with $n-1$ pairs of columns distinct from itself. Any pair of columns determines ω links. Hence the number of the links in question is $\omega(n-1)$.

By comparing these two enumerations we have

(5.4.8) $$\pi_j = \frac{\omega(n-1)}{v-1}.$$

Hence the statement $\pi_1 = \pi_2 = \ldots = \pi_n (=\mu)$ is correct.

So the inversive geometries each furnish one extremal table like this. Namely according to M_1, M_2, M_5 we have in turn

$$n = q^2+1, \quad m = q(q^2+1), \quad \omega = Q+1.$$

By substituting these into (5.4.7) the right hand side takes on the following value:

$$2q(q+1)(q^2+1).$$

Consequently, $v = q+1$, $\mu = q(q+1)$ in properties M_3, M_4.

5.5 Incidence structure and the t-block design

1° The block design t-(v, k, λ) can be regarded as a special incidence structure on an incidence table. The specializing assumptions being

(1) k incidence signs occur in every row of the table

$$\lambda_1 = \lambda_2 = \ldots = \lambda_m = k.$$

(2) Any t columns are crossed by exactly λ rows such that any of the squares of intersection contains an incidence sign.

From this definition of the t-block design follows property (3): Every column contains the same number of incidence signs.

A proof of (3) will be given later, first we make some remarks.

The example of Fig. 102 represents a block design 3-(10, 4, 1). We mentioned in the preceding section that this can also be considered as a block design 2-(10, 4, 4). — Namely any pair of points of the Möbius plane of order three is incident with exactly four circles.

In fact we can prove the following

Theorem: *Every t-block design is also a t'-block design for each integer t' such that* $2 \leq t' \leq t$.

Let the table Θ be the incidence table of a block design t-(v, k, λ); then the number of the rows crossing the columns of a subset consisting of a number t of columns is always λ, independent of which subset of t columns is chosen. Let us now choose arbitrarily $t-1$ columns from the table Θ. Let us consider the rows which are crossed by every one of the aforesaid columns in a square occupied by an incidence sign, suppose the number of these rows is m. If we delete the other rows and the chosen $t-1$ columns there remains an incidence table Θ' which has $v'=v+1-t$ columns and m' rows. Now, any column of the table Θ' is crossed by exactly λ rows in a square containing an incidence sign. Hence the number of the signs in Θ is given by $v'\lambda = \lambda(v+1-k)$. This number could also have been determined by means of the rows. Originally, every row contained k signs but after the deletions there remained in every row of Θ' only $k+1-t$ incidence signs. Hence the total number of the signs of the table Θ is $m'(k+1-t)$. By comparing the two kinds of enumerating we have

$$m' = \lambda \frac{v+1-t}{k+1-t}.$$

This number does not depend on the choice of the $t-1$ columns that we deleted; it depends only on the parameters t, v, k, λ of the original block design. It is now clear with the number $\lambda^* = m'$ that the original block design t-(v, k, λ) can also be considered as a block design t'-(v, k, λ^*) if $t' = t-1$. Repetition of this reasoning proves the theorem.

Now let us return to the proof of the property (3) of the t-block design.

The Θ-table of the block design t-(v, k, λ) is, according to the theorem, the table of a block design 2-(v, k, λ^*). As was proved at the end of the preceding section, every column of a block design 2-(v, k, λ^*) contains the same number of incidence signs, and hence (3) is proved.

2° We shall understand by an *automorphism* of a t-block design a pointwise one-to-one mapping which assigns blocks to blocks and only to blocks. In this sense we use the names *block-preserving mapping* and *image block*.

The automorphisms of a t-block design form a group.

If the elements of a *t*-block design are briefly denoted by $1, 2, \ldots, v$ then an automorphism of the *t*-block design can be expressed as a permutation of $\{1, 2, \ldots, v\}$. Thus the group formed by the automorphisms of the *t*-block design is a subgroup of the complete permutation group S_v.

In recent years the study of permutation groups has greatly assisted in the study of *t*-block designs.

The group formed by all automorphisms of the *t*-block design is called the *complete automorphism group* and any subgroup of its is called *automorphism group*.

A permutation group **P** on the set $H = \{1, 2, \ldots, n\}$ is said to be *k-transitive* if given any two subsets of k elements of **H**

$$X = \{x_1, x_2, \ldots, x_k\} \quad \text{and} \quad Y = \{y_1, y_2, \ldots, y_k\}$$

there exists a permutation $\pi \in P_n$ such that

$$y_j = x_j^\pi \quad (j = 1, 2, \ldots, k).$$

Clearly, *if a group is k-transitive on a set* **H** *then it is h-transitive for every* $h \leq k$.

We mention (without proof), a theorem from the theory of permutation groups.

Theorem: (due to Cofman): *If* **M** *is a Möbius plane of order q having an automorphism group* **A** *which maps a subset* **H** *of the plane* **M** *consisting of* $k(>q+1)$ *points onto itself and if it induces on the elements of* **H** *a 3-transitive permutation group, then* **H** *is identical with the point set* **M** *and the Theorem of Miguel holds on the plane* **M**.

5.6 Problems and exercises to Chapter 5

38. Determine the complete automorphism group of the Petersen graph. (The interrelation with the complete automorphism group of the D-configuration makes the solution of this problem easier.)

39. Prove that the order of a group transitive on the set **H** is divisible by the degree of the group.

40. Prove that a transitive Abelian group is always regular.

41. If $q = 2^r$, then we use the quadric

$$x_1 x_2 + x_3 x_4 + \varrho x_3^2 + \varrho x_4^2 = 0$$

defined over the coordinate field $K = GF(2^r)$ to construct the Möbius plane, where $\varrho \in K$ and has the property that $\varrho x^2 + x + \varrho$ is irreducible in the field **K**. Using this construct the incidence table of the Möbius plane of order four.

42. Construct the incidence table of the Möbius plane of order two and determine the complete automorphism group of this plane.

CHAPTER 6

SOME ADDITIONAL THEMES IN THE THEORY OF FINITE GEOMETRIES

In this final chapter we shall present a selection of the more interesting recent developments in the theory of finite geometries.

6.1 The Fano plane and the theorem of Gleason

It was first recognized by Fano that an important role is played in the theory of finite projective planes by the planes, the so-called Fano planes, on which the diagonal points of every complete quadrangle are collinear. More than a half century elapsed until the fundamental theorem of Gleason displayed the structural properties of this class of the finite projective planes. Since 1956 we have known that any Fano plane is isomorphic with a Galois plane of even order. Thus the study of the projective spaces in which every plane quadrangle is a Fano quadrangle does not present any fundamentally new problem; the study of these spaces can be considered, in the above sense, as closed.

1° First we shall prove a lemma concerning the Fano planes from which it follows that the order of every Fano plane is even.

Lemma: *A Fano plane has a coordinate structure such that the order of every element of the Abelian group established by the addition of points is equal to* 2.

Proof: We discussed the point addition in connection with Fig. 62, p.165. Everything derived there without the use of the D-theorem, i.e. merely assuming the validity of the axiom system **I**, holds also for every Fano plane. Hence we have only to prove independently of the D-theorem that for any three points A, B, C of the line e other than the point Z, i.e. for any three points A, B, C of the point set e' we have

$$A+A = 0, \quad A+B = B+A, \quad (A+B)+C = A+(B+C).$$

Firstly, we shall determine the sum point $A+A$. According to our assumption the diagonal points of the quadrangle $OAVU$ i.e. the points $OU \cap AV = Y$, $OA \cap UV = Z$ and $UA \cap OV = R$ lie on a single line (in Fig. 62 this is the ideal line). Now if $B=A$, it follows that AU and UX are coincident, which implies

$R = X$. Hence O, V, X are collinear, thus the role of the sum point $OZ \cap VX = S$ is taken over by $OZ \cap OV = O$. Hence the order of every element is 2.

Secondly we shall prove the property $A + B = B + A$ (we refer to Fig. 103). Consider the following quadrangles named by the inhomogeneous coordinates of the vertices:

α: $\{(b, b), (a, a), X, Y\}$,

β: $\{(a, a+b), (0, b), X, Y\}$,

γ: $\{(b, b+a), (0, a), X, Y\}$.

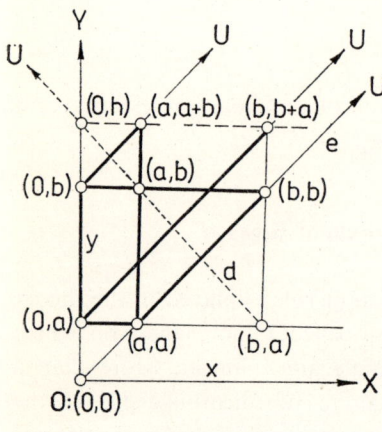

Figure 103

The diagonal points of these are in turn the following collinear triples of points:

$\{U, (a, b), (b, a)\}$, $\{U, (a, b), (0, a+b)\}$,

$\{U, (b, a), (0, b+a)\}$.

— Here the point $(0, a+b)$ is the point of intersection of line y with the line joining the points X and $(b, b+a)$. — The collinearity of the triple of points implies the collinearity of the five points

δ: $\{U, (a, b), (b, a), (0, a+b)$

and $(0, b+a)\}$.

Since the two latter points lie both on the line y and on the line d joining the points U and (a, b), they coincide with the point $y \cap d$. And this signifies that, in fact, $A + B = B + A$.

We have from the above as an additional result that the line d joining the points (a, b) and U is incident with the point (x, y), if and only if

$$x + y = h \quad (h = a+b = b+a).$$

Finally we prove the associativity of the addition (we refer to Fig. 104). Consider the quadrangles

α: $\{(c, b+c), (a, a+b), X, Y\}$,

β: $\{(a, a+(b+c)), (0, b+c), X, Y\}$,

γ: $\{(c, c+(a+b)), (0, a+b), X, Y\}$.

They have collinear triples of diagonal points, which are in turn:

$\{(c, a+b), (a, b+c), U\}$, $\{(a, b+c), (0, a+(b+c)), U\}$,

$\{(c, a+b), (0, (a+b)+c), U\}$

and from this it follows that the five points

$$\delta: \{U, (c, a+b), (a, b+c), (0, a+(b+c)) \text{ and } (0, (a+b)+c)\}$$

are collinear.

The line r joining the points X and $(a, a+(b+c))$ and the line s joining to points X and $(c, (a+b)+c)$ cut the line y in the fourth and fifth points above. But these two points are also on the line d hence these two points coincide in the point $y \cap d$. Thus

$$a+(b+c) = (a+b)+c$$

that is the associativity is valid.

Summarizing, the addition of points establishes an Abelian group in which the order of every element is 2. And, according to a well-known theorem of group theory, if every element of an Abelian group has the period 2, then the order of the group is $q=2^r$.

We emphasize the fact that in the proof of the theorem above we did not refer to the Desargues' theorem.

2° A Fano plane has the R-property (cf. Section 3.4) for any choice of three collinear points X, Y and Z. Consider the set formed by the pencil of lines through X, where we remove the point X from each of the lines of the pencil; let this set be denoted by x. Similarly we obtain the sets y and z from the pencils

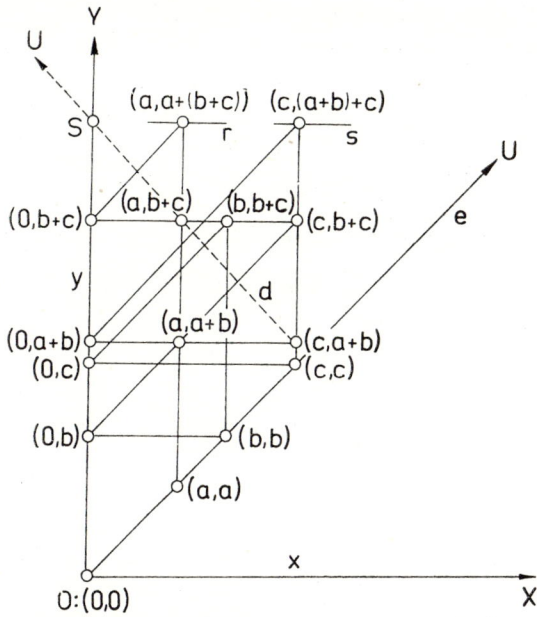

Figure 104

through Y and Z, respectively. The families of lines $\{x\}$ $\{y\}$ $\{z\}$ form an R-net, this is equivalent with the associativity of the operation established by the net and associativity, as we have seen, follows from the assumption that we are dealing with a Fano plane. Gleason was the first to realize the far-reaching consequences of this fact, he gave a new formulation of the MR-configuration, which will be referred to as *G-condition*. Firstly we introduce some new notations.

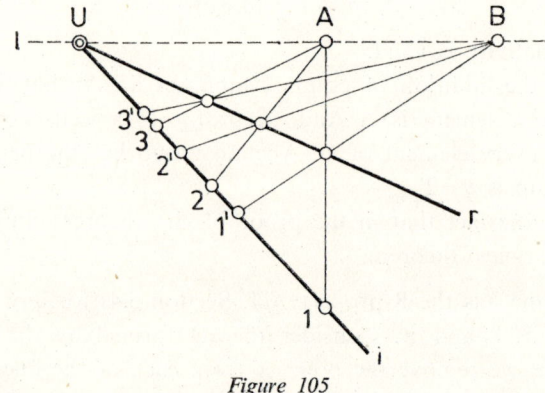

Figure 105

Let l be an arbitrary line of a Fano plane and let U, A and B be three distinct points of the line (Fig. 105). Let the lines i and r pass through the point U. If our plane is of order q, then the points of the line i other than the point U can be indexed by the set $1, 2, ..., q$. Let $1'', 2'', ..., q''$ denote a permutation of $1, 2, ..., q$. We shall now produce certain permutations by means of projections. Let the line r be one of the lines passing through the point U and distinct from the line l. The symbol $\begin{bmatrix} \alpha \\ ir \end{bmatrix}$ indicates that we are projecting the points of the range $i-U$ from the point A onto the line r. The symbol

(6.1.1) $$\begin{bmatrix} \alpha \\ ir \end{bmatrix} \begin{bmatrix} \beta \\ ri \end{bmatrix}$$

indicates that the image-range obtained by the aforesaid projection is then projected from the point B upon the line i. The mapping obtained as the composition of the two projections induces a permutation $1', 2', ..., q'$ of the points $1, 2, ..., q$ of the line i. Thus for each position of the line r we obtain a permutation. The set of q permutations so obtained will be denoted by the symbol

(6.1.2) $$\begin{pmatrix} \alpha\beta \\ i \end{pmatrix}.$$

Obviously, *the set* $\begin{pmatrix} \alpha\beta \\ i \end{pmatrix}$ *of permutations is simply transitive on the range of points* $i-U$, that is given any two points P and P' of the range $i-U$, there exists a unique line r such that the permutation (6.1.1) carries P into P'. This line is, in fact, the line joining the points $AP \cap BP'$ and U.

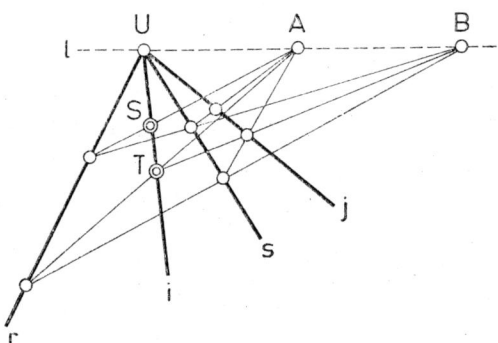

Figure 106

Similarly, the product of projections

(6.1.3)
$$\begin{bmatrix} \alpha \\ ir \end{bmatrix} \begin{bmatrix} \beta \\ rs \end{bmatrix} \begin{bmatrix} \alpha \\ sj \end{bmatrix} \begin{bmatrix} \beta \\ ji \end{bmatrix} = \pi$$

is a permutation of the set $i-U$ (Fig. 106). Naturally, this permutation may have the property that — as in the case of our figure — the point S of the range $i-U$ is a fixed point. Now we are ready to formulate

Condition G: *If, for any choice of distinct points* U, A, B *on the line* l *and of the lines* i, j, r, s *passing through* U, *the permutation* π *induced by them has a fixed point on the range* $i-U$ *then every point of the range is a fixed point.*

The fixed point S of the figure implies that every point T of the range $i-U$ is a fixed point, too. Compare the points U, A and B of Fig. 106 with the points Z, X and Y, respectively, of Fig. 90 p. 193. It is then seen readily that *the condition G and the condition R are equivalent.*

3° Let us return to the study of set of the permutations $\begin{pmatrix} \alpha\beta \\ i \end{pmatrix}$. In this set there occurs one permutation which leaves every element unchanged; namely if we put $r=i$ in (6.1.1).

Every permutation in the set has an inverse permutation in the set (Fig. 107). In order to prove this, it is enough to consider the effect $S \to S'$ of the permutations $\begin{bmatrix} \alpha \\ ir \end{bmatrix} \begin{bmatrix} \beta \\ ri \end{bmatrix}$ induced by the line r upon a single point S of the range $i-U$. The line joining the points $S'A \cap SB$ and U, the line j together with the lines $s=i$ and r form a quadruple (i, r, s, j) such that the permutation π induced by them makes the point S a fixed point of the range $i-U$. But then the validity of the condition G implies that every point of the range is a fixed point with respect to the permutation π, i.e. the permutation induced by the line j is the inverse of that induced by the line r. In the figure this is expressed by writing $j=r^{-1}$.

The product of any two permutations of the set is also a permutation of the set (Fig. 108). In order to prove this statement, consider the permutations induced by the lines r and s, respectively, where r and s are arbitrary lines passing through

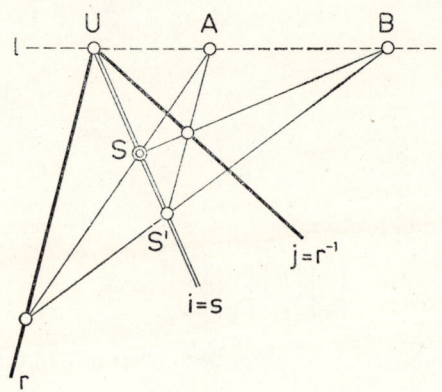

Figure 107

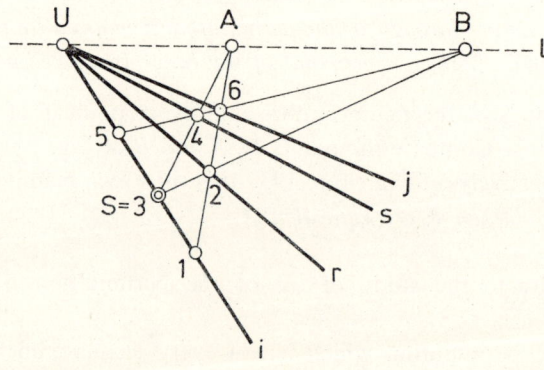

Figure 108

the point U. Their effect is considered by starting from the point labelled by 1 of the range $i-U$. The permutation

$$\begin{bmatrix} \alpha \\ ir \end{bmatrix} \begin{bmatrix} s \\ ri \end{bmatrix} = [r]$$

belonging to r carries the point 1 into the point $S=3$. The permutation

$$\begin{bmatrix} \alpha \\ is \end{bmatrix} \begin{bmatrix} \beta \\ si \end{bmatrix} = [s]$$

belonging to s carries the point $S=3$ into the point 5. Hence the product of the permutations $[r]$ and $[s]$ carries the point 1 into the point 5. There exists, however, a line j such that the permutation

$$\begin{bmatrix} \alpha \\ ij \end{bmatrix} \begin{bmatrix} \beta \\ ji \end{bmatrix} = [j]$$

induced by it, carries the point 1 into the point 5. Thus the permutations $[r][s]$ and $[j]$ have the same effect upon the point 1. But the validity of the condition **G** then implies (cf. Fig. 106) that the effect of $[r][s]$ is the same as that of $[j]$, hence $[r][s]=[j]$ and $[j] \in \begin{pmatrix} \alpha\beta \\ i \end{pmatrix}$.

The permutation set in question contains the identity, the inverse of any permutation, and the product of any two permutations, hence *it forms a group*.

Conversely, if we start from the assumption that the set $\begin{pmatrix} \alpha\beta \\ i \end{pmatrix}$ of the permutations is a group, then it follows that the group has no permutation, other than the identity, which fixes an element of the range $i-U$. From this follows immediately (as can be seen from Figure 108 showing the construction of the permutation product $[r][s]=[j]$) that if the permutation corresponding to (6.1.3) fixes one point, then it fixes every point and hence the condition **G** is satisfied.

Let the *condition* **G*** signify that the permutation set $\begin{pmatrix} \alpha\beta \\ i \end{pmatrix}$ forms a group with respect to the composition of the permutations. Thus our results are summed up in

Gleason's lemma: *The conditions* **G** *(and consequently* **R***) and* **G*** *are equivalent.*

We emphasize again that our statements hold for any line l of the plane.

4° We shall now study the structure of the plane of order q satisfying the condition **G** in connection with the elations of the plane. Let l be the axis, and the point U of the line l the centre of an elation. There exists always a collineation satisfying these conditions, since the identity — which assigns to every point the point itself as an image — fulfils the definition of an elation. If the pair (U, l) determine more than one elation, the number of elations is at most q. The elations in question form a group; this group will be denoted by $\Gamma(U, l)$.

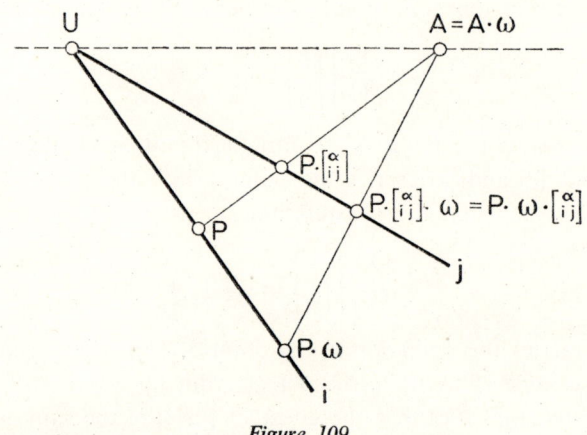

Figure 109

Any element of this group induces on any line i passing through the point U a permutation leaving the point U unchanged and this permutation, provided that it is the identity, has no further fixed point on the line i. Since any element of $\Gamma(U, l)$ is uniquely determined by a pair of points (P, P') — distinct from U — of the line i (where P' is the image of P), the permutations induced by the elations on the line i represent in a one-to-one manner the elements of $\Gamma(U, l)$. The image of a point P of the line i under ω is also on the line i, and let the mapping be expressed by writing $P' = P \cdot \omega$. The product $\begin{bmatrix} \alpha \\ ij \end{bmatrix} \begin{bmatrix} \beta \\ ji \end{bmatrix}$ of projections assigns also an image lying on the line i to a point P of the line i. This mapping is expressed by the symbol $P' = P \cdot \begin{bmatrix} \alpha \\ ij \end{bmatrix} \begin{bmatrix} \beta \\ ji \end{bmatrix}$, or by $P' = P \cdot \begin{bmatrix} \alpha \\ ij \end{bmatrix} \cdot \begin{bmatrix} \beta \\ ji \end{bmatrix}$.

The commutation rule

(6.1.4)
$$\begin{bmatrix} \alpha \\ ij \end{bmatrix} \omega = \omega \begin{bmatrix} \alpha \\ ij \end{bmatrix}$$

is explained by Fig. 109: the point P, the projection point $Q = P \cdot \begin{bmatrix} \alpha \\ ij \end{bmatrix}$ and the point A are collinear. Thus the points P', Q' and A' ($=A$) are collinear. These image-points can also be expressed in the form $P \cdot \omega$, $P \cdot \begin{bmatrix} \alpha \\ ij \end{bmatrix} \cdot \omega$, A. From this we can see that the point $P \cdot \begin{bmatrix} \alpha \\ ij \end{bmatrix} \cdot \omega$ is none other than the projection of the point $P \cdot \omega$ from A onto the line j, and this is the same as (6.1.4), From this follows also the commutation rule:

(6.1.5) $\qquad \omega \begin{bmatrix} \alpha \\ ir \end{bmatrix} \begin{bmatrix} \beta \\ ri \end{bmatrix} = \begin{bmatrix} \alpha \\ ir \end{bmatrix} \begin{bmatrix} \beta \\ ri \end{bmatrix} \omega \quad (U \in r).$

Consider the following

(6.1.6) $\qquad \begin{bmatrix} \alpha \\ ij \end{bmatrix} \begin{bmatrix} \alpha \\ jk \end{bmatrix} = \begin{bmatrix} \alpha \\ ik \end{bmatrix}$ and $\begin{bmatrix} \alpha \\ ij \end{bmatrix} \begin{bmatrix} \alpha \\ ji \end{bmatrix} = 1,$

where i, j, k are lines passing through the point U and 1 denotes the identity mapping.

Consider now a permutation σ of the points of the line i which commutes with any one of the permutations $\begin{bmatrix} \alpha \\ ij \end{bmatrix} \begin{bmatrix} \beta \\ ji \end{bmatrix}$ for any choice of $(U \in) j \neq l$, $A \in l$, $B \in l$, $A \neq U \neq B \neq A$.

We shall show that this mapping σ can be extended into a mapping σ^* of the entire plane such that $\sigma^* \in \Gamma(U, l)$ should hold.

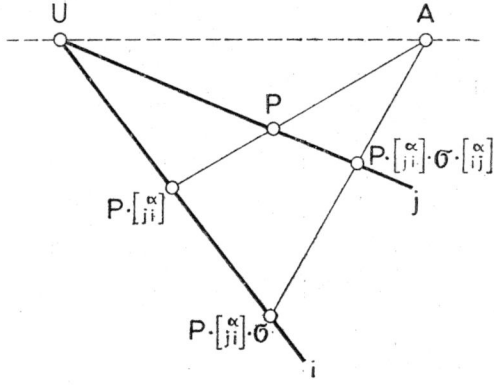

Figure 110

In order to do this consider the point P not incident with the line i and the line $PU = j$ ($\neq l$) (Fig. 110). The extension of the permutation σ over every point P of the plane is performed by means of a point $A(\in l - U)$ of the line l. If $P \in l$, then for the mappings σ^* this is expressed by

$$P \cdot \sigma^* = P,$$

however, $P \in l$, then
$$P \cdot \sigma^* = P \cdot \begin{bmatrix} \alpha \\ ji \end{bmatrix} \cdot \sigma \cdot \begin{bmatrix} \beta \\ ij \end{bmatrix}.$$

The mapping σ^* so defined is a collineation. In order to prove this, it suffices to consider special triples of points which have one of their points on the fixed line l. — Since, if both X, Y, V and Y, Z, V are collinear ($V \in l$), then also X, Y, Z are collinear. Let us therefore take the points P and Q of the lines j and k respectively, passing through the point U and distinct from each other as well as from the lines l and i, further let $PQ \cap l = B \neq A$. Put $P' \cdot \sigma^*$ and $Q' = Q \cdot \sigma^*$; and we have clearly $B' = B$. We have to prove now the collinearity of $BP'Q'$.

From the collinearity of BPQ we have
$$P \cdot \begin{bmatrix} \beta \\ jk \end{bmatrix} = Q.$$

Furthermore, the projection of the point P' (lying, together with the point P, on the line j) from the point B onto the line k can be expressed in the form
$$P' \cdot \begin{bmatrix} \beta \\ jk \end{bmatrix} = P \cdot \sigma^* \cdot \begin{bmatrix} \beta \\ jk \end{bmatrix}$$

and from this, by applying the definition of σ^*, and the relations (6.1.5), (6.1.6) we have

$$P' \cdot \begin{bmatrix} \beta \\ jk \end{bmatrix} = P \cdot \begin{bmatrix} \alpha \\ ji \end{bmatrix} \cdot \sigma \cdot \begin{bmatrix} \alpha \\ ij \end{bmatrix} \cdot \begin{bmatrix} \beta \\ jk \end{bmatrix} =$$
$$= P \cdot \begin{bmatrix} \alpha \\ ji \end{bmatrix} \cdot \sigma \cdot \begin{bmatrix} \alpha \\ ij \end{bmatrix} \begin{bmatrix} \beta \\ ji \end{bmatrix} \cdot \begin{bmatrix} \beta \\ ik \end{bmatrix} \begin{bmatrix} \alpha \\ ki \end{bmatrix} \cdot \begin{bmatrix} \alpha \\ ik \end{bmatrix} =$$
$$= P \cdot \begin{bmatrix} \alpha \\ ji \end{bmatrix} \cdot \begin{bmatrix} \alpha \\ ij \end{bmatrix} \begin{bmatrix} \beta \\ ji \end{bmatrix} \cdot \begin{bmatrix} \beta \\ ik \end{bmatrix} \begin{bmatrix} \alpha \\ ki \end{bmatrix} \cdot \sigma \cdot \begin{bmatrix} \alpha \\ ik \end{bmatrix} =$$
$$= P \cdot \begin{bmatrix} \beta \\ jk \end{bmatrix} \cdot \begin{bmatrix} \alpha \\ ki \end{bmatrix} \cdot \sigma \cdot \begin{bmatrix} \alpha \\ ik \end{bmatrix}.$$

Thus we have
$$P' \cdot \begin{bmatrix} \beta \\ jk \end{bmatrix} = Q \cdot \sigma^* = Q'$$

and this states that by projecting the point P' of the line j from B onto k we obtain precisely the point Q'. Thus the collinearity of $BP'Q'$ is proved. If we consider further the behaviour of the point U and the line l it is readily seen that σ^* is an elation, hence $\sigma^* \in \Gamma(U, l)$.

The results obtained in paragraph 4 are summarized by the

Second lemma of Gleason: There is a *one-to-one correspondence between the elements of the elation group* $\Gamma(U, l)$ *and the permutations of the line i commuting with the elements of any choice of the points* A *and* B. *The permutations induced by the elements of* $\begin{pmatrix} \alpha\beta \\ i \end{pmatrix}$ *on the range* $i - U$ *are the permutations which represent the elations in the correspondence in question.*

5. In this paragraph we shall state and outline the proof of Gleason's fundamental theorem, which gives a sufficient condition for a finite plane to be Desarguesian. Gleason proved this theorem by exploiting some strong results concerning finite groups. We mention firstly its consequence for the Fano plane under the name: "Gleason's theorem". Namely on the Fano plane we have the associativity of the addition of points, as we have already proved. Consequently, we have the

Gleason's theorem: *every Fano plane is a Desarguesian plane.*

Gleason's fundamental theorem: *If the condition* **G*** *is valid on a projective plane whose order is a power of a prime, then the plane is Desarguesian.*

Proof — the proof relies heavily on theorems which are essentially algebraic and thus we shall only quote these here and shall return to them in more detail later. First of all, assume the validity of the condition **G***, i.e. the fact that the permutations given $\begin{bmatrix} \alpha \\ ij \end{bmatrix} \begin{bmatrix} \beta \\ ji \end{bmatrix}$ constitute a group, denoted by the symbol $\begin{pmatrix} \alpha\beta \\ i \end{pmatrix}$. The permutation $\begin{bmatrix} \beta \\ ij \end{bmatrix} \begin{bmatrix} \alpha \\ ji \end{bmatrix}$ is clearly the inverse of $\begin{bmatrix} \alpha \\ ij \end{bmatrix} \begin{bmatrix} \beta \\ ji \end{bmatrix}$. Obviously, the inverses of the elements of a group themselves form a group, the original group itself. This is the meaning of the formula

(6.1.7)
$$\begin{pmatrix} \alpha\beta \\ i \end{pmatrix} = \begin{pmatrix} \beta\alpha \\ i \end{pmatrix}.$$

By returning to Fig. 106 we can see that the order of the group $\begin{pmatrix} \alpha\beta \\ i \end{pmatrix}$ is equal to the order of the plane, i.e. p^r where p is prime. In fact, if we specify the image k' of a point k of the line i, then the line r inducing this permutation is uniquely obtained: it is the line joining the points U and $kA \cap k'B'$, and the image k' of a fixed point k can be chosen in $q = p^r$ ways.

Consider now the complete permutation group of the p^r points of the range $i - U$; let this group now be denoted by $\mathbf{Q}_i$. Further let $A \neq D \in l - U$ and let the line j range over p^r lines passing through the poin U but distinct from the line j. Express the permutations induced by the line j of the groups

$$\begin{pmatrix} \alpha\beta \\ i \end{pmatrix} = \begin{pmatrix} \beta\alpha \\ i \end{pmatrix} \text{ and } \begin{pmatrix} \alpha\delta \\ i \end{pmatrix} = \begin{pmatrix} \delta\alpha \\ i \end{pmatrix}$$

in the form

$$\begin{bmatrix} \beta \\ ij \end{bmatrix} \begin{bmatrix} \alpha \\ ji \end{bmatrix} \text{ and } \begin{bmatrix} \alpha \\ ij \end{bmatrix} \begin{bmatrix} \delta \\ ji \end{bmatrix}, \text{ respectively.}$$

The product of the two permutations so paired is clearly

$$\begin{bmatrix} \beta \\ ij \end{bmatrix} \begin{bmatrix} \delta \\ ji \end{bmatrix} \in \begin{pmatrix} \alpha\delta \\ i \end{pmatrix}.$$

While the line j ranges over the lines passing through U (but distinct from l), this product ranges over the elements of the group $\begin{pmatrix} \alpha\delta \\ i \end{pmatrix}$.

Write instead of the permutations

$$\begin{bmatrix} \alpha \\ ij \end{bmatrix} \begin{bmatrix} \beta \\ ji \end{bmatrix} \text{ and } \begin{bmatrix} \alpha \\ ij \end{bmatrix} \begin{bmatrix} \delta \\ ji \end{bmatrix}$$

the more concise symbols B_j and D_j, further instead of the subgroups

$$\begin{pmatrix} \alpha\beta \\ i \end{pmatrix}, \begin{pmatrix} \alpha\delta \\ i \end{pmatrix}, \begin{pmatrix} \beta\delta \\ i \end{pmatrix}$$

of the group $\mathbf{Q}_i$ the symbols $\mathbf{B}, \mathbf{D}, \mathbf{G}$. Thus we can say that every element of the group $\mathbf{G}$ can be expressed as a product $B_j \cdot B$ if $B_j \in \mathbf{B}$, $D_j \in \mathbf{D}$; and every $B_j \cdot D_j \in \mathbf{G}$.

But then by choosing an arbitrary pair of lines r and s (passing through the point U), $B_r \cdot D_r \in \mathbf{G}$, $B_s \cdot D_s \in \mathbf{G}$ and from this follows $(B_r D_r)(B_s D_s) \in \mathbf{G}$. But the latter permutation can be expressed in the form $B_k \cdot D_k$, because it is an element of $\mathbf{G}$:

$$(B_r D_r)(B_s D_s) = B_k D_k,$$

and from this it follows that

$$D_r B_s = (B_r^{-1} B_k)(D_k D_s^{-1}),$$

which implies that

$$\mathbf{D} \cdot \mathbf{B} \subseteq \mathbf{B} \cdot \mathbf{D}.$$

However, if we deduce first the relation

$$B_s^{-1} D_r^{-1} = (D_k D_s^{-1})^{-1} (B_r^{-1} B_k)^{-1},$$

then we obtain from it

$$\mathbf{B} \cdot \mathbf{D} \subseteq \mathbf{D} \cdot \mathbf{B}.$$

By comparing the two kinds of result we have $\mathbf{B}\cdot\mathbf{D} = \mathbf{D}\cdot\mathbf{B}$, that is

(6.1.8) $$\binom{\alpha\beta}{i}\binom{\alpha\delta}{i} = \binom{\alpha\delta}{i}\binom{\alpha\beta}{i}.$$

Therefore, if we consider the pairs A, X of points of the point set $l-U = \{A, B, ..., L, ..., Z\}$ where A is fixed and X ranges over the elements of the point set, then each group $\binom{\alpha\xi}{i}$ has the order q and the product (6.1.8) of the groups is pairwise commutative. —

And from this by a well-known theorem in the theory of groups it follows that the *product*

(6.1.9) $$\mathbf{K} = \binom{\alpha\beta}{i}\cdots\binom{\alpha\lambda}{i}\cdots\binom{\alpha\zeta}{i}$$

is a group, namely a p-group, and thus its centre contains other permutations besides the identity. ($\mathbf{T}_1$-**theorem**.)

Let the permutation B be an element of the centre of the group $\mathbf{K}$, distinct from the identity. We have seen that $\binom{\alpha\beta}{i}\binom{\alpha\delta}{i}$ contains $\binom{\beta\delta}{i}$ hence B can be interchanged by every subgroup $\binom{\beta\delta}{i}$. But B is a permutation of the points of the line i, by returning to the former theorem we may say that the group $\Gamma(U, l)$ is not reduced to the identity. This implies, according to theorem $\mathbf{T}_2$ formulated below (and due also to Gleason), that on the plane in question the theorem of Desargues holds without restriction and therefore we have finished the proof of Gleason's theorem.

We return now to the deep theorem used above, namely:

Theorem $\mathbf{T}_2$: *If on a finite plane the elation $\Gamma(U, l)$ does not consist of the identity only, then for any line l of the plane and for any point U of the line l the theorem of Desargues holds with unrestriced validity.*

The proof of the theorem will not be given, but essentially it falls into the following two parts.

Firstly we deduce from the hypotheses of $\mathbf{T}_2$ that *the micro-Desargues theorem holds on the plane* (i.e. the case when the centre of perspective lies on the axis); then we appeal to the famous

Zorn—Levi theorem: *Every finite micro-Desarguesian plane is Desarguesian.* This classical theorem really belongs to the realm of algebra, its proof is reduced to the proof of the algebraic theorem stating that *every finite skew field is a finite field.*

6° Several attempts at simplifying the proofs of the Wedderburn theorem and of the Zorn—Levi theorem have been made, as well as of the Gleason theorem.

6.2 The derivation of new planes from the Galois plane

When at the end of the last century it was proved by Hilbert that the Theorem of Desargues does not depend on the axioms of incidence, he used a method which was simplified almost to the level of a triviality in 1902 by Moulton through the construction of an ingenious model. As an introduction, we shall consider Moulton's model, its significance appearing at the introduction of a coordinate system. Next we shall show how this idea of Hilbert and Moulton can be applied to the theory of finite projective planes.

1° Consider the classical projective plane and let us change the definition of the *"line"* but in such a way that the axiom system **I** should remain valid on the new "plane" (Fig. 111). For this purpose let us decompose the plane by an *x*-axis into two half-planes Σ and Σ'. Every line which has a positive slope (as the line c) is changed as follows: the part lying in the half-plane Σ' is replaced by a half-line having a slope half as steep as the original line, i.e. tg $\omega = 2$ tg ω'. That is, the original line is transformed into a broken line and this line broken at the point T will be considered as a "line". The other lines (those of negative

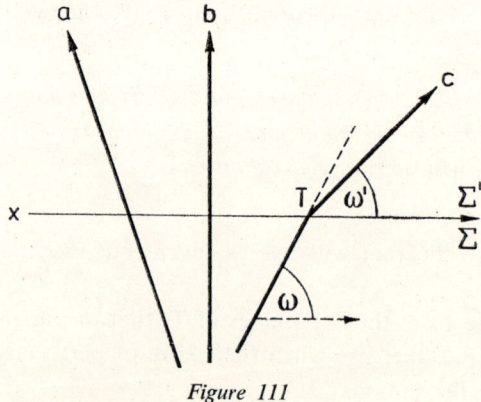

Figure 111

slope as the line *a*, the vertical line *b* and the lines parallel to the *x*-axis) are not altered. It is easy to see that on this Moulton plane the axioms l_1, l_2, l_3, l_4 are valid. This is left to the reader.

Now consider on the classical projective plane the Desargues configuration formed by the pair of triangles *ABC* and *A′B′C′* in perspective from the centre

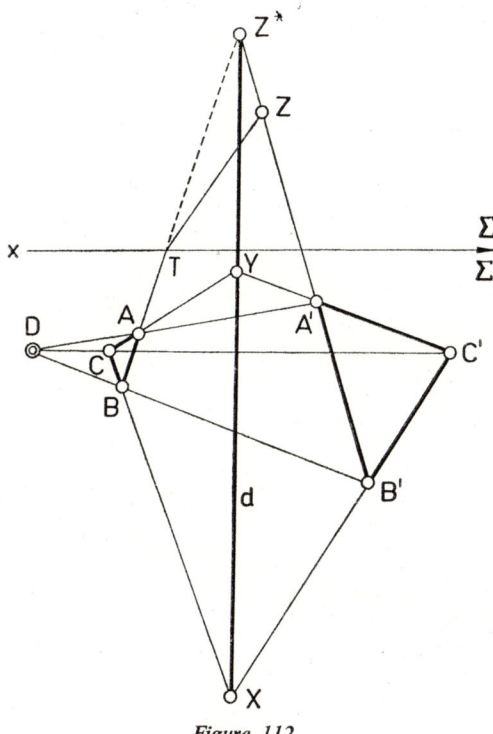

Figure 112

D and from the axis *d*. If we deform our plane into a Moulton plane determined by the line *x* perpendicular to line *d*, then the structure of our configuration remains unaltered, except for the side *AB* of the triangle *ABC*. On the new plane the "line" joining the points *A* and *B* will be the broken line *BATZ*. We may say that in the triangles *ABC* and *A′B′C′* of the Moulton plane are in perspective from the point *D*. The points of intersection *X*, *Y*, *Z* of the corresponding sides, however, do not lie on the same line: $Z \notin d = XY$. — We cannot say that our plane is anti-Desarguesian, since we can find a configuration for which the Theorem of Desargues is valid. Namely, by moving the figure downward till the configuration of 10 points falls entirely in the half-plane Σ.

Hence the axiom system **I** *is independent of the theorem of Desargues.*

The axioms of incidence suffice for the introduction of the notion of the parallelogram, by nominating a line and its points as ideal elements, even on a finite plane. Naturally the notions of translation, of the vector, and of the addition of vectors can also be introduced, if opposite sides of a parallelogram are interpreted as equal vectors. However, we find that to ensure the transitivity of vector-equality an extension of the axiom system I is necessary.

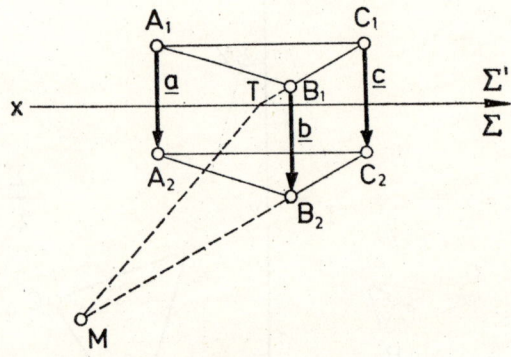

Figure 113

On the classical affine plane suppose the lines supporting the vectors $\overrightarrow{A_1A_2}=\mathbf{a}$, $\overrightarrow{B_1B_2}=\mathbf{b}$, $\overrightarrow{C_1C_2}=\mathbf{c}$ are distinct and parallel (Fig. 113); because the MD-theorem holds the conditions $\mathbf{a}=\mathbf{b}$, $\mathbf{a}=\mathbf{c}$ imply $\mathbf{b}=\mathbf{c}$, since $A_1A_2B_2B_1$ and $A_1A_2C_2C_1$ are parallelograms, it follows by the MD-theorem that the quadrangle $B_1B_2C_2C_1$ is also a parallelogram. However, if the micro-Desargues theorem (MD-theorem) does not hold on a plane then the transitivity of vector equality may also not hold.

For example consider the plane deformed by the line x of the figure into a Moulton plane. Though on the Moulton plane we have $A_1B_1 \parallel A_2B_2$ and $A_1C_1 \parallel A_2C_2$, the lines B_1C_1 and B_2C_2 are, however, not parallel, meeting as they do in the point M. Consequently $\mathbf{b} \neq \mathbf{c}$ in spite of $\mathbf{a}=\mathbf{b}$ and $\mathbf{a}=\mathbf{c}$.

Of course for vectors introduced on the Galois plane the equality of the vectors is a transitive relation, since the Theorem of Desargues holds without restriction. If one could find a procedure analogous to the Moulton construction, then the Galois plane could also be deformed into a non-Desarguesian plane. This is an interesting problem since the Galois planes are well known to us, but as yet we have seen only one finite non-Desarguesian plane, a plane of order nine. Of course, the search for an analogous procedure to the Moulton construction is made more difficult by the fact that on a Galois plane we have no metrical process corresponding to "the replacing of a line by one of half the slope" in the Moulton construction.

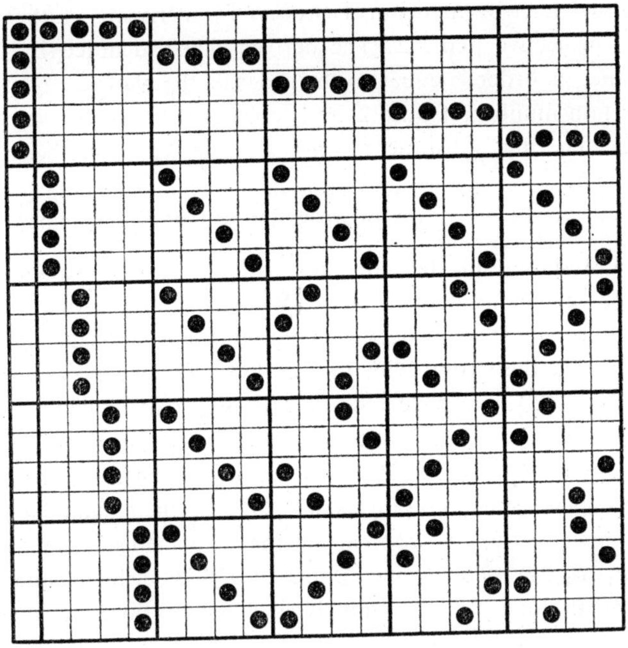

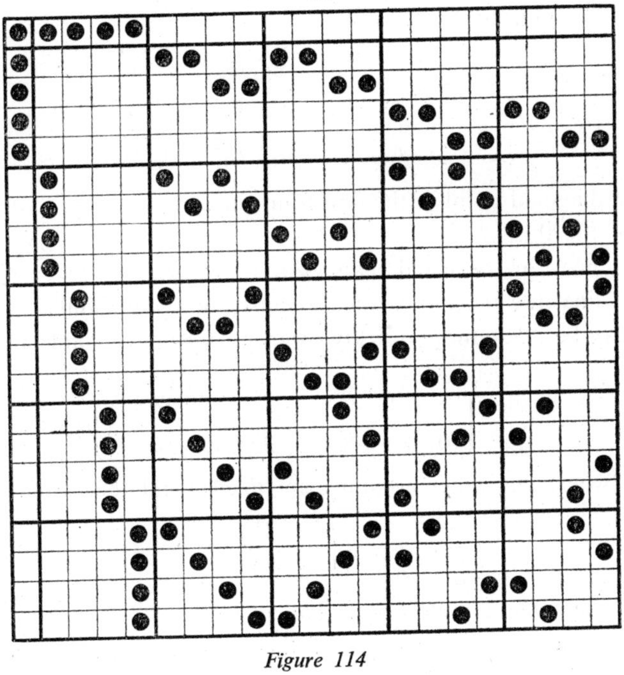

THE DERIVATION OF NEW PLANES

Σ

Σ'

Figure 114

We shall now describe in detail a process for deriving another plane from a given finite projective plane Σ. The derived plane, denoted by Σ', is also a projective plane, its order is equal to the order of the plane Σ. The construction removes a certain number of lines Σ and replaces them by the same number of *line-substitutes*. The line substituent consists of the same number of points as a line of the plane Σ.

Now we shall verify by means of an example that this can be done.

2° Figure 114 shows side by side the incidence tables of a Galois plane of order four and that of its derived plane. We identify the k-th column (a point of the plane) of the table with the k-th column of the table Σ ($k=1, 2, \ldots, 21$); further, we identify their rows with indices 1, 14, 15, ..., 21 as well; in the rows with indices 2, 3, ..., 13 we have modified the sign pattern of the original table. The modified rows are to be understood to replace the lines $l_2, l_3, \ldots, l_{13}$ of the plane Σ by the subsets $l_2, l_3, \ldots, l_{13}$ of five points, which are not lines on the original plane but quadrangles with one of their diagonal points. Thus for instance we can see by comparing the two tables that the point set $l'_2 = \{P_1, P_6, P_7, P_{10}, P_{11}\}$ contains the points P_1, P_6, P_7 of line l_2 and of the points P_1, P_{10}, P_{11} of the line l_3; P_1 point is a diagonal point of the quadrangle $P_6 P_7 P_{10} P_{11}$.

It is easy to check by Fig. 114 that the point sets $l'_2, l'_3, \ldots, l'_{13}$ of five elements are given by the quintuples of points:

$$l_2, l_3; \quad l_3, l_2; \quad l_4, l_5; \quad l_5, l_4;$$
$$l_6, l_8; \quad l_7, l_9; \quad l_8, l_6; \quad l_9, l_7;$$
$$l_{10}, l_{13}; \quad l_{11}, l_{12}; \quad l_{12}, l_{11}; \quad l_{13}, l_{10};$$

The common diagonal point of the first four quadrangles is P_1, that of the second is P_2, and of the third P_3.

We could check in the usual manner that by considering every row of the table Σ' as a line, the three incidence axioms are again fulfilled. However, this verification can be made more quickly by transforming the table Σ by a permutation

$$\omega = \begin{pmatrix} 8 & 9 & 10 & 11 & 16 & 17 & 18 & 19 \\ 10 & 11 & 8 & 9 & 18 & 19 & 16 & 17 \end{pmatrix}$$

of columns (where the other columns are fixed), and then by a transposition

$$\sigma = \begin{pmatrix} 20 & 21 \\ 21 & 20 \end{pmatrix}$$

of rows into the table Σ'. We know that an elementary transformation $\omega\sigma = \pi$ like this does not alter the validity of the axioms $\mathsf{l}_1, \mathsf{l}_2, \mathsf{l}_3$. Consequently, Σ'

is also an incidence table of a finite projective plane of order four and herewith the proof is completed.

We know, however, that there exists only one kind of a plane of order four up to isomorphism, i.e. the Galois plane of order four, so the case of our construction could not produce a non-Desarguesian plane. (This, however, is not a general fact, witness the case of $q=9$).

As we have mentioned, the line-substitute consists of quadrangles with one of their diagonal points, four of them with the diagonal point P_1, four with the diagonal point P_2, and four with the diagonal point P_3; the remaining two diagonal points being the ideal points P_2 and P_3, P_3 and P_1 and P_1 and P_2^* respectively.*

If we delete the ideal points of the line-substitute, we obtain 12 set of vertices of quadrangles of the original plane, these, of course, are Fano quadrangles having a common triple of (collinear) diagonal points. We can see immediately from table Σ' that any two of the quadrangles in question have at most a single vertex in common. Two quadrangles which do not have a common vertex and which determine line-substitues common (ideal) diagonal points could be called a *parallel* pair of quadrangles. — Any non-ideal point of the plane Σ is the common vertex of exactly three of these 12 quadrangles.

We see further from our figure that any of the lines $l_{14}=l'_{14}$, $l_{15}=l'_{15}$, ..., $l_{21}=l'_{21}$ of the plane Σ' preserved from the lines of the original plane contains a single vertex of each of the 12 quadrangles, but of course none of their diagonal points.

An essential result of our analysis is that *the line-substitute of the plane of order four are subplanes of order two (Fano planes) having for their ideal points three fixed ideal points of the original plane.*

In the possession of these results we are now able to generalize the construction somewhat.

3° We shall deal only with planes (not necessarily Galois planes) having for their order a square number: $n=q^2$ (where q is not necessarily a power of a prime number). By considering a line l_1 of a plane Σ of this kind as an ideal line, the points $P_1, P_2, ..., P_{q^2}, ..., P_{q^2+1}$ of line l_1 as ideal points, we delete these ideal elements to obtain an affine plane consisting of $n^2=q^4$ points. We require now the plane Σ so satisfy the following.

Condition: *We choose* $q+1$ *points of the line* l_1 *of plane* Σ; *let these be denoted by* $P_1, P_2, ..., P_{q+1}$. *Suppose* $A \notin l$, $B \notin l$, $A \neq B$, *and* $AB \cap l_1 = P_r$ $(r=1, 2, ...,$

---

* We consider an arbitrarily chosen line of the plane in the case of our example the line l_1 as the ideal line and the points of this, the points P_1, P_2, P_3, P_4, P_5 as the ideal points.

$q+1$). Then we shall assume that for every pair of points A, B there exists a proper subplane Σ^* containing each of the points A, B, $P_1, P_2, \ldots, P_{q+1}$ but containing no other point of the line.

We shall denote the subplane satisfying the conditions by the symbol $\Sigma^* = (\mathbf{H}; A, B)$, where $\mathbf{H} = \{P_1, P_2, \ldots, P_{q+1}\}$. We shall construct a subplane Σ^* and we shall determine the properties of the subplanes satisfying the conditions (Fig. 115).

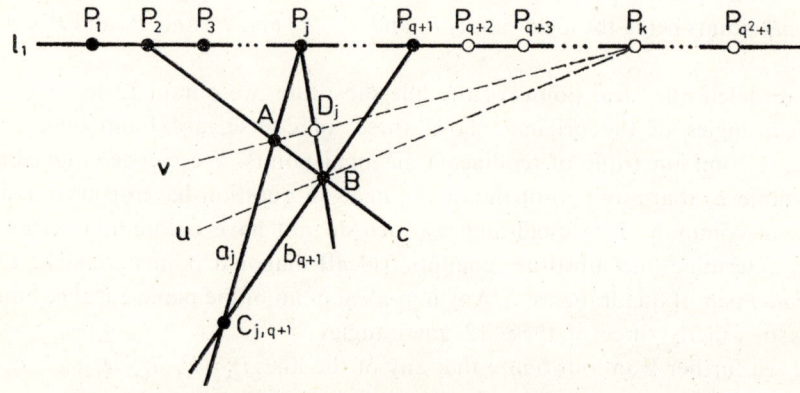

Figure 115

S_1 *The subplane* $(\mathbf{H}; A, B) = \Sigma^*$ *is of order* q.

It follows from the definition of a subplane that if we consider on the plane Σ $AB \cap l_1 = P_r$, $AP_j = a_j$, $BP_h = b_h$ (where P_j, P_h, P_r are distinct elements of the set $\mathbf{H}$) then $a_j \cap b_h = C_{jh} \in \Sigma^*$. We obtain thus, however, only $q(q-1)$ points, which, together with $(q+1)+2$ points defining the plane give q^2+3 points in all. Further points of the lines ABC belong to the subplane Σ^*, other than the points A, B, P_r. These $(q-2)$ points are obtained if the elements of the point set $\mathbf{H} - \{P_j, P_h, P_r\}$ are projected from C_{jh} onto AB. If we add these to the former points, we have q^2+q+1 points of the subplane Σ^* and a plane of order q consists of precisely this many points. If subplane Σ^* were of order greater than q then the line $P_1 P_2$ belonging to Σ^* would contain more than $h+1$ points of plane Σ, which would be a contradiction. Thus we have completed not only the proof S_1, but also the construction of the points of Σ^*.

S_2 *The existence of the subplane* $(\mathbf{H}; A, B) = \Sigma^*$ *implies its uniqueness*.

Namely the existence of the subplane Σ implies that the number of its points is q^2+q+1. The points obtained by the above construction are points of each plane Σ^* defined by $(\mathbf{H}; A, B)$; thus Σ^* is unique.

In our figure, the points of the original plane Σ forming the plane Σ^* are darkly coloured (the other points are lightly coloured).

S$_3$ *Any subplane Σ^* of the original plane Σ is cut in a single point by any line passing through any point of the point set $\mathsf{l}_1-\mathbf{H}$ and distinct from l_1.*

Namely, let $P_k \in \mathsf{l}_1$ but $P_k \notin \mathbf{H}$, further let B be a non-ideal point of an arbitrarily chosen subplane Σ^*. If $BP_k = u$ met Σ^* in a second point A, then we would have $AB = BP_k = u$. Since $A, B \notin \Sigma^*$, the point of intersection of the lines u and l_1 would belong to $\mathbf{H}$, in contradiction to our assumption. If Σ^* and P_k are fixed, then the number of lines u is equal to the number of the non-ideal points of Σ^*, i.e. q^2. Consequently, each of the non-ideal lines passing through the point P_k cuts the subplane Σ^* in a single point.

S$_4$ *If we fix the point set $\mathbf{H}$ of the plane Σ and a non-ideal point B of it, then the number of subplanes $(\mathbf{H}; A, B) = \Sigma^*$ is $q+1$.*

In the proof of the statement we shall refer to the lines $u = BP_k$ and $v = AP_k$, $P_k \in \mathbf{H}$ (Fig. 115). Neither u nor v have a point in common with subplane Σ^* other than points B and A, respectively. Through every non-ideal point of line v there passes a subplane which belongs to $\mathbf{H}$; in fact, if we consider a point $X \in v$ of this kind and a non-ideal point Y for which $Y \in XP_j$ and $P_j \in \mathbf{H}$, then $(\mathbf{H}; X, Y)$ satisfies the condition. These two statements imply that by projecting the points of $\mathbf{H}$ from B onto v we obtain every point D_j ($j = 1, 2, \ldots, q+1$) in which the subplanes belonging to the fixed B and $\mathbf{H}$ cut the line v and different subplanes $(H; B, D_j) = \Sigma^*$ belong to different points D_j. Consequently the number of subplanes in question is $q+1$.

S$_5$ *If we fix the point set $\mathbf{H}$ of the plane Σ, then the number of subplanes $(\mathbf{H}; A, B) = \Sigma^*$ is $q^3 + q$.*

Every one of the q^2 non-ideal points of line u discussed above is contained in $q+1$ subplanes and every subplane intersects the line in a non-ideal point, hence the number of subplanes in question is $q^3 + q$.

S$_6$ *Given a non-ideal point P which is not a point of a fixed subplane Σ_0^* belonging to $\mathbf{H}$, there exists a single subplane Σ^* incident with the point P such that $\Sigma^* \cap \Sigma^* = \mathbf{H}$. Two subplanes like these are said to be parallel to each other and we use the symmetric symbol $\Sigma_0^* \| \Sigma^*$ to denote this relationship.*

The clue of the construction of the derived plane is just this statement to be proved. — In this statement, if we substitute the word "line" for the word "subplane", it states that given a line and a point not on the line, there is one and only one line containing the given point and parallel to the given line. Of course, we have to scrutinize the truth of the statement even if the role of the given Σ^* is taken over by a line u for which $u \cap l_1 \in \mathbf{H}$. This, however, is obvious, since according to S_3 the line u is cut in an ideal point only by an "ordinary" line joining the points P and $u \cap l_1$. —

After this incidental remark we shall give the proof of S_6. The condition $u \cap l_1 \in \mathbf{H}$ is satisfied by $q^2 - q$ u-lines passing through the point P, which each cut Σ_0^* in one non-ideal point X. Thus we exhaust only $q^2 - q$ of the non-ideal points of Σ_0^*, let the remaining $q^2 - (q^2 - q) = q$ points be called Y-points. Clearly any line PY of the plane Σ can intersect the ideal line l_1 only in a point of $\mathbf{H}$ and it cannot contain two distinct Y-points, since otherwise P would lie on the subplane Σ_0^* in contradiction to the fact that the subplane Σ_0^*, according to S_2 is uniquely determined by its points Y_1 and X_2. However, the q subplanes $(\mathbf{H}; P, Y) = \Sigma^*$ do not exhaust the whole set of the planes passing through point P, the number of which is, according to S_4, equal to $q+1$. Thus we have still a subplane Σ^*, which in fact does not have a non-ideal common point with subplane Σ_0^*.

4° In the possession of statements $S_1 - S_6$, we can construct the derived plane Σ, i.e. of the plane Σ'.

We preserve all the points of the original plane, except the points of l_1, that is we pass first to the affine plane Σ obtained from the projective plane Σ by deleting l_1. We preserve all the lines of the original plane which do not have ideal points lying in $\mathbf{H}$, but the ideal points of these lines are deleted. We shall call these lines r-lines.

We delete all the other lines and instead of them we introduce as "lines" the subplanes Σ^* determined by $\mathbf{H}$ but we delete the points lying in the point set $\mathbf{H}$. We shall call these lines s-lines.

Thus, if we suppress the distinction between r-lines and s-lines, i.e. if we speak only of *lines* we arrive at an affine plane $\bar{\Sigma}'$. The following structural properties of the figure follows readily from the construction:

a) The figure $\bar{\Sigma}'$ consists of $q^4 = n^2$ points and of $q^4 + q^2 = n^2 + n$ lines.

b) Any line of the figure consists of $q^2 = n$ points and $q^2 + 1 = n + 1$ lines pass through each of its points.

This is the number of the elements necessary to build up an affine plane of order q^2. And these elements in fact form an affine plane, since the following conditions are satisfied:

A_1' There is one and only one line passing through any two distinct points.
(Cf. l_1 and S_2).

A'_2 Given a line and a point not on the line, there is one and only one line containing the given point and parallel to the given line (C. S_6).

A'_3 There exists a real triangle. (Since there exist at least two distinct r-lines.)

And these are the axioms defining an affine plane.

If we extend the affine plane — in the classical manner — by the ideal points defined by equivalence classes of parallel lines, we obtain a projective plane of the same order. Let this be denoted by $\overline{\Sigma'} \to \Sigma'$.

The minimal example leading to a non-trivial result is the construction by means of the incidence table of the Galois plane of order nine. The derived plane so obtained is non-Desarguesian. Note that the construction may start not only of a Desarguesian plane, but from any finite projective plane, and it may lead to new kinds of planes.

Other kinds of derivation have also been developed; these, however, will not be discussed here.

6.3 A generalization of the concept of the affine plane

As we have seen, the axioms of incidence for the projective plane are not sufficient to ensure that the plane is Desarguesian. However, as is well known, the axioms of incidence for projective spaces of dimension greater than two are so "strong" that they imply the validity of the theorem of Desargues even if the two triangles in question lie in the same plane of the space.

Naturally we may ask, can the system of axioms of projective space be "weakened" to such an extent that spaces exist satisfying the new system; but at the same time these spaces need not be Desarguesian?

Such a program is the mapping system of descriptive geometry called "central representation". — Several programs of this kind were developed e.g. by L. Lombardo Radice (1951), E. Sperner (1960), A. Barlotti (1962); of these we shall mention only the ideas of Barlotti.

1° From the method of the central representation we mention only the particular parts which suggested the notion of the three-dimensional affine space to Barlotti.

Consider classical projective space of 3 dimensions and select a plane as the ideal plane; let us take any other plane in the space as a plane of projection and let the centre of projection be a (non-ideal) point K not incident with π. We shall understand by the image of any arbitrary point $P \neq K$ of the space the point $\Pi \cap KP = P'$. Obviously, this mapping is well defined but is not one-to-one. This can be remedied by introducing other elements determining the plane graphically. Namely we shall prove that if we require that the mapping of the lines as well as of the planes should be one-to-one, then we arrive at a one-to-one representation of the points by means of a construction with lines and planes.

The one-to-one representation of a line or of a plane is not established by a central projection but by *the (non-interchangeable) pair formed by a trace element and a direction element* as follows: We shall understand by the *trace* point of a line l of the space the point $\Pi \cap l = N$, and by the *trace* line of a plane σ of the space the line $\Pi \cap \sigma = n$. The line passing through the points K and parallel to the line l meets the plane Π of projection in the *direction* point M_l, the plane passing through the point K and parallel to σ intersects the plane Π of projection in the *direction* line $m_\sigma(-)$. We shall not discuss the cases when the spatial elements are of special position with respect to the mapping system K, Π, the complete discussion is left to the reader. — We may say that a given pair (N_l, M_l) of points in the *model* of a line of the space and a given pair (n_σ, m_σ) of lines is the model of a plane of the space.

This model is extremely simple from the point of view of expressing the relations of incidence and parallelism. In fact, $l \in \sigma$ holds, if and only if $N_l \in n_\sigma$, $M_l \in m_\sigma$ (in general). Similarly, we can speak of parallelism in the order corresponding to the case $l \| g, l \| \sigma, \sigma \| \tau$ if and only if $M_l = M_g$ $M_l \in m_\sigma$, $m_\sigma = m_\tau$. In addition we mention the fact, obvious from the definition, that the trace line and the direction line of a plane are parallel to each other. In other words, if we denote the ideal line of u, we have $n_\sigma + m_\sigma \in u$.

Now, if we submit the space S_3 to a polarity, then we can consider the elements (N_l, M_l) and (n_σ, m_σ) as the elements representing the lines and points of the image-space and we can consider incidence before.

Note that everything said about the lines and planes of the space (respectively, in view of the duality, about the points of the space) with respect to the notions of incidence and of parallelism of lines can be translated into statement valid on plane Π. Now the idea of Barlotti was to construct a model like this from a *not necessarily Desarguesian plane*. We shall now discuss this in more detail.

2° Let Π be any projective plane of order q and let u be an arbitrarily chosen line on it and U an arbitrary point lying on the line u. We fix the pair $U \in u$ and further we fix two lines n and m distinct from u and from each other but satisfying $n \cap m = U$. Line u is said to be the *fundamental line and the point* U is said to be the *fundamental point*.

We shall define now the *affine space* S realized by a model of the plane π in the following manner. We shall understand by a point P a pair of lines on the plane Π which fulfils the conditions

$$n_p \neq u \neq m_p \quad \text{and} \quad U \neq n_p \cap m_p \in u.$$

This is expressed by the symbol $P=(n_p, m_p)$.

We shall speak of the *lines of first kind* (r-lines) and the *lines of second kind* (s-lines) of the space **S**. By an r-line we shall mean a pair of points (N_r, M_r) of the plane π which satisfies the conditions

$$N_r \notin u, \quad M_r \notin u.$$

Similarly, we shall understand by an s-line a pair of points (N_s, M_s) of the plane Π satisfying the conditions

$$U \neq N_s \in u \quad \text{and} \quad M_s \neq U.$$

These are expressed by the symbols $r=(N_r, M_r)$ and $s=(N_s, M_s)$, respectively.

We have still to define the incidence relation of points and lines, and also the parallelism of lines of same kind. (We shall exclude the parallelism of lines of different kinds.)

We shall understand by the incidence of a point $A=(n_A, m_A)$ and an r-line, $r=[N_r, M_r]$ the following: $N_r \in n_A, M_r \in m_A$ and of a point A and an s line, $s==[N_s, M_s]$ the following:

$$M_s, n_A \cap n, m_A \cap m$$

are collinear and $N_s = n_A \cap u$.

By the parallelism of two lines of the first kind $r_1=[N_{r_1}, M_{r_1}]$ and $r_2==[N_{r_2}, M_{r_2}]$ we shall understand that the triples of points

$$N_{r_1}, N_{r_2}, U \quad \text{and} \quad M_{r_1}, M_{r_2}, U$$

are collinear.

Similarly we shall understand by the parallelism of the two lines of the second kind $s_1=[N_{s_1}, M_{s_1}]$ and $s_2=[N_{s_2}, M_{s_2}]$ that the triplet M_{s_1}, M_{s_2}, U of points is collinear.

It is easy to see — the verification is left to the reader — if Π is a Desarguesian plane, then the affine space (model) so constructed is an ordinary affine space; the ideal points of the space are introduced by the assumption that "the set of the lines parallel to a given line has one ideal point common to all lines of the set".

However, if the plane Π is a non-Desarguesian projective plane obtained by the derivation discussed in Section **6.2** we do not obtain ordinary affine space. Of course, in compiling a list of the properties of this space we arrive at a system of axioms which is weaker than the system of axioms serving to the definition

17 Introduct on

of an affine Galois space. We have not introduced thus far the notion of a plane in the affine space S. However, we shall not develop this program, we only claim without proof that our model satisfies the following system of axioms compiled by Sperner:

SA$_1$ *One and only one line is incident with two distinct points.*

SA$_2$ *The same number of points are incident with every line; the number of points incident with the same line is called order of the space.*

SA$_3$ *Parallelism is an equivalence relation.*

SA$_4$ *Given a point P and a line t, one and only one line passes through the point P and is parallel to the line t.*

6.4 Problems and exercises to Chapter 6

43. Let Pascal's Theorem be given in the following special form:

"Let the quadruples of points U, A_1, A_3, A_5 and U, A_2, A_4, A_6 lie on a line a and on a line b respectively, and $a \neq b$, further let the triple of points $U, X = A_1 A_2 \cap A_4 A_5$, $Y = A_2 A_3 \cap A_5 A_6$ lie on the line u". In this special form Pascal's Theorem is called the MP-*theorem* (Fig. 116). Prove the validity of the MP-theorem on the Fano plane.

44. If a plane is of order at least three and the conditions of the MP-theorem are fulfilled for the points $A_1, \ldots, A_6$, then the lines $A_1 A_4, A_2 A_5, A_3 A_6$ meet

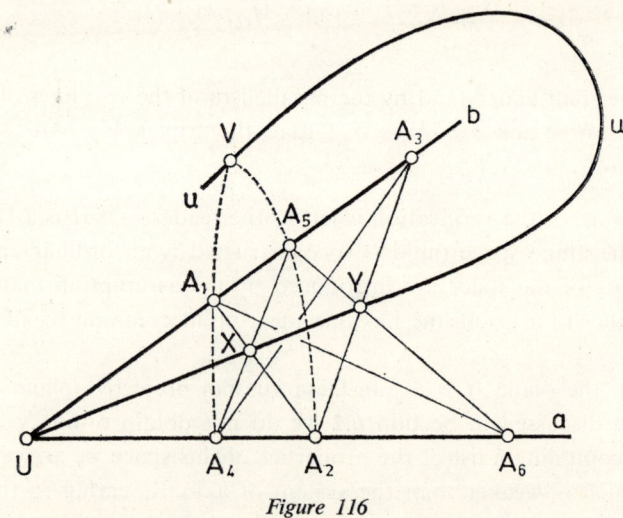

Figure 116

in a point V. Prove this statement and in addition the statement that in the case of a Fano plane V also lies on the line u.

45. If $a \neq b$ are fixed lines, further if A_1 and A_2 are points of the line a distinct from each other and from the point $a \cap b = U$, also if B_1 and B_2 are points of the line b distinct from each other and from U, then the points

$$A_1 B_1 \cap A_2 B_2 = P, \quad A_1 B_2 \cap A_2 B_1 = P'$$

are said to be *conjugate* pair of points with respect to the pair $\{a, b\}$ of the lines. Prove that if the MP-theorem holds on the plane, then the conjugates with respect to $\{a, b\}$ of an arbitrary fixed point P satisfying the conditions $P \notin a, p \in b$ exhaust all the points of the line l_p, other than U. — The line l_p assigned thus to the point P is called the polar P with respect to $\{a, b\}$.

46. If Q is an arbitrary point as defined in Exercise 45 and $Q \neq U$, then prove that the polar of Q with respect to $\{a, b\}$: $l_Q = UP$. A pair $l(=l_p), l'(=l_Q)$ of lines so related is said to be *conjugate* with respect to $\{a, b\}$.

47. If l and l', occurring in the previous exercise, form a conjugate pair of lines with respect to $\{a, b\}$, then a and b also form a conjugate pair of lines with respect to $\{l, l'\}$. Prove the validity of this statement.

48. On a Fano plane every line l is self-conjugate with respect to $\{a, b\}$, if $a \neq b \neq l \neq a$ and $a \cap b \in l$. Prove the validity of this statement. If the MP-theorem holds on a plane unrestrictedly, the plane is called an MP-plane. We shall understand by a conjugate pair of lines with respect to $\{a, b\}$ without stating explicitly that we are considering the pair l, l' of lines assigned to each other in the above manner in the case of $a \neq b \neq l \neq a$ and $a \cap b \in l$. If for the conjugate pair l, l' of lines we have l, l', then l is said to be *self-conjugate* with respect to $\{a, b\}$.

49. Prove that on an MP-plane of even order there exists a self-polar line with respect to $\{a, b\}$.

50. Let the line l on an MP-plane be self-conjugate with respect to $\{a, b\}$. Prove that l is self-conjugate with respect to every $\{u, v\}$ for which $u \neq l \neq v \neq u$ and $u \cap v \in l$ is fulfilled.

51. If on an MP-plane l is self-conjugate with respect to an $\{a, b\}$ then prove that a is self-conjugate with respect to $\{b, l\}$.

52. Prove, on an MP-plane of even order, that if any three distinct lines a, b, c pass through a point then c is self-conjugate with respect to $\{a, b\}$.

53. Prove that every MP-plane of even order is a Fano plane.

In possession of the solutions of these ten problems it follows from Gleason's Theorem that the following theorem is true.

Luneburg's Theorem: *Every* MP-*plane of even order is a Desarguesian plane.*

The following problems require much work but are nevertheless quite instructive:

54. We have already seen the incidence table of the Galois plane of order nine (in connection with Figs 44, 45, 46, p. 93, 94—96) and we know that the sub-table $\Gamma(3)$ of the table $\Gamma(9)$ formed by the intersection of the rows and columns of indices

$$1, \quad 2, 3, 4, \quad 11, 12, 13, \quad 20, 21, 22, \quad 29, 30, 31$$

furnishes an incidence table representing a subplane of order three of the original plane. Let the Galois plane given by the table $\Gamma(9)$ be called Σ, and consider the set $H = \{P_1, P_2, P_3, P_4\}$. It is easy to check that, for instance, the subplane $(H; P_{13}, P_{20})$ has the aforesaid incidence table $\Gamma(3)$. Construct the incidence table of the derived plane of the plane Σ in question induced by H. — The table of 91 squares × 91 squares gives a clear picture of this comparatively complex structure.

55. Try to determine some properties of the new plane Σ' of order nine by means of the incidence table. For instance the property that it is non-Desarguesian.

56. Prove the validity of Sperner's axiom system for the Barlotti model.

A further instructive exercise is given, by trying to introduce the notion of the plane into the space dealt with in Exercise 56. A plane is a point set which contains all the points of at least two lines and all points which satisfy the following conditions:

- B_1 *The points of the line joining two distinct points of the plane are also points of the plane.*
- B_2 *If every point of two distinct lines belong to the same plane, then they are either parallel or meet each other in a point of the plane.*

If the plane Π serving as a basis of the model is non-Desarguesian, we do not know as yet that a plane satisfying B_1, B_2 exists corresponding to two lines intersecting each other.

57. Prove that the set of points $P = (n_p, m_p)$ satisfying the condition $n_p = m_p$ forms a plane.

58. Define the lines which, in the case of the model built upon any Π, determine a plane, but not a plane of the type mentioned in Exercise 57.

Exercises 43—58 continued the themes of Chapter 6. Exercises following the former chapters were of the same character. Now at the end of our book we give some exercises concerning the whole of the material of the book. We include such questions as to test the reader's ability to view certain of the topics discussed in this book from different points of view.

59. The following is a system of axioms for a three-dimensional space.

PT$_0$ *The space S is the set of abstract elements called points* (A, B, ...,) *and lines* (a, b, ...) *satisfying the following axioms*:

PT$_1$ *If* A, B $\in$ S *and* A $\neq$ B, *then there exists exactly one l such that* A, B $\in$ l. — *It is denoted by* AB = l. —

PT$_2$ *If* A, B, C $\in$ S *and there exists no l for which* A, B, C $\in$ l, *further if* P $\in$ BC *and* Q $\in$ CA *but* P $\neq$ Q, *then there exists a point* R *such that* R $\in$ PQ *and* R $\in$ AB.

— We can define the notion of a *plane* in terms of this axiom system as follows: If there exist three distinct point U, V, W $\in$ S not lying on the same line, then consider the set consisting of the points of each line UQ where Q $\in$ VW; this set will be called a plane and will be denoted by (U, VW). —

PT$_3$ Every line contains three distinct points.

PT$_4$ There exists a line.

PT$_5$ Not all the points of the space belong to the same line.

PT$_6$ Not all the points of the space belong to the same plane.

Now if we delete a plane, together with its points and lines of the space S satisfying the axiom system **PT**, we arrive at the notion of the affine space. Conversely show how this affine space can be extended into a projective space S, satisfying the axiom system **PT**, by the addition of ideal elements.

60. Compare the axiom system of the affine space defined in Exercise 59 with the axiom system of Sperner.

The following problems concern plane geometry where we assume only the validity of the axioms l_1, l_2, l_3 unless otherwise stated.

An automorphism of a plane is called a *perspectivity*, if there exists a fixed point K, called the centre, such that any point on a line through K is transformed into a point on the same line; and if there exists a line t, called the axis, such that every point of t is fixed.

61. Prove that the only perspectivity with more than one centre is the identity; similarly, that the only perspectivity with more than one axis is the identity.

62. Given any line t, any point K and two points P and P', distinct from t and k, satisfying the requirements

$$K \in PP', \quad K \neq P \neq P' \neq K, \quad P \in t, \quad P' \notin t,$$

then there exists uniquely a perspectivity with centre K and axis t, such that the image of P under this perspectivity is P'.

63. Determine the maximal number of the perspectivities with centre K and with axis t and $K \in t$.

64. Prove that if an automorphism of a projective plane has an axis, then it has a centre, too and conversely, if an automorphism has a centre then it has an axis.

65. Consider the following two assumptions: 1) For each pair of points (P, P') a perspectivity with centre K and axis t exists such that P' is the image of P under this perspectivity (Exercise 62.1). 2). Given any pair of triangles in perspective from the point K, such that two pairs of corresponding sides intersect in points of the line t, then the third pair of corresponding sides also intersect in a point of the line t. Prove that each of 1) and 2) implies the other.

66. Prove that the validity of statement 1) of exercise 65, for every choice of centre K and axis t is equivalent to the unrestricted validity of the Theorem of Desargues.

67. Given $K_1 \neq K_2$, $K_1 \in t$, $K_2 \in t$. Let π_1 and π_2 be two perspectivities with centres K_1 and K_2, respectively, the same axis t. Consider the mapping $\pi_1 \pi_2 = \omega$. Prove that the mapping π is again a perspectivity with the axis t and centre K where $K_1 \neq K \neq K_2$ and $K \in t$.

68. Let both the centres $K_1 \neq K_2$ lie on the axis t and let $\Gamma(K_j, t) = \Gamma_j$ ($j = 1, 2$) denote the groups formed by all perspectivities with centres K_1 and K_2 respectively, and the same axis t of the plane, furthermore let $\Gamma(t) = \Gamma$ denote the group formed by all perspectivities which have their centre on the axis t. Assume that both Γ_1 and Γ_2 have more than one element. Prove that Γ is a commutative group. (Theorem of Baer.)

NB. It is known that if $\Gamma_1 = 1$ and $\Gamma_2 \neq 1$ then the group Γ is not necessarily commutative.

69. Let the conditions of the former exercise be augmented by the assumption that the plane should be finite. Prove that every element other than the unit element of the group Γ has for its order the same prime number p. (Theorem of Baer.)

70. Consider the extended axiom system of incidence in which the fourth axiom D is some version of restricted validity of the Theorem of Desargues. Determine the coordinate structure of a plane with respect to axiom D, if the plane is finite.

Exercise 59—70 have been included because the discussions in this book were not purely axiomatic. However, after these exercises even a beginner will understand the necessity of a strict axiomatic treatment and thus willingly read a work based upon this principle.

7. APPENDIX

In this appendix we give, first of all, a concise compendium of algebraic material. Not all of this material is necessary for the reader's understanding of this book, but is given in order that the reader need not refer to another book if he wishes to supplement his algebraic knowledge. In addition there are some brief supplements in connection with the main topics of this book which are considered to be both interesting and useful.

7.1 Notes concerning algebraic structures in general

1° The *product* $\mathbf{A} \times \mathbf{B} = \mathbf{C}$ of the sets $\mathbf{A}$ and $\mathbf{B}$ is the set of the ordered pairs (a, b) where $a \in \mathbf{A}$ and $b \in \mathbf{B}$.

Given a set $\mathbf{H} = \{1, 2, \ldots, n\}$ and a permutation group Γ over the set. Let $x \in \mathbf{H}$, the *stabilizer* of the element x is the subgroup of the set

$$\Gamma_x = \{\pi \in \Gamma | x^\pi = x\}.$$

The *orbit* of the element x under the group Γ is the following set $x^\Gamma = \{x^\pi | \pi \in \Gamma\}$. In both cases x^π denotes the element upon which the element x is mapped by the permutation π.

In order to visualize these two important concepts, consider the examples given in Figs 97 and 97* (p. 216). In these figures the elements of $\mathbf{H}$ were denoted by $1, 2, \ldots, 13$ and the elements of Γ by $\sigma_1, \sigma_2, \ldots, \sigma_6$ (σ_1 is the unit element). It is easy to check that e.g. $\Gamma_{11} = \{\sigma_1, \sigma_3\}$, and these two permutations form a subgroup of Γ; $\Gamma_3 = \{\sigma_1\}$, the identity is a trivial subgroup; Γ_8 is the set of all elements of Γ, a trivial subgroup. It is further easy to check that $3^\Gamma = \{1, 3, 4, 5, 7, 10\}$; $2^\Gamma = \{2, 9, 11\}$; $8^\Gamma = \{8\}$.

A permutation group Γ over a set $\mathbf{H}$ is said to be *regular*, if for every $x \in \mathbf{H}$ holds $\Gamma_x = 1$ (identity).

2° Let $\mathbf{A}$ be an arbitrary non-empty set, then the set of all ordered pairs

$$(x, y) \quad (x, y \in \mathbf{A})$$

can be denoted — as it was already mentioned — by $\mathbf{A} \times \mathbf{A}$. A mapping of the set $\mathbf{A} \times \mathbf{A}$ into the set $\mathbf{A}$ is said to be a *binary operation*, the image of (x, y) being denoted by $x \circ y$. The non-empty set $\mathbf{A}$ is said to be *fundamental set* of operation.

Let us now compile the hierarchy of sets with binary operations.

A **groupoid** denoted by $(\mathbf{A}, \circ)$, is set $\mathbf{A}$ with a binary operation denoted by the sign $\circ$. If the set $\mathbf{A}$ of the groupoid is finite, say $|\mathbf{A}|=n$, then the operation table of the groupoid is a table of $n \times n$ squares having the element $x \circ y \in \mathbf{A}$ in its square of intersection of the row of index x and the column of index y. We may say that a groupoid is given, once we write down its operation table.

A **quasigroup** is a groupoid $(\mathbf{A}, \circ)$ in which for any pair $(a, b) \in \mathbf{A}$ of elements both the equations $a \circ x = b$ and $y \circ a = b$ have one and only one solution. — The operation table of a finite quasigroup is a Latin square. A quasigroup with a unit element is also called a *loop*.

A **semigroup** is a groupoid $(\mathbf{A}, \circ)$ in which for any triple $a, b, c \in \mathbf{A}$ we have $a \circ (b \circ c) = (a \circ b) \circ c$.

A **group** is a loop which is at the same time a semigroup. The operation table of a finite group was already introduced in connection with the notion of an R-net.

An **Abelian group** is a group $(\mathbf{A}, \circ)$ in which for every pair of elements $(a, b) \in \mathbf{A}$ we have $a \circ b = b \circ a$. The operation table of a finite Abelian group is symmetrical with respect to the main diagonal, when the rows and columns are indexed in the usual way.

The order of derivation of these notions is visualized by the directed graph in Fig. 117. The labelling of the points of the graph signifies: G=groupoid, K=quasigroup, F=semigroup, C=group, A=Abelian group.

3° Now we quote some group theoretical notions and theorems which are particularly important from the point of view of finite geometries.

Cyclic group: If the group contains an element whose powers exhaust the group then the group is said to be cyclic.

If a group G contains two subgroups A and B such that for any element $g \in G$ there exists a unique element $a \in A$ and a unique element $b \in B$ where $g = ab$;

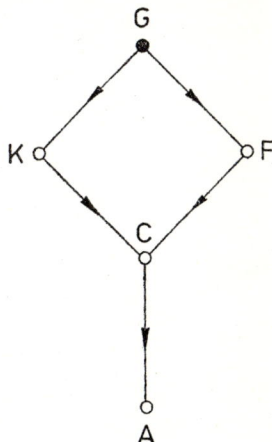

Figure 117

and for each $a \in A$ and $b \in B$ we have $ab = ba$. Then **G** is said to be the direct product of **A** and **B**, denoted by $\mathbf{G} = \mathbf{A} \times \mathbf{B}$. Similarly we can define the direct product of several subgroups of a group.

Theorem of Lagrange: *If* **A** *is a subgroup of the finite group,* **G**, *then* $|\mathbf{G}|:|\mathbf{A}|$ *is an integer.*

A consequence of the theorem is that the order of any element of a finite group is the divisor of the order of the group. Another consequence of the theorem is that a group of prime order is always cyclic.

A **primary group** (p-group) is a group for which there exists a prime p such that the order of every element of the group is a power of prime number p.

An **elementary group** is a group in which the order of every element other than the unit element is a prime number. — A p-group is clearly an elementary group.

Central element: An element c of a group **G** is said to be a central element, if it commutes with every element of **G**, i.e. $cx = xc$ for all $x \in \mathbf{G}$.

Normal subgroup: A subgroup **N** of group **G** is said to be *normal* if for every element $x \in \mathbf{G}$ the set $\mathbf{N}x = \{nx | n \in \mathbf{N}\}$ is the same as the set $x\mathbf{N} = \{xn | n \in \mathbf{N}\}$. This is denoted by $\mathbf{N} \triangleleft \mathbf{G}$.

Centre of a group: The central elements of a group constitute a normal subgroup $Z(\mathbf{G})$ called the centre of the group.

Theorem: *If the order of the group* **A** *is a prime power, then the centre contains elements other than the unit element of the group.*

Conjugates of an element of a group: Two elements a and b of group **G** are said to be conjugate if there exists an element g such that $ag = gb$, i.e. if $g^{-1}ag = b$. In other words, b is the *transform* of the elements a by the element g. If $g^{-1}ag = a$, we say that element a is *invariant* under the transformation by element g.

Conjugate of a subgroup: If the subgroups **A** and **B** of a group **G** are such that there exists $g \in \mathbf{G}$ for which $\mathbf{A}g = g\mathbf{B}$, i.e. $g^{-1}\mathbf{A}g = \mathbf{B}$, then subgroup **B** is a conjugate of subgroup **A** in **G**. In other words, **B** is the *transform* of the subgroup **A** by the element g. If $g^{-1}\mathbf{A}g = \mathbf{A}$, we say that the subgroup **A** is invariant under the transformation by the element g. — Since the normal subgroup of a group is invariant under a transformation with any of its elements, it is also called *invariant subgroup* of the group. —

Now we shall demonstrate by an example how the notions of group theory can be applied to the concise expression of certain geometric facts. The picture on the left hand side of Fig. 118 (where P is a fixed point, t is a fixed line, X is a variable point ranging over all points of the plane) leads to the following theorem. Reflect the points of the plane in the line t, then reflect the images so obtained in the point P, and then reflect these new image-points in line t again; the same result is obtained, if the points are reflected in the point P', the mirror-image of the point P under the reflection in line t. Formulate this fact by means of the group constituted by the (congruency) mappings. Let the reflection with respect to point P; be denoted by π. Obviously $\pi^2=1$; i.e. $\pi=\pi^{-1}$. Similarly, by denoting the reflection in the line t we have that $\tau^2=1$, hence $\tau=\tau^{-1}$. Now

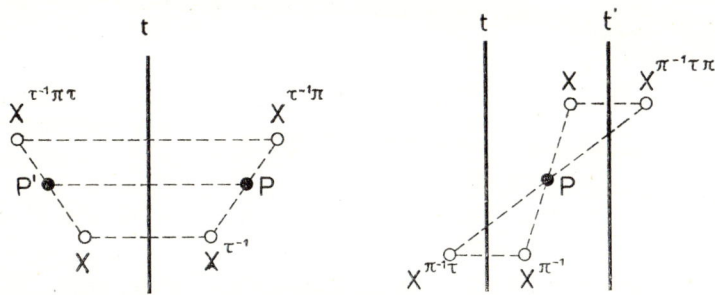

Figure 118

our geometric statement concerning the three reflections above can be expressed by the following formula: The right hand picture of our figure shows that of a reflection in P, then in t, and finally again P is the same as reflection in the line t' which is the mirror-image of t with respect to the point P. This can be expressed by the following formula: $\tau^1 = \pi^{-1}\tau\pi$.

Furthermore it is easy to see that the incidence of point P and line t and the validity of the equality $\pi\tau=\tau\pi$ are equivalent conditions. In order to see this it is enough to consider either the centre of the reflection $\tau^{1-}\pi\tau = \pi$ invariant under the transformation corresponding to the one formula, or the axis of the reflection $\pi^{-1}\tau\pi = \tau$ invariant under the transformation corresponding to the other.

The advantage of the group theoretic language is particularly striking, if we try to express the same geometric notions and relations in the language of coordinate geometry, too. It suffices perhaps to compare the following:

$$\left.\begin{array}{l} P: (x, y) \\ t: [m, b] \end{array}\right\} P \in t: y = mx+b \qquad \left.\begin{array}{l} P: \pi \\ t: \tau \end{array}\right\} P \in t: \pi\tau = \tau\pi.$$

After this incidental example let us continue our review of the group theoretical material.

Sylow's Theorem: *If the order of a finite group* **G** *is divisible by a power* $\mathfrak{p}^r$ *of a prime number, then the group has a subgroup of order* $\mathfrak{p}^r$.

Sylow subgroup: *If* $|\mathbf{G}|=p^k \cdot s$, *where s is an integer not divisible by p, then the subgroups of order* p^k *of group* **G** *are said to be Sylow p-subgroups.*

Consequence i: If the order of finite group is divisible by a prime number p, then the group has an element of order p.

Consequence ii: The order of every finite p-group is a p-power.

Fundamental theorem of Abelian groups: Every finite Abelian group can be decomposed into a direct product of cyclic groups having for their orders powers of primes; the sequence of the powers of primes constituting the orders of these cyclic groups follows uniquely from the order of the group, the sequence of the factors disregarded. —

This is to be understood in the following manner: If $|\mathbf{G}| = p_1^{r_1} \cdot p_2^{r_2} ... p_k^{r_k}$, then there exist cyclic groups $\mathbf{A}_1, \mathbf{A}_2, ..., \mathbf{A}_k$ having for their order $p_1^{r_1}, p_2^{r_2}, ..., p_k^{r_k}$ and $\mathbf{G} = \mathbf{A}_1 \times \mathbf{A}_2 \times ... \times \mathbf{A}_k$. Hence finite Abelian groups can completely be characterized up to isomorphism by finite sets of powers of primes.

4° With respect to sets with two binary operations we restrict ourselves to the most important facts. The fundamental set will be denoted by $\mathbf{A}$, the two operations by $\perp$ and $\curlywedge$ (or, where no misunderstanding can occur, by $+$ and $\cdot$). Structures like these will be denoted by $(\mathbf{A}, \perp, \curlywedge)$ or by $(\mathbf{A}, +, \cdot)$.

A **ring** is a structure $(\mathbf{A}, \perp, \curlywedge)$ such that $(\mathbf{A}, \perp)$ is an Abelian group, $(\mathbf{A}, \curlywedge)$ is a semigroup and the operation $\curlywedge$ is distributive over the operation $\perp$, i.e.

$$(a \perp b) \curlywedge c = (a \curlywedge c) \perp (b \curlywedge c),$$
$$c \curlywedge (a \perp b) = (c \curlywedge a) \perp (c \curlywedge b).$$

This notion is a generalization of the structure $(\mathbf{Z}, +, \cdot)$ of the integers, where (1) $ab \neq ba$ can occur and: (2) $ab = 0$ may hold even for $a \neq 0 \neq b$.

A **field** is a structure $(\mathbf{A}, \perp, \curlywedge)$ such that $(\mathbf{A}, \perp)$ is an Abelian group and $(\mathbf{A}', \curlywedge)$ is a group, where $\mathbf{A}'$ is the set obtained from $\mathbf{A}$ by omitting the zero element (i.e. identity element of the Abelian group), further the operation $\perp$ is distributive over the operation $\curlywedge$. — If the structure $(\mathbf{A}', \curlywedge)$ is also an Abelian group, then we speak of a commutative field. —

Theorem: *A ring is a field if and only if it consists of at least two elements, and further that both the equations $a \curlywedge x = b$ and $y \curlywedge a = b$ can be solved for any pair of elements $a, b \in \mathbf{A}$ where $a \neq 0$.*

— The notion of the field is developed as a generalization of the structure $(\mathbf{Q}, +, \cdot)$ of rational numbers by omitting the requirement that the multiplication be commutative.

Theorem of Wedderburn: *Every finite field is commutative.* —

This profound theorem was of great importance in the development of finite geometries.

A quasifield: is a structure $(A, \perp, \curlywedge)$ having the properties that $(A, \perp)$ is an Abelian group and $(A, \curlywedge)$ is a quasigroup with unit element (i.e. a loop) and we have left distributivity, i.e. for any three elements $a, b, c \in A$ we have

$$a \curlywedge (b \perp c) = (a \curlywedge b) \perp (a \curlywedge c).$$

The quasifield is *distributive* if it is also right distributive.

A quasifield is *associative*, if the operation $\curlywedge$ is associative.

An **alternative quasifield** is a distributive quasifield in which the following two rules hold:

$$a(a'x) = x \quad \text{if} \quad a \neq 0 \quad \text{and} \quad aa' = 1;$$
$$(xa')a'' = x \quad \text{if} \quad a'' \neq 0 \quad \text{and} \quad a'a'' = 1.$$

Theorem: *If a quasifield is both distributive and associative then it is a field.*

As an example let us construct a non-associative quasifield. Let $\mathbf{K} = GF(q)$, $q = p^r$, $p \neq 2$, $r > 2$. Consider the mapping

$$\Theta(x): x \to x^r \quad (x \in \mathbf{K}).$$

We can derive now from $\mathbf{K} = (A, +, \cdot)$ a new structure $\mathbf{K}^* = (A, \perp, \curlywedge)$ in the following manner: Let $(A, +) = (A, \perp)$ and let $(A, \curlywedge)$ defined as follows: if there exists a $c \in A$ for which $a = c^2$, then $a \curlywedge b = a \cdot b$; otherwise (i.e. if a is not a square) $a \curlywedge b = a \cdot b^p$. It is not difficult to prove that $\mathbf{K}^*$ is a quasifield, but not an associative one. From our point of view the example is of interest, since we can establish a non-Desarguesian finite projective plane by the coordinate structure $\mathbf{K}^*$.

The hierarchy of the binary structures with two operations is shown in Fig 119. The meaning of the labelling is the following: G = ring, Q = quasifield, D = distributive quasifield, A = associative quasifield, T = field, K = commutative field.

5° A *vector space (linear space)* over a commutative field $\mathbf{K}$ consists of an Abelian group $\mathbf{M}$, whose elements are called vectors, such that given any element $\alpha \in \mathbf{K}$ and any vector $\mathbf{a} \in \mathbf{M}$ determine a vector $\alpha \cdot \mathbf{a} \in \mathbf{M}$, called the *scalar product* of α and $\mathbf{a}$, subject to the following conditions:

If $\mathbf{a}, \mathbf{b} \in \mathbf{M}$ and $\alpha, \beta \in \mathbf{K}$ then

$$\alpha \cdot (\mathbf{a} + \mathbf{b}) = \alpha \cdot \mathbf{a} + \alpha \cdot \mathbf{b}$$
$$(\alpha + \beta) \cdot \mathbf{a} = \alpha \cdot \mathbf{a} + \beta \cdot \mathbf{a}$$
$$(\alpha \cdot \beta) \cdot \mathbf{a} = \alpha \cdot (\beta \cdot \mathbf{a}) \quad \text{and} \quad 1 \cdot \mathbf{a} = \mathbf{a}$$

where 1 is the identity element of the field $\mathbf{K}$.

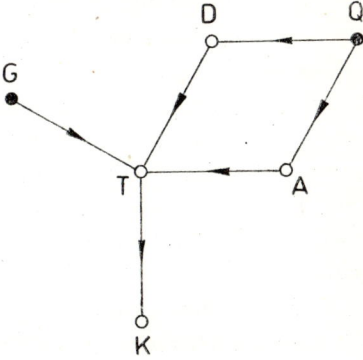

Figure 119

Linear independence: A system $\{\mathbf{a}_1, \mathbf{a}_2, \ldots, \mathbf{a}_k\}$ vectors in a linear space **M** is said to be linearly independent if the relation

$$\lambda_1 \mathbf{a}_1 + \lambda_2 \mathbf{a}_2 + \ldots + \lambda_k \mathbf{a}_k = 0$$

implies that

$$\lambda_1 = \lambda_2 = \ldots = \lambda_k = 0.$$

Basis, dimension, coordinates: If **M** does not contain a linearly independent system of $n+1$ vectors but $\{\mathbf{e}_1, \mathbf{e}_2, \ldots, \mathbf{e}_n\}$ is a linearly independent system, then this system of n elements is said to be a *basis* of vector space and the number n is the *dimension* of the vector space. If $(\mathbf{e}_1, \mathbf{e}_2, \ldots, \mathbf{e}_n)$ is a basis and if every element of the ordered n-tuple $(\xi_1, \xi_2, \ldots, \xi_n)$ of elements belongs to the field **K**, further if

$$\mathbf{x} = \xi_1 \mathbf{e}_1 + \xi_2 \mathbf{e}_2 + \ldots + \xi_n \mathbf{e}_n,$$

then we say that $(\xi_1, \xi_2, \ldots, \xi_n)$ is a *coordinate sequence* of the vector **x** relative to the basis $(\mathbf{e}_1, \mathbf{e}_2, \ldots, \mathbf{e}_n)$.

Theorem: *Any element of a vector space can be expressed uniquely in the form*

$$\mathbf{x} = \xi_1 \mathbf{e}_1 + \xi_2 \mathbf{e}_2 + \ldots + \xi_n \mathbf{e}_n$$

in terms of a given basis; the zero vector is expressed by the n-tuple $(0, 0, \ldots, 0)$ of coordinates.

6° The final note of this section is given as an illustration of the strong connections between the development of projective geometry and that of algebra. In 1942 Baer investigated the projective planes for which every one of the perspectivities $P \to P'$ induced by the ideal line l of the plane and by a fixed ideal point Y of it ($P \neq Y \neq P'$; $Y \in PP'$; $P, P' \notin l$) exists. These planes are the so-called (Y, l)-*transitive* planes. Later, *translation planes* were investigated. There are projective planes that are (Y, l)-*transitive* for every point Y of l. It turns out that the notion of the *algebraic plane over a quasifield* and the notion of the translation plane are equivalent concepts.

If we impose stronger conditions on the set of the translations of the plane, then the coordinate structure satisfies stronger conditions.

(1) If we require, on the translation plane, the existence of all perspectivities having a fixed ideal point X as centre and a line passing through the fixed ideal point $Y(X \neq Y)$ as axis; and moreover we require the existence of all perspectivities with centre Y and with an axis passing through point X, then the corresponding coordinates form an associative quasifield.

(2) If we require further, on the translation plane, the existence of every elation having a fixed ideal point Y as centre and an arbitrary line passing through point Y as axis, then the corresponding coordinates form a *distributive quasifield*.

We always understand by distributivity of a quasifield, the left distributivity $a(b+c) = ab+ac$.

Dual of a translation plane: This is a plane on which every elation exists having a fixed ideal point Y as centre and having a line passing through point Y as axis. — These planes do not have a generally accepted name; the German word for a plane like this is "*Scherungsebene*". —

A **right quasifield** is a structure $(\mathbf{Q}, \perp, \curlywedge)$ in which $(\mathbf{Q}, \perp)$ is an Abelian group (its neutral element is denoted by 0); $(\mathbf{Q}', \curlywedge)$ is a quasigroup with unit element, $\mathbf{Q}' = \mathbf{Q} - 0$ (the unit element is denoted by 1) such that for every element a of $\mathbf{Q}$ we have $a \cdot 0 = 0 \cdot a = 0$; for every triple a, b, c of elements of $\mathbf{Q}$ hold $(a \perp b) \curlywedge c = (a \curlywedge c) \perp (b \curlywedge c)$; if $a \neq b$ then the equation $(-x \curlywedge b) \perp (x \curlywedge a) = c$ has one and only one solution in $\mathbf{Q}$, where $-x$ denotes the inverse of the element x in the Abelian group $(\mathbf{Q}, \perp)$.

It is known that in the case of a finite right quasifield the last requirement is already implied by the preceding four. In 1965 F. Bartalozzi showed that finite *proper* left quasifields exist having p^{2r} elements where $p \neq 2$. By the word proper we understand that the operation is neither commutative nor associative and that right distributivity is not fulfilled. — We mention here that R. Moufang as early as 1931 had investigated planes which satisfy the axioms I_1, I_2, I_3 and the MD-theorem, and he arrived at the notion of an *alternative quasifield*. However, we have dealt in this book with finite planes and we know that, according to the Zorn—Levi theorem, finite alternative quasifields do not exist.

7.2 Notes concerning finite fields and number theory

The most simple example of a finite field is a residue class field mod p (p is a prime), denoted here by Π_p (or more briefly by Π).

Isomorphic fields: If a field $\mathbf{K}$ has a one-to-one mapping φ onto a field $\mathbf{K}'$ such that

$$a \to a' \quad \text{and} \quad b \to b' \quad (a, b \in \mathbf{K}, \ a', b' \in \mathbf{K}')$$

implies $a+b \to a'b'$, $ab \to a'b'$ (φ is an operation preserving mapping), then the two fields are said to be isomorphic and this is denoted by $\mathbf{K} \cong \mathbf{K}'$.

Theorem: *Isomorphism is a reflexive, symmetric and transitive relation; an isomorphism $\mathbf{K} \cong \mathbf{K}'$ assigns to the zero element and the unit element of $\mathbf{K}$ to the zero element and the unit element respectively of $\mathbf{K}'$.*

Subfield, prime field: If a subset $\mathbf{K}$ of a field $\mathbf{L}$ is its own field (with the operations defined in $\mathbf{L}$), then $\mathbf{K}$ is said to be a *subfield* of $\mathbf{L}$ and $\mathbf{L}$ is said to be an *extension* of $\mathbf{K}$. This is denoted by $\mathbf{K} \subseteq \mathbf{L}$. If a field has no (proper) subfield, we call it a *prime* field.

Theorem: *Any finite field has one and only one prime subfield which is isomorphic with Π_p.*

Characteristic: Consider the sum with the least number of terms formed with the unit element e of a finite field Λ for which $e+e\ldots +e=0$. The number of terms of this sum is said to be the characteristic of the field Λ.

Theorem: *The characteristic of a finite field is a prime number and the residue class field belonging to this prime is isomorphic to the prime field of the field.*

Independent elements of Λ with respect to Π: If $\alpha_k \in \Lambda$ and $c_k \in \Pi$ ($k=1, 2, \ldots, r$) and $c_1\alpha_1+c_2\alpha_2+\ldots+c_r\alpha_r=0$ ($0\in\Lambda$) imply $c_1=c_2=\ldots=c_r=0$, then the elements $\alpha_1, \alpha_2, \ldots, \alpha_r$ of Λ constitue an independent system with respect to Π.

Rank of a field: If the maximal system of independent elements with respect to a field Π of the field Λ is $\alpha_1, \alpha_2, \ldots, \alpha_n$, then the number of the elements of this system is said to be the rank of Λ with respect to Π; (i.e. $n=\Lambda:\Pi$).

Degree of extension: The number n in the former definition with respect to the extension of Π into Λ is said to be the degree of the extension.

Theorem: *If $\Pi \subset \Gamma \subset \Lambda$, then $(\Lambda:\Pi) = (\Lambda:\Gamma)\cdot(\Gamma:\Pi)$.*

The **notation** $\Lambda|x|$ indicates the set of the polynomials in an indeterminate x with coefficients in Λ, where addition and multiplication of two polynomials can be performed in the usual manner, but we compute with the coefficients in the manner defined in Λ. — As an example let us consider the finite field Λ given by the operation tables of Fig. 44 (p. 93); consider the polynomials

$$a(x) = 2x^3+x+1 \quad \text{and} \quad b(x) = 5x^4+3x^3+7x+6.$$

By using the operation table it is easy to check that

$$a(x)+b(x) = c(x) = 5x^4+5x^3+8x+7,$$
$$a(x)\cdot b(x) = 7x^4+6x^6+5x^5+x^4+6x^3+7x^2+4x+6.$$

Obviously, the degree of the product polynomial is the sum of the degrees of the factors, while the degree of the sum of two polynomials can be smaller than the degree of each of them.

Now in the possession of the above notions and theorems we can formulate the theorems, which are necessary in the discussion of Galois geometries as follows:

Fundamental Theorem: *The number of the elements of a finite field is a power of prime: There exists a field of $\mathfrak{q}$ elements for every prime power $\mathfrak{q}=\mathfrak{p}^r$; two fields of $\mathfrak{q}$ elements are isomorphic.* — Obviously, the characteristic of the field of $\mathfrak{q}=p^r$ elements is p. —

1° The p^r-th power of any element of the finite field of p^r elements is equal to the element itself. For a prime field this is the Theorem of Fermat. —

2° Every element of a finite field of characteristic p is a p-th root, more exactly, while x ranges over the elements of Λ, x^p generates a permutation of the elements.

3° We have for any two elements a, b of the field

$$(a+b)^p = a^p + b^p, \quad (ab)^p = a^p b^p.$$

4° The mapping $x \to x^p$ is an automorphism of the field, called the Frobenius automorphism. All automorphisms of the field can be produced by the iteration of this mapping.

5° The multiplicative group of a finite field is a cyclic group.

6° Let $a_1, a_2, \ldots, a_{q-1}$ denote the elements other than zero of a finite field, then

$$a_1 a_2 \ldots a_{q-1} = -1,$$

where -1 signifies the additive inverse of the unit element. — If we formulate this theorem for a field of residue classes we obtain Wilson's Theorem: $(p-1)! \equiv -1 \pmod p$.

We insert a number theoretical remark:

Wilson's theorem gives a characterization of the prime numbers. Namely if we disregard

$$(4-1)! \equiv 2 \pmod 4,$$

we have for all other composite numbers

$$(n-1)! \equiv 0 \pmod n$$

and for all primes

$$(p-1)! \equiv -1 \pmod p.$$

7° Given a set $\Pi_p[x]$ defined over the prime field Π_p of the field Λ of p^r of elements (i.e. over the field of residue classes), we choose any polynomial

$$a(x) = x^r + a_1 x^{r-1} + \ldots + a_r$$

of the set $\Pi_p[x]$, irreducible over the field Π_p. If we consider the set of the polynomials

$$c_\nu(x) = c_{r-1} x^{r-1} + c_{r-2} x^{r-2} + \ldots + c_0$$

where $c_j \in \Pi_p$ ($j = 0, 1, \ldots, r-1$) then the set $c_\nu(x)$ consists of p^r polynomials — this is the number of the r-tuples of the ordered elements $(c_{r-1}, c_{r-2}, \ldots, c_0)$ in Π_p. These polynomials constitute a field with respect to addition and multiplication, which is isomorphic to the field Λ, if we reduce modulo p and modula $a(x)$ the polynomial obtained by the operation.

Let us remember that this theorem was already applied in the construction of finite fields. The meaning of the reduction of the sum and product were also discussed there, but we repeat it here. We replace the coefficients of the result by their smallest non-negative residues mod p. In the product the degree of the polynomial can be greater than $(r-1)$, but is less than $2r-1$. In this case the result is replaced by the residue obtained by dividing by the polynomial $a(x)$.

8° An element of a finite field can have 0, 1 or 2 square roots. If the field has characteristic $p=2$, then each of its elements has just one square root; if $p \neq 2$, then the set of the non-zero elements can be decomposed into two subsets of equal number of elements such that no element of the one subset has a square root and every element of the other has two square roots. The latter subset forms a subgroup of the multiplicative group of the field.

9° Consider an arbitrary element d of a field Λ of 2^r elements and consider the element
$$D = d2^{r-1} + d2^{r-2} + \ldots + d.$$
While d ranges over every element of Λ, $D=0$ and $D=1$ occurs in 2^{r-1} cases, each.

10° The quadratic equation $z^2+z+d=0$ either cannot be solved in a field of even characteristic, or has two distinct solutions, according as $D=1$, or $D=0$ for the element d in question.

11° In a field of even characteristic the number of the solutions of the equation $x^2+bx+c=0$ is 1 in the case $b=0$ (cf. 8°), and if $b \neq 0$, then we reduce it by the substitution
$$z = xb^{-1}, \quad d = cb^{-2}$$
to the equation $z^2+z+d = 0$.

12° In a field of odd characteristic the number of the solutions of the equation $x^2+bx+c=0$ is 0, 1 or 2 according as l^2-4c is non-square, zero, or square, respectively.

Note that a knowledge of theorems 8°–12° is necessary when discussing curves of the second order of the Galois plane.

We finish our notes regarding the theory of finite fields by remarking that the number theoretical theme known under the name of *cyclic difference bases* can also be considered as an application of the theory of finite groups. The first prominent result occurred in a paper of I. Singer which appeared in 1938. In this paper Singer proved the existence of the simple cyclic difference bases by their construction. And this can directly be used for the establishment of the incidence tables of Galois planes. (See the text relevant to Figs 5 and 6, pp. 23, 26.)

7.3 Notes concerning planar ternary structures

The gradual generalization of the notion of the projective plane starting from the classical projective plane went hand-in-hand with the gradual generalization of the notion of coordinates suitable for the algebraic representation of the plane. When the algebraic representation of non-Desarguesian planes came into prominence, the role of the coordinate field was taken over by a new kind of structure, called a *ternary structure,* having a fundamental set with a *ternary operation* with respect to the triples of elements of the fundamental set. By fixing of one of the elements of the triples we obtain a binary operation and by determining the essential relations of these operations we can determine the essential properties of the ternary structure.

Consider a fundamental set S of at least two elements. We assign to an ordered triple (m, x, b) of elements of the set S an element of S, this element will be denoted by the operational symbol $\langle xmb \rangle$ and the unique mapping is called a *ternary operation*. If this operation is defined over every ordered triples of elements, i.e. if

$$(m, x, b) \rightarrow \langle xmb \rangle \in S,$$

we speak of a *ternary structure* $(S, \langle \ \rangle)$.

The different kinds of ternary structures can be defined by axioms with respect to the ternary operation. Their names are more a "historical tradition", than an expression of their contents.

Planar ternary ring (denoted by PTR). This is defined by the following axioms:

Θ_1: There exists a single element $u \in S$ for which $\langle abu \rangle = c$, where a, b, c are arbitrarily chosen elements of S.

Θ_2: If a, b, c, d are arbitrary elements of set S, but $a \neq c$, then there exists a single element u in S for which

$$\langle uab \rangle = \langle ucd \rangle.$$

Θ_3: If a, b, c, d are arbitrary elements of set S, but $a \neq c$ then there exists a pair of elements in S, for which $\langle auv \rangle = b$ and $\langle cuo \rangle = d$.

Intermediate ternary ring (denoted by ITR). This is a PTR satisfying also the following axioms:

Θ_4: If a, b, c, d are arbitrarily chosen elements of set S, but $a \neq c$ and t is an element of the set such that $\langle atb \rangle = \langle ctb \rangle = d$, then $\langle xtb \rangle = d$ for every element x of the set S.

Θ_5: If a, b, c, d are arbitrarily chosen elements of S, but $a \neq c$, and t is an element of the set such that $\langle tab \rangle = \langle tcb \rangle = d$, then $\langle txb \rangle = d$ for every element x of the set S.

Hall ternary ring (denoted by HTR). This is a PTR for which the following axioms are also satisfied:

Θ_0: The set S has elements 0, 1 such that $0 \neq 1$ and for any pair a, b of its elements

$$\langle a0b \rangle = \langle 0ab \rangle = b, \quad \langle a10 \rangle = \langle 1a0 \rangle = a.$$

Θ_2^*: Given the elements a, b, c of set S, where $a \neq 0$, there is one and only one element of S which satisfies the equation $\langle xab \rangle = c$.

Then we have the following theorems:

1° With respect to PTR, the existence of the pair (u, v) of elements defined in Θ_3 implies its uniqueness.

2° With respect to ITR, axioms Θ_4 and Θ_5 are independent from the preceding axioms.

3° If the fundamental set of a PTR is finite, then the subsystem $\Theta_0 \wedge \Theta_1$ of the system of axioms

$$\Theta_1 \wedge \Theta_2 \wedge \Theta_3 \wedge \Theta_0 \wedge \Theta_2^\circ$$

imply that the other axioms are also fulfilled if and only if Θ_3 is also valid.

4° There can be assigned to any quadrangle OXYE of a plane defined by the axioms system I a coordinate structure (**K**, $\langle \rangle$), namely a HTR so that there exists a one-to-one correspondence between the points of the plane excluding the line XY (and of the points of this line) and the ordered pairs (x, y) of the set **K**; the lines of the plane as points sets are sets of pairs (x, y) corresponding either to a condition $x = a$, or to a condition $y = \langle xmb \rangle$, where a, b, m are fixed elements of the set **K**.

It is important to note that the coordinate structure HTR will not be the same (in general) for different choices of quadrangle.

5° Given an HTR; let it be denoted by (**K**, $\langle \rangle$). Consider as a "point" an ordered pair (x, y) of elements of set **K**, and consider as lines the sets of the points so defined by either the equation $x = a$, or the equation $y = \langle xmb \rangle$. Thus we can define an affine plane on which the vertices of the fundamental quadrangle are in turn: O: (0, 0), E: (1, 1), Y: the ideal point of the line of equation $x = 0$, X: the ideal point of the line of equation $y = 0$.

BIBLIOGRAPHICAL NOTES

We cannot hope to compile the complete literature of finite geometries up to the present day. The excellent monograph of Dembowski, recommended in the following short bibliography, has performed this task, for the period 1782—1968. At the end of his book Dembowski published a bibliography of 48 pages.

As to the bibliographical notes to be found below we mention books and papers suitable for a more thorough study of the topics merely sketched in this book.

Books

BLUMENTHAL, L. M. (1961): *A Modern View of Geometry.* Freeman, San Francisco
BUSACKER, R. G.—SAATY, T. L. (1966): *Finite graphs and Networks.* McGraw-Hill, New York
DEMBOWSKI, P. (1968): *Finite Geometries,* Springer, Berlin
HALL M. JR. (1959): *The Theory of Groups.* The Macmillan Company, New York
KERÉKJÁRTO, B. (1966): *Les Fondements de la Géométrie,* Tome 2. Géométrie projective. Akadémiai Kiadó, Budapest
RÉDEI, L. (1967): *Algebra. Vol.* 1. Akadémiai Kiadó, Budapest
RIORDAN, J. (1958): *An Introduction to Combinatorial Analysis.* Wiley, New York
SEGRE, B. (1961): *Lectures on Modern Geometry.* Cremonese, Roma. This book contains as an appendix the following work:
LOMBARDO RADICE, L. *Non-Desarguesian Finite Graphic Planes*

Papers

ACZÉL, J. (1965): Quasigroups, net and nomograms. *Advances in Math.,* **1**, 383—450
ACZÉL, J—KÁRTESZI, F. (1967): Funktionskomposition, vertauschbare Funktionen und Iterationsgruppen vom geometrischen Standpunkt aus. *Math.-Phys. Semesterberichte (Göttingen),* **14/1**, 79—88
BARLOTTI, A. (1962): Una costruzione di una classe di spazi affini generalizzati. *Boll. Un. Mat. Ital.,* **17**, 394—398
BARTALOZZI, F. (1965): Su una classe di quasicorpi (sinistri) finiti. *Rendic. Mat.,* **24**, 165—173
BOSE, R. C. (1938): On the application of the properties of Galois fields to the problem of the construction of hyper Greco-Latin squares. Sankya, *Ind. J. Statistics,* **3**, 323—338
BRUCK, R. H.—RYSER, J. H. (1949): The non-existence of certain finite projective planes. *Can. J. Math.,* **1**, 88—93
COFMAN, J. (1967): Triple transitivity in finite Möbius Planes. *Atti Accad. Naz. Lincei Rendic.,* **42**, 616—620
CROME, D. W. (1965): The construction of finite regular hyperbolic planes from inversive planes of even order. *Col. Math.,* **13**, 247—250

Fano, G. (1892): Sui postulati fondamentali della geometria proiettiva. *Giornale di Matematiche*, **30**, 106—132

Gleason, A. M. (1956): Finite Fano planes. *Amer. J. Math.*, **78**, 797—887

Hall, M. (1943): Projective planes. *Trans. Amer. Soc.*, **54**, 229—277

Kárteszi, F. (1960): Piano finiti ciclici come risoluzioni di un certo problema di menimo. *Boll. Un. Mat. Ital.*, **15**, 522—528

Kárteszi, F. (1963): Alcuni problemi della geometria d'incidenza. *Confer. Sem. Mat. Univ. Bari*, **88**, 1—14

Lüneburg, H. (1967): Gruppentheoretische Methoden in der Geometrie. *Jahresbericht d. DMV.* **70**, 16—51

Lüneburg, H. (1969): *Transitive Erweiterungen endlicher Permutationsgruppen.* Springer, Berlin, 1—119. (Lecture Notes in Math. 84)

Qvist, B. (1952): Some remarks concerning curves of the second degree in a finite plane. *Ann. Acad. Fenn.*, **134**, 1—27

Reiman, I (1968): Su una proprietà dei 2-disegni. *Rend. di Mat.*, 1—7

Rosati, L. A. (1957): Piani proiettivi desarguesiani non ciclici. *Boll. Un. Mat. Ital.*, **12**, 230—240

Segre, B. (1955): Ovals in a finite projective plane. *Can. T. Math.*, **7**, 414—416

Segre, B. (1967): Introduction to Galois geometries. *Atti della Accad. Naz. dei Lincei, Memorie*, **8**, 135—236

Singer, J. (1938): A theorem in finite projective geometry and some applications to number theory. *Trans. Amer. Math. Soc.*, **43**, 377—386

Sperner, E. (1960): Affine Räume mit schwacher Inzidenz und zugehörige algebraische Strukturen. *J. reine angew. Math.*, **204**, 205—215

Szele, T. (1956): Elemi bizonyítás a véges testek elméletének alaptételére. *Mat. Lapok*, **VII**, 249—254

Tallini, G. (1960): Le geometrie di Galois e le loro applicazioni alla statistica e alla teoria dell'informazione. *Rendiconti di Matematica*, **19**, 379—400

Turán, P.—Sós, V.—Kővári, T. (1954): On a problem of K. Zarenkiewicz. *Coll. Math,*. **3**, 50—57

Zappa, G. (1960): Piano grafici a caratteristica 3. *Ann. Math. Pura Appl.*, **49**, 157—166

INDEX

N.B. Items in the Appendix have not been included in the index

affine plane 30—32
affinity 54—59
arithmetic space 84—85
articulation point of the graph 188
associated plane 31
associated point-octet 205
automorphism 138, 141

basis 89
bipartite graph 188
block 203
block design 203

canonical form 123
Cayley table 181
central-axial collineation 51
centre of the collineation 50
class of the graph 188
classical projective geometry 1
classical projective plane 1
closed plane 205
collineation 49
combinatorial structure 4
complete graph 187
complete orthogonal system of the Latin
  squares 58—67
complete Γ-hook 14
complete permutation group 51, 201
configurations 164
configuration table 165
cross ratio 94—98
curve 124
cycle (graph) 188
cyclic form 31
cyclic incidence table (Ω-table) 12
cyclic subgroups 51

derivation 232—241
Desargues' plane (space) 43
Desargues' theorem 41
Desarguesian configuration 170—177
diagonal line 54
diagonal pattern 14
diagonal point 53
diagonal triangle 54
dimension of the space 74, 89
duality 3, 94

edge of graph 187
elation 51
elementary transformation 164
empty graph 187
equivalence classes 201
exterior line 126
exterior point 112

Fano figure 1
Fano plane 2, 219
Fano quadrangle 97
fixed lines 49
fixed points 49
fundamental line 102
fundamental point 99
fundamental triangle 101, 104

Galois plane (space) 36—39, 84—85
gamma (Γ)-hook 13
gamma (Γ)-table 14
girth of a graph 188
Gleason's theorem 229
graph 187
group of collineations 51

Hall's model 43
Hamilton circuit 188
Hamilton graph 188
heteromorphic graph 189
heteromorphic incidence table 165
homogeneous coordinates 85, 93
homography 142
homologous elements 17
homology 51
homothetic 55
hyperbolic plane 32
hyperoval 113

ideal point 32
imprimitive group 202
incidence 1
incidence table 1, 4, 11, 19
independent point system 86
independent subspaces 90
interior point 112
intersection space 90
intransitive permutation group 201
invariant line 49

invariant point 49
inversive geometry 205—213
inversive plane 209
involution 52
isomorphic graph 189
isomorphic incidence table 165
isomorphic plane 4
isomorphism 6, 138, 141

join (of two subspaces) 90

k-transitive permutation group 218

Latin squares 60
line 1, 37
linear combination 39
linear transformations 106

maximal subspace 90
Möbius plane 209

net 177
non-Desarguesian plane 43
non-point 85
norming of coordinate sequence 85
N-theorem 206
nucleus of the oval 113

omega- (Ω-) table 12
order of the graph 188
ordinary Desargues configuration 171
oval 110
ovoid 210

parallelism 32
parcel 13
parcel rectangle 17
path of a graph 188
pentagon theorem 170
permutation group 48, 201
Petersen graph 185
plane of order q 4
plane phase 43
Poincare's spherical model 185
point 1, 37
point of graph 187
polar of the point 126
polar triangle 127
pole of a line 127

primitive group 202
projective geometry 2, 36
projective plane 4

quadric 210
quasigroup 59

regular graph 188
regular hyperbolic plane 32
Reidemeister configuration 180

secant 126
Segre's theorem 133
self-conjugate point 121, 126
sigma- (Σ-) table 164
sign-rectangle 15
skew line 126
skew subspaces 90
special Desargues configuration 172
stereographic mapping 205
subphase 43
subplane 29
subspace of the smallest dimension 90
system of axioms for a Möbius plane 209
system of axioms for a non-Desarguesian plane 232
system of axioms for a projective plane 2, 3, 4
system of axioms for a regular hyperbolic plane 34, 35
system of axioms for a three-dimensional space 247
system of axioms for an affine plane 32
system of axioms for Galois space 84—86
system of coordinate 99

tangent 124, 126
ternary operation 19
theorem of Crowe 35
theorem of Miguel 206
theorem of Singer 31
tract 34
transformation (ω-, σ-, τ-) 5, 6
transitive permutation group 201
translation plane 184
tree (graph) 188

unit point 99

QA
167
K23
1976

JAN 19 1977